AQA BIOLOGY Specification A

A New Introduction to BIOLOGY

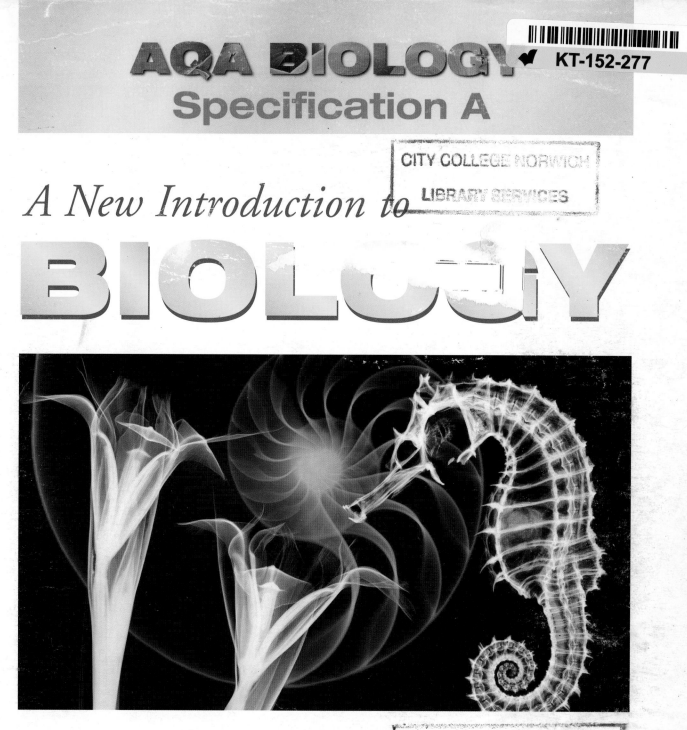

Bill Indge
Martin Rowland
Margaret Baker

Hodder & Stoughton
A MEMBER OF THE HODDER HEADLINE GROUP

Orders: please contact Bookpoint Ltd, 78 Milton Park, Abingdon, Oxon OX14 4TD. Telephone: (44) 01235 827720, Fax: (44) 01235 400454. Lines are open from 9.00–6.00, Monday to Saturday, with a 24 hour message answering service. Email address: orders@bookpoint.co.uk

A catalogue record for this title is available from The British Library

ISBN 0 340 78167 X

First published 2000

Impression number 10 9 8 7 6 5 4 3 2
Year 2005 2004 2003 2002 2001 2000

AQA Examination questions reproduced or adapted by permission of the Assessment and Qualifications Alliance.

Cover photo from: Montage of images from Photodisc.

Photo acknowledgements
Biophoto Associates (Figs 1.3, 1.5, 1.6, 1.7, 1.10, 3.11, 5.18, 5.27, 6.1, 9.1, 9.2, 9.3, 9.4, 9.5, 9.6, 10.4)
Bourne Hall Clinic (Fig. 13.12)
John Birdsall (Fig. 3.10, 4.27, 7.1, 11.2)
Holt Studios International (Figs 3.3, 3.8, 3.30, 4.25, 10.5a, 12.4b, 12.14, 12.15, 12.22a, 12.22b, 12.25, 12.26, 12.29, 13.7b)
Hulton Getty (Fig. 13.15)
Bill Indge (Figs 12.4a, 12.4c, 12.6, 12.16, 12.18, 12.21, 12.22c, 13.9, 13.10)
Natural History Picture Agency (Fig. 3.12, 3.26, 3.36, 7.10, 7.11, 10.5c, 10.10, 13.8)
Planet Earth Pictures (Figs 3.1, 3.32, 4.1, 5.25, 5.26, 6.3, 6.5, 10.5b, 12.1, 12.2, 12.31, 13.1)
Science Photo Library (Figs 1.4, 1.11, 2.1, 2.13, 3.22 3.24, 4.13, 6.10, 6.23, 8.1, 9.8, 9.13, 12.24, 13.2, 13.11, 13.18)
SECOL (Fig. 5.1)
Tropix (Fig. 6.9)
Wellcome Picture Library (Fig. 11.1)

Designed by Allan Sommerville, Cambridge Publishing Management.
Typeset by Cambridge Publishing Management.

Printed in Italy for Hodder & Stoughton Educational, a division of Hodder Headline Plc, 338 Euston Road, London NW1 3BH by Printer Trento.

Contents

Introduction

Biology is never far from the headlines. While this book was being written, the human genome has been sequenced and we now know the complete arrangement of the three thousand million bases that make up human DNA. In Africa, one in four Kenyans is now estimated to be HIV-positive and 350 people die every day from AIDS. Smoke is once again blanketing Southeast Asia from fires burning out of control in Indonesia. Biologists are concerned with all these issues. They work in the fields of cell biology, medicine, food production and ecology and the work they do is vital to us all.

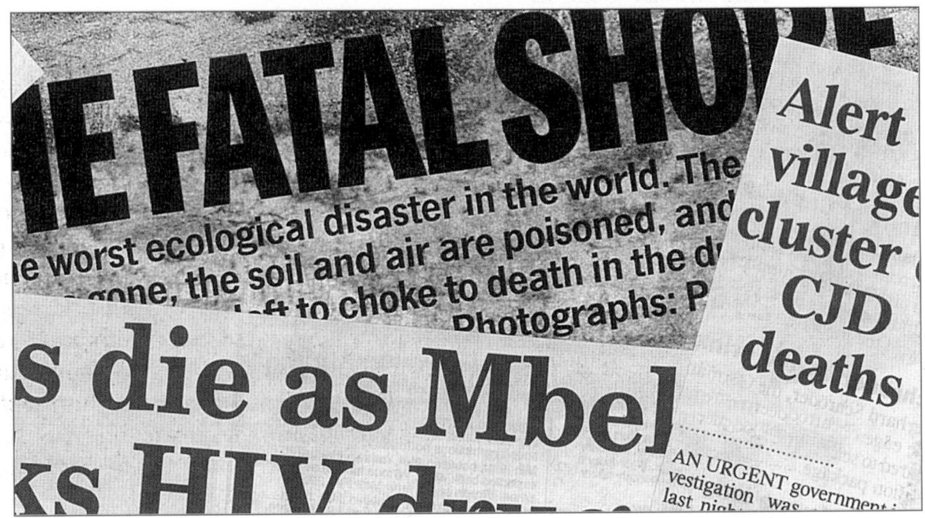

This book is an introduction to the study of biology. It is based on the AQA Specification A Biology AS course. The first four chapters are concerned with the basic biological principles that underpin nearly all the biology that you will study later. They concentrate on providing you with an understanding of cell biology. You will look at proteins and other biological molecules, enzymes, cell structure and the way in which substances get in and out of cells. All these are topics that you will encounter many times in the rest of your A level course.

Physiology is the study of how living organisms work. In Chapters 5 and 6 you will look at the physiology of the blood and gas-exchange systems. These first six chapters cover Module 1 of the specification.

Module 2 emphasises the ways in which humans make use of biology and is covered by the remainder of the book. The different chapters show how different aspects of biology play an important part in our lives. We consider the use of enzymes in biotechnology; genes and gene technology; the ways in which our knowledge of ecology has allowed us to increase crop yield and, finally, how our understanding of physiology has enabled us to control aspects of animal reproduction.

About this book

Before you begin working from this book, it would be a good idea to have a look at the way it has been written. Each chapter has been presented in the same way and shares a number of features.

The chapter opening

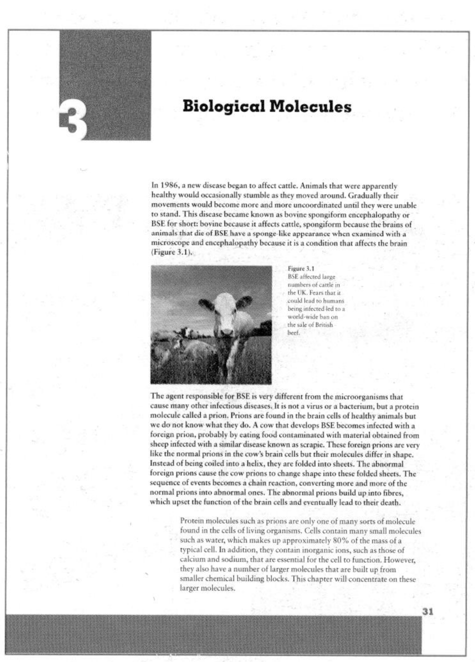

3

Biological Molecules

In 1986, a new disease began to affect cattle. Animals that were apparently healthy would occasionally stumble as they moved around. Gradually their movements would become more and more uncoordinated until they were unable to stand. This disease became known as bovine spongiform encephalopathy or BSE for short: bovine because it affects cattle, spongiform because the brains of animals that die of BSE have a sponge-like appearance when examined with a microscope and encephalopathy because it is a condition that affects the brain (Figure 3.1).

Figure 3.1
BSE affected large numbers of cattle in the UK. Fears that it could lead to humans being infected led to a world-wide ban on the sale of British beef.

The agent responsible for BSE is very different from the microorganisms that cause many other infectious diseases. It is not a virus or a bacterium, but a protein molecule called a prion. Prions are found in the brain cells of healthy animals but we do not know what they do. A cow that develops BSE becomes infected with a foreign prion, probably by eating food contaminated with material obtained from sheep infected with a similar disease known as scrapie. These foreign prions are very like the normal prions in the cow's brain cells but their molecules differ in shape. Instead of being coiled into a helix, they are folded into sheets. The abnormal foreign prions cause the cow prions to change shape into these folded sheets. The sequence of events becomes a chain reaction, converting more and more of the normal prions into abnormal ones. The abnormal prions build up into fibres, which upset the function of the brain cells and eventually lead to their death.

Protein molecules such as prions are only one of many sorts of molecule found in the cells of living organisms. Cells contain many small molecules such as water, which makes up approximately 80% of the mass of a typical cell. In addition, they contain inorganic ions, such as those of calcium and sodium, that are essential for the cell to function. However, they also have a number of larger molecules that are built up from smaller chemical building blocks. This chapter will concentrate on these larger molecules.

31

We have chosen something topical to start each chapter. This is usually an interesting application of biology related to the contents of the chapter. It will help to show you how biology plays an important part in the real world.

The text

The most important word in the study of AS Biology is 'understanding'. If you understand the basic ideas, you will always be able to learn the necessary details later. But if you don't understand the underlying biology, you will find it very difficult to remember the facts. In writing this book, we have kept this in mind and have tried to explain the basic ideas as clearly as possible. Most of you will have come to AS biology from a background of GCSE science so we have avoided introducing unnecessary technical terms. Where these terms are essential, they have been explained carefully in the 'Key Terms' section. The book has been illustrated in colour throughout. The drawings and the information in the captions should provide you with further help in understanding the basic biology.

Text questions

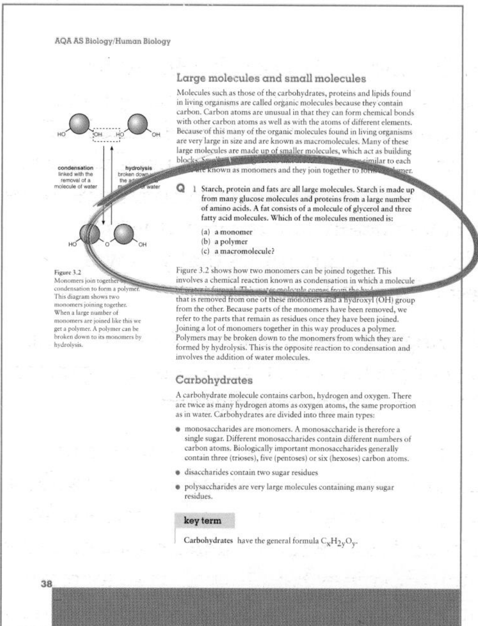

AQA AS Biology/Human Biology

Large molecules and small molecules

Molecules such as those of the carbohydrates, proteins and lipids found in living organisms are called organic molecules because they contain carbon. Carbon atoms are unusual in that they can form chemical bonds with other carbon atoms as well as with the atoms of different elements. Because of this many of the organic molecules found in living organisms are very large in size and are known as macromolecules. Many of these large molecules are made up of smaller molecules, which act as building blocks. [illegible] similar to each [illegible] are known as monomers and they join together to [illegible].

Q 1 **Starch, protein and fats are all large molecules. Starch is made up from many glucose molecules and proteins from a large number of amino acids. A fat consists of a molecule of glycerol and three fatty acid molecules. Which of the molecules mentioned is:**

(a) **a monomer**
(b) **a polymer**
(c) **a macromolecule?**

Figure 3.2
Monomers join together [illegible] condensation to form a polymer. This diagram shows two monomers joining together. When a large number of monomers are joined like this we get a polymer. A polymer can be broken down to its monomers by hydrolysis.

Figure 3.2 shows how two monomers can be joined together. This involves a chemical reaction known as condensation in which a molecule [illegible] that is removed from one of these monomers and a hydroxyl (OH) group from the other. Because parts of the monomers have been removed, we refer to the parts that remain as residues once they have been joined. Joining a lot of monomers together in this way produces a polymer. Polymers may be broken down to the monomers from which they are formed by hydrolysis. This is the opposite reaction to condensation and involves the addition of water molecules.

Carbohydrates

A carbohydrate molecule contains carbon, hydrogen and oxygen. There are twice as many hydrogen atoms as oxygen atoms, the same proportion as in water. Carbohydrates are divided into three main types:

- monosaccharides are monomers. A monosaccharide is therefore a single sugar. Different monosaccharides contain different numbers of carbon atoms. Biologically important monosaccharides generally contain three (trioses), five (pentoses) or six (hexoses) carbon atoms.
- disaccharides contain two sugar residues
- polysaccharides are very large molecules containing many sugar residues.

key term

Carbohydrates have the general formula $C_xH_{2y}O_y$.

38

The text questions should help you to understand what you have just been reading and they are meant to be answered as you go along. Most of them are very straightforward and can be attempted from the information in the paragraph or two which come immediately before. We haven't included any answers. If you get stuck, read the last paragraph again. If you still have problems, make a note of the question and get some help.

Extension boxes

If you restrict yourself to learning the absolute minimum, you should be successful in your AS biology examinations, providing that all you want is a pass. If you want to do better or go on to study the subject at A level, you will need to get a wider understanding of the subject. This is where the Extension boxes should help. They are intended to do two things. Not only will they tell you a little more about topics mentioned in the main text, but they will also help you to begin to understand the links between different aspects of biology.

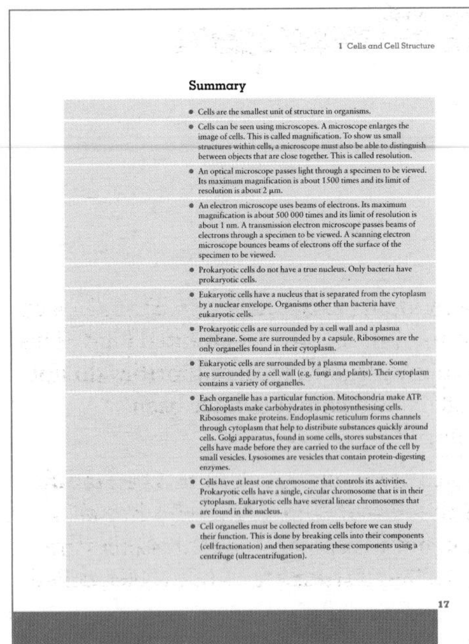

1 Cells and Cell Structure

Summary

- Cells are the smallest unit of structure in organisms.
- Cells can be seen using microscopes. A microscope enlarges the image of cells. This is called magnification. To show us small structures within cells, a microscope must also be able to distinguish between objects that are close together. This is called resolution.
- An optical microscope passes light through a specimen to be viewed. Its maximum magnification is about 1500 times and its limit of resolution is about 2 μm.
- An electron microscope uses beams of electrons. Its maximum magnification is about 500 000 times and its limit of resolution is about 1 nm. A transmission electron microscope passes beams of electrons through a specimen to be viewed. A scanning electron microscope bounces beams of electrons off the surface of the specimen to be viewed.
- Prokaryotic cells do not have a true nucleus. Only bacteria have prokaryotic cells.
- Eukaryotic cells have a nucleus that is separated from the cytoplasm by a nuclear envelope. Organisms other than bacteria have eukaryotic cells.
- Prokaryotic cells are surrounded by a cell wall and a plasma membrane. Some are surrounded by a capsule. Ribosomes are the only organelles found in their cytoplasm.
- Eukaryotic cells are surrounded by a plasma membrane. Some are surrounded by a cell wall (e.g. fungi and plants). Their cytoplasm contains a variety of organelles.
- Each organelle has a particular function. Mitochondria make ATP. Chloroplasts make carbohydrates in photosynthesising cells. Ribosomes make proteins. Endoplasmic reticulum forms channels through cytoplasm that help to distribute substances quickly around cells. Golgi apparatus, found in some cells, stores substances that cells have made before they are carried to the surface of the cell by small vesicles. Lysosomes are vesicles that contain protein-digesting enzymes.
- Cells have at least one chromosome that controls its activities. Prokaryotic cells have a single, circular chromosome that is in their cytoplasm. Eukaryotic cells have several linear chromosomes that are found in the nucleus.
- Cell organelles must be collected from cells before we can study their function. This is done by breaking cells into their components (cell fractionation) and then separating these components using a centrifuge (ultracentrifugation).

17

Summary

At the end of each chapter is a summary. This tells you everything you ought to know when you have finished the work. It will enable you to check that you have done everything that the specification requires.

Examination questions

The AQA examination questions are reproduced or adapted by permission of the Assessment and Qualifications Alliance.

Assignment

Studying biology involves much more than learning facts. It also involves acquiring a range of skills. A biologist should be able to apply facts to new situations; interpret photographs and drawings, graphs and tables, and design experiments. As a biologist, you should also be able to use mathematical skills to carry out a range of simple calculations, and you must be able to use scientific English to communicate your knowledge and ideas effectively. Success in your unit tests will therefore mean that not only will you have understood the basic ideas and learnt the necessary facts, but you will have developed a range of skills. Each chapter in this book finishes with an assignment. The purpose of the text in the chapter is to provide you with the factual information you require. The purpose of the assignment is to help you to master the skills you need.

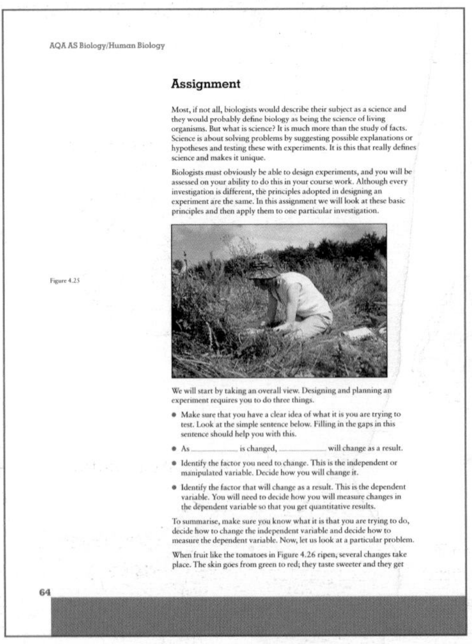

AQA AS Biology/Human Biology

Assignment

Most, if not all, biologists would describe their subject as a science and they would probably define biology as being the science of living organisms. But what is science? It is much more than the study of facts. Science is about solving problems by suggesting possible explanations or hypotheses and testing these with experiments. It is this that really defines science and makes it unique.

Biologists must obviously be able to design experiments, and you will be assessed on your ability to do this in your course work. Although every investigation is different, the principles adopted in designing an experiment are the same. In this assignment we will look at these basic principles and then apply them to one particular investigation.

Figure 4.25

We will start by taking an overall view. Designing and planning an experiment requires you to do three things.

- Make sure that you have a clear idea of what it is you are trying to test. Look at the simple sentence below. Filling in the gaps in this sentence should help you with this.
- As .... is changed, ... will change as a result.
- Identify the factor you need to change. This is the independent or manipulated variable. Decide how you will change it.
- Identify the factor that will change as a result. This is the dependent variable. You will need to decide how you will measure changes in the dependent variable so that you get quantitative results.

To summarise, make sure you know what it is that you are trying to do, decide how to change the independent variable and decide how to measure the dependent variable. Now, let us look at a particular problem.

When fruit like the tomatoes in Figure 4.26 ripen, several changes take place. The skin goes from green to red; they taste sweeter and they get

64

Cells and Cell Structure

key term

Cells
The cell is the basic unit of structure in almost all organisms.

Prokaryotic cells
do not have a nucleus.

Eukaryotic cells
do have a nucleus.

Soil becomes polluted when land is used for dumping chemical waste. From March 2000, local authorities in the UK must now publish the sites of polluted land. This poses a problem. Because some dumping is done illegally, local authorities do not always know when soil has become polluted. New techniques have been developed to identify land that has been polluted.

One of these new techniques uses cells from earthworms. Earthworms live in soil. They eat soil and digest the organic matter that is in it. Unfortunately, they also absorb many of the harmful substances from polluted soil. These substances are destroyed in the cells of an earthworm's body. Small structures within cells, called lysosomes, are involved in this destruction.

In the new technique, a small sample of cells is taken from earthworms. Done properly, this does not harm the earthworms. These cells are then injected with a red dye. Healthy cells are not stained by the red dye because it is removed by the lysosomes. If an earthworm has passed through polluted soil, its lysosomes are not able to remove the injected dye. The dye spills into the rest of the cell, which then turns red.

Scientists can dig up worms at sites to be tested for pollution, remove a few cells from the worms and inject them with red dye. Cells that quickly turn red show that a worm has been burrowing through polluted soil.

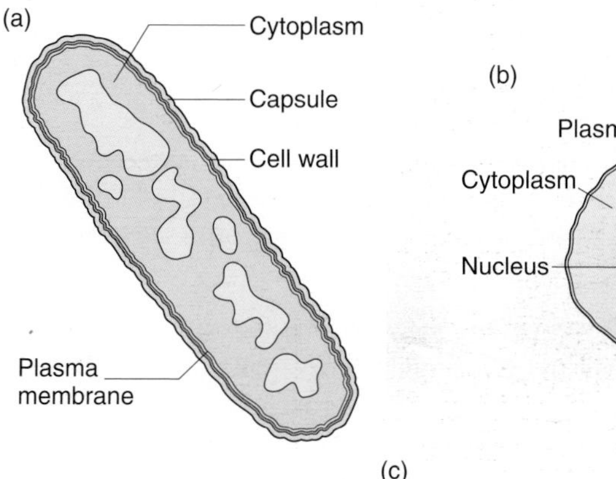

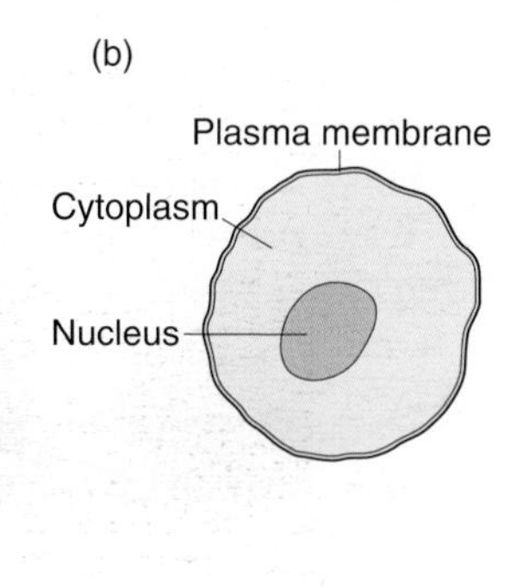

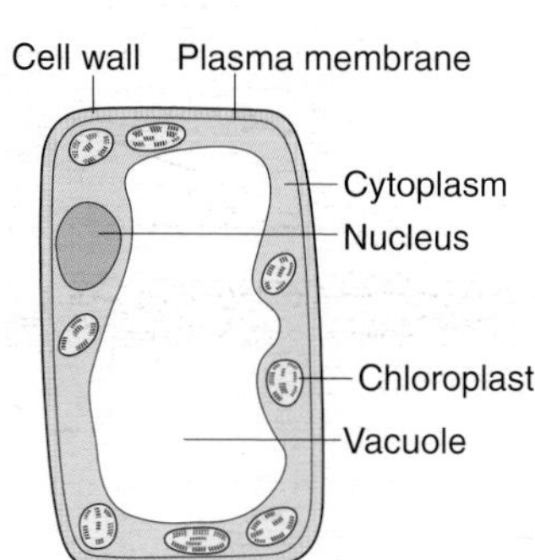

Figure 1.1
(a) A bacterial cell; (b) an animal cell; (c) a plant cell. The bacterial cell does not have a nucleus: it is called a prokaryotic cell. The animal and plant cells, although different from each other, both have a nucleus. They are called eukaryotic cells.

Two types of cell

Living things are called organisms. Almost without exception, they are made of small units called cells. Some organisms are made from only one cell. Bacteria, yeast and the malarial parasite *Plasmodium* are examples of single-celled organisms. Most organisms are made from many cells. From a single fertilised egg cell, your body grew to about one trillion cells by the time you were born.

Figure 1.1 shows three types of cell: a bacterial cell, an animal cell and a plant cell. These cells differ from each other in a number of ways. One of these is crucial: bacterial cells have no nucleus. Because of this, bacterial cells are called **prokaryotic cells** (*pro* means 'before' and *karyon* means 'nucleus'). Animal and plant cells do have a nucleus. They are called **eukaryotic cells** (*eu* means 'true' and *karyon* means 'nucleus').

The main features shown in each of the cells in Figure 1.1 are summarised in Table 1.1.

Feature	Function	Bacterial cell	Animal cell	Plant cell
Capsule	Protects cell from attack, e.g. by human granulocytes (see Chapter 6).	Present	Absent	Absent
Cell wall	Provides strength and stops cell bursting in dilute solutions.	Present (chitin)	Absent	Present (cellulose)
Plasma membrane	Barrier between cell and its surroundings. Controls movement of substances in to, and out of, cell.	Present	Present	Present
Cytoplasm	Performs most of the 'work' of the cell.	Present	Present	Present
Nucleus	Contains the genetic code that controls the cell's activities.	Absent	Present	Present

Table 1.1 The main features of bacterial cells, animal cells and plant cells. Bacterial cells are prokaryotic. Animal and plant cells are eukaryotic.

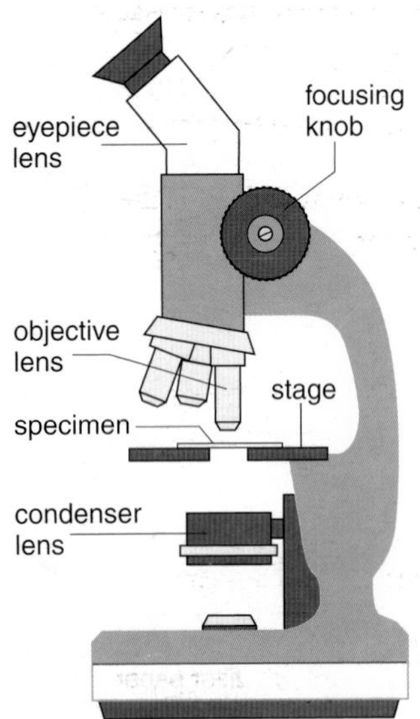

Figure 1.2
An optical microscope. This uses lenses to focus light through a specimen on its stage. This type of microscope can be carried around and is relatively cheap.

Q 1 (a) **Name *one* way in which a plant cell is similar to a prokaryotic cell but different from an animal cell.**

(b) **Name *one* way in which a plant cell is similar to an animal cell but different from a prokaryotic cell.**

The optical microscope

You could not see the detail shown in Figure 1.1 with your naked eye. You would need to use an optical microscope to see the detail of animal and plant cells. Even with an optical microscope, you would see only the shape of a bacterial cell.

You will use an optical microscope during your course of study. Like any other skill, your ability to use an optical microscope will improve with practice. Figure 1.2 shows the sort of optical microscope you are likely to use. It has three groups of lenses: the condenser, the objective and the eyepiece. These magnify the object placed on the stage of the microscope. We can control the magnification using different objective lenses. The coarse and fine focus knobs enable us to focus the microscope.

Preparing temporary mounts

Figure 1.3 shows a thin film of plant cells. This film of cells is easy to obtain. To view such a film of cells with an optical microscope, you need to put it on a glass slide. This is called a mount. Because this mount quickly dries up, it does not last for more than a few hours. It is therefore called a **temporary mount.** Table 1.2 shows the steps you need to follow to prepare a temporary mount in a school or college laboratory.

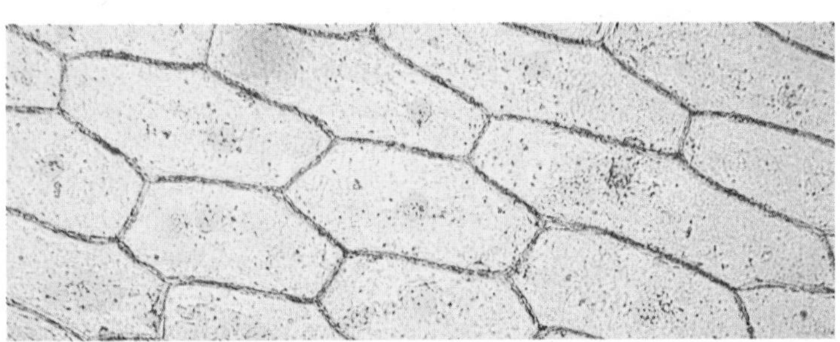

Figure 1.3
A thin film of cells from the epidermis ('skin') of an onion leaf. A stain has been used to show up the cell walls and the nuclei. Without the stain, we would not be able to see these structures clearly.

Procedure	Explanation
Place a drop of water on a glass slide	
Peel a film of cells from the inner surface of the fleshy leaf of an onion Cut a small piece from this film and float it on the drop of water on the slide	Floating flattens the film of cells. Done properly, this ensures that we see a layer of cells that is only one cell thick Float film of cells on drop of water Slide
Use a mounted needle to lower a thin cover slip of glass over the layer of cells Coverslip mounted needle	Covers the temporary mount and slows the evaporation of the water
Place a drop of iodine solution at one edge of the cover slip Put a piece of filter paper at the opposite edge of the cover slip Allow the filter paper to soak up the water from under the cover slip This will draw iodine solution under the cover slip	Iodine will stain parts of the cell Some of the cell's structure will now be easier to see filter paper Iodine solution Iodine solution drawn under coverslip
Remove any surplus water or iodine solution from the slide View the slide under the microscope	You will now see cells with cell walls, cytoplasm and nuclei

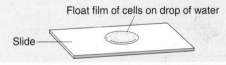

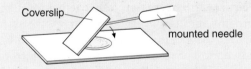

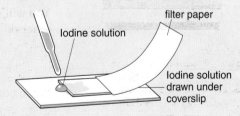

Table 1.2 Stages in the production of a temporary mount. Many different stains can be used to show cell structure. Iodine is just one of these stains.

The electron microscope

Microscopes are used to make objects look bigger. This is called **magnification** and lets us see details of cell structure that cannot be seen with the naked eye. The optical microscope in Figure 1.2 has a low-power objective lens and a high-power objective lens. A typical low-power objective lens magnifies objects to 10 times their real size. The image is further magnified by the eyepiece lens, which usually magnifies 10 times. Using a $10\times$ objective lens and a $10\times$ eyepiece lens means that we see cells 10×10, i.e. 100 times, their real size.

Q 2 How much bigger than their real size would we see cells if we used a $10\times$ eyepiece lens and a $40\times$ objective lens?

key terms

Magnification describes the extent to which an image has been enlarged.

Resolution describes the ability to distinguish between points that are close together. The resolving power of light is about 2 μm (pronounced 'two micrometres'; 1μm is one millionth of a metre, i.e. 10^{-6} m). The resolving power of electrons is about 1 nm (1 nm, pronounced 'one nanometre', is one thousand-millionth of a metre, i.e. 10^{-9} m).

You might expect that a high-power lens would let us see more detail than a low-power lens. Up to a point this is true. However, there comes a point at which increasing the magnification of an optical microscope does not show more detail. To understand this, we need to know something about the physical nature of light.

Light is made from particles (called photons) that travel in a wave. We can only see that two objects are separate from each other if light can pass between them. Light can pass between the letters on this page, so we see the individual letters. When objects are very close together, light cannot pass between them. We cannot see them as separate objects; they appear as a single object. This ability to distinguish between objects is called **resolution**. The size of the wavelength of light limits its **resolution** to about 2 μm.

Electrons have a much smaller wavelength than light. This means that they can pass between objects that are too close together to let light pass between them. In turn, this means that electrons can resolve objects that light cannot resolve. Microscopes that use electrons instead of light are called **electron microscopes**. The resolution of a modern electron microscope is about 1 nm. A **transmission electron microscope** passes beams of electrons through a very thin object. A **scanning electron microscope** bounces beams of electrons off the surface of an object.

Q 3 Explain the difference between magnification and resolution.

Figure 1.4 shows an electron microscope in use. It is much bigger than an optical microscope and so cannot be moved around. Its size is partly caused by:

- the power supply to its electromagnets, which focus beams of electrons
- the pump, which is used to make a vacuum inside its specimen tube.

An electron microscope is also much more expensive than an optical microscope. You are likely to see one only in a university or a scientific research organisation.

Table 1.3 summarises the major differences between optical microscopes and electron microscopes.

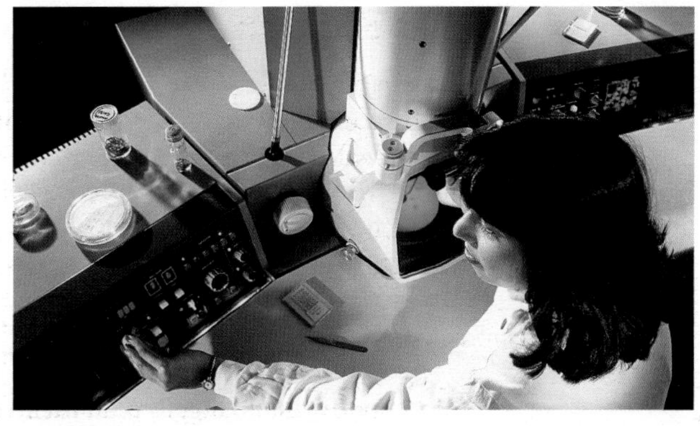

Figure 1.4
A transmission electron microscope. This type of microscope is very expensive and, unlike an optical microscope, cannot be moved. The tube containing the specimen also contains a vacuum to stop electrons being scattered by air molecules.

Feature	Optical microscope	Electron microscope
Radiation used	Beams of light	Beams of electrons
Method of focusing	Glass lenses	Electromagnets
Maximum magnification of specimen being observed	Typical student microscopes magnify up to 400 × the real size of the specimen The maximum magnification of a research microscope is about 1500 ×	About 500 000 × the real size of the specimen
Resolution	About 2 μm	About 1 nm
Presence of vacuum within microscope	Absent Scattering of light by air molecules does not affect image production	Present Scattering of electrons by air molecules interferes with image production
Specimens that can be viewed	Specimens can be alive or dead; If dead, they can be stained Specimens must be relatively thin so that light can pass through them	Specimens are always dead since they cannot survive inside a vacuum **Transmission microscope:** specimens must be extremely thin so that electrons can pass through them **Scanning microscope:** specimens need not be thin since beams of electrons are bounced off their surface

Table 1.3 A comparison of optical and electron microscopes.

Q 4 **Why must electron microscopes contain a vacuum?**

Ultrastructure of eukaryotic cells

We can see cell structure using an optical microscope. An electron microscope shows us a cell's ultrastructure. Figure 1.5 shows the ultrastructure of a epithelial cell. The photograph was taken using the image from a transmission electron microscope. Such a photograph is called an **electron micrograph**. Figure 1.5 shows us much more about a eukaryotic cell than Figure 1.3 does. For example, in Figure 1.5 we can see:

- the plasma membrane, which separates the cell from its surroundings
- a nuclear envelope, which separates the nucleus from the cytoplasm
- many different structures, called **organelles**, within the cytoplasm.

Figure 1.5
Electron micrograph of a epithelial cell.

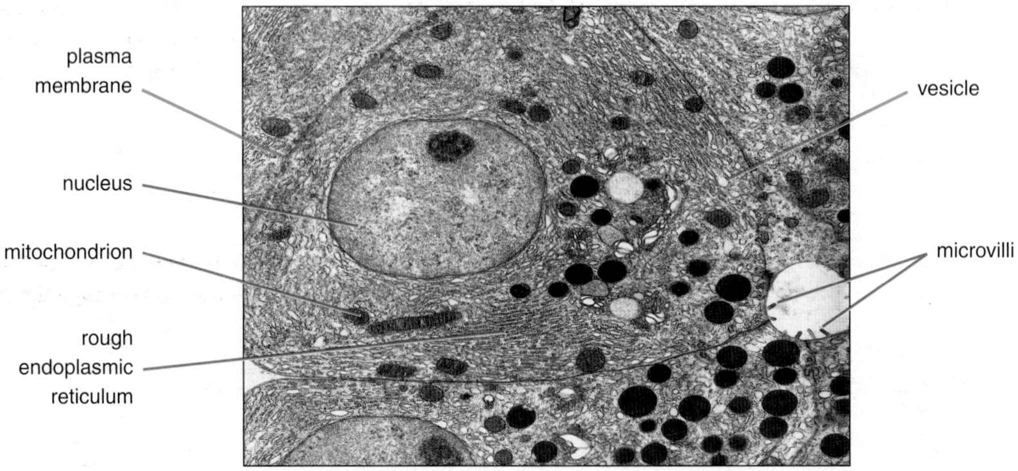

plasma membrane
nucleus
mitochondrion
rough endoplasmic reticulum
vesicle
microvilli

Figure 1.6 is an electron micrograph of a plant cell. It shows many of the same structures you can see in Figure 1.5. It also shows the cell wall and chloroplasts that are not found in animal cells. Table 1.4 summarises the function of the structures shown in Figures 1.5 and 1.6.

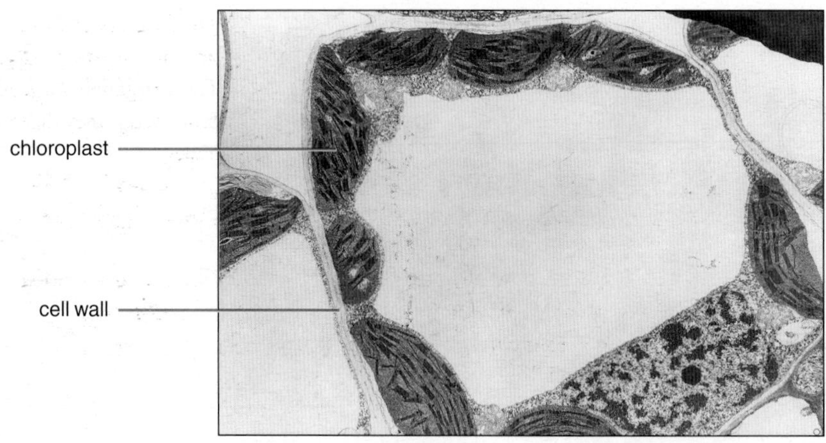

Figure 1.6
Electron micrograph of a mesophyll cell.

Q 5 **Explain why we cannot see cell organelles in Figure 1.3.**

Feature	Description	Function
Plasma membrane	Appears as two dark lines These represent two layers of phospholipids (see Chapter 2)	Separates cell from its surroundings Controls movement of substances in to, and out of, cell Allows cell identification and cell adhesion
Cell wall	Layers of cellulose fibres around plant cells and of chitin around fungal cells Absent from animal cells	Provides mechanical strength and support Stops cell bursting in dilute solutions
Chloroplast	Cigar-shaped organelle found in photosynthesising cells Its inner membrane has folds, called lamellae, which surround a fluid, called the stroma Chlorophyll is found in the lamellae	Traps light energy and uses it to produce carbohydrates from carbon dioxide and water, i.e. photosynthesis
Endoplasmic reticulum (ER)	A series of tubes running through the cytoplasm The tubes are surrounded by membrane Rough ER has ribosomes studded into its membranes; smooth ER has no ribosomes	Speeds up the distribution of substances through the cytoplasm
Golgi apparatus	A stack of flattened sacs surrounded by membrane Small vesicles bud off its edge Not found in plant cells	Stores (and chemically modifies) substances produced in the cell These substances are later secreted using vesicles
Lysosome	A small sphere of liquid surrounded by membrane	The fluid contains powerful protein-digesting enzymes (lysozymes) These organelles are used to digest protein, e.g. they digest the cell after it dies In some white blood cells, they are used to digest bacteria (see Chapter 6)
Microvilli	Microscopic folds in the plasma membrane Only found in some types of epithelial cells	Increase surface area of membrane for absorption

Erratum:the statement “Golgi apparatus is not found in plant cells” should read, Golgi apparatus found in both animal and plant cells.

Table 1.4 The major organelles found in eukaryotic cells (see Figures 1.5 and 1.6). *Continued on next page*

Continued from previous page

Mitochondrion	Cigar-shaped organelle found in all cells Its inner membrane has folds, called cristae, which surround a fluid matrix	Produces adenosine triphosphate (ATP), which cells use as an energy source This is the process of respiration
Nucleus	Contains chromosomes and stores of nucleic acid (in one nucleolus or in several nucleoli) Separated from cytoplasm by a nuclear envelope made of two layers of plasma membrane containing small holes, called nuclear pores	Chromosomes contain genes made of deoxyribonucleic acid (DNA) These genes control the cell's activities
Ribosome	This can be seen only as a small black sphere Sometimes ribosomes are free in the cytoplasm; sometimes they are bound to ER, forming rough ER	Using chemical instructions from the nucleus, ribosomes assemble amino acids to make proteins
Vesicle	A small sphere of liquid surrounded by membrane	See Golgi apparatus

Table 1.4 The major organelles found in eukaryotic cells (see Figures 1.5 and 1.6).

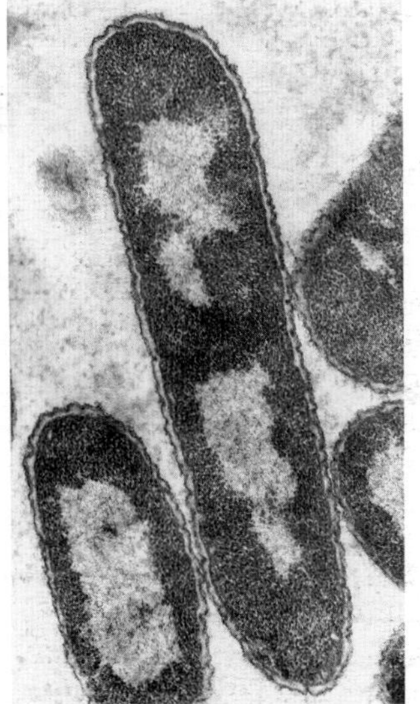

Figure 1.7
Electron micrograph of a bacterial cell. Notice that the genetic material is free in the cytoplasm and there are no cell organelles.

Ultrastructure of prokaryotic cells

Figure 1.7 is an electron micrograph of a bacterial cell. In it, you can see the capsule, cell wall and cytoplasm shown diagrammatically in Figure 1.19. You can also see the plasma membrane and the genetic material. Notice in the prokaryotic cell in Figure 1.7 that:

- the genetic material is not surrounded by a nuclear envelope, as it is in the eukaryotic cells shown in Figures 1.5 and 1.6. Instead, the single, circular chromosome is free in a special region of the cytoplasm.
- the cytoplasm does not contain most of the cell organelles shown in Figures 1.5 and 1.6. The only organelle present is the ribosome. The ribosomes in prokaryotic cells have the same function as, but are smaller than, the ribosomes of eukaryotic cells.

Now that we know the ultrastructure of prokaryotic and eukaryotic cells, we can revisit Table 1.1. At the start of the chapter, we had not learned about cell ultrastructure. We can use our knowledge of cell ultrastructure to improve the comparison of prokaryotic cells and eukaryotic cells given in Table 1.1. This has been done in Table 1.5.
You should remember the information in Table 1.5 if asked to compare prokaryotic and eukaryotic cells in a Unit Test.

Feature	Prokaryotic cell	Eukaryotic cell
Capsule	Present	Absent
Cell wall	Present (chitin)	Present in fungi (chitin) and in plants (cellulose) Absent in animals
Plasma membrane	Present	Present
Cytoplasm	Present	Present
• ribosomes	• small ribosomes are present – they are always free in the cytoplasm	• larger ribosomes are present – some are free in the cytoplasm and some are attached to rough ER
• other organelles	• absent	• present – they include chloroplasts, ER, Golgi apparatus, lysosomes and mitochondria
Nucleus	Absent	Present
• nuclear envelope	• absent	• present
• nucleoli	• absent	• present
• chromosomes	• single and circular	• many and linear

Table 1.5 A fuller comparison of prokaryotic and eukaryotic cells than that given in Table 1.1

Cell fractionation and ultracentrifugation: separating cell organelles

You might wonder how biologists have been able to work out the function of organelles, given their size. This is done by collecting a sample containing large numbers of only one type of organelle. The function of the organelles in such a sample can then be studied. Samples of only one organelle can be obtained by the process of cell fractionation followed by the process of cell ultracentrifugation.

Cell fractionation breaks up (homogenises) cells. This can be done using a pestle and mortar. (You might use a pestle and mortar at home, for example to grind spices together.) Electric blenders and homogenisers are more likely to be used in research laboratories for breaking up cells. After breaking up the cells, a fluid mixture called a **homogenate** is obtained. This homogenate is filtered to remove bits of cells that have not been broken up properly. The filtrate from this process contains a mixture of cell organelles. We now need to separate the different organelles in this mixture.

Q 6 Whilst being homogenised, cells are kept in a solution that:
(a) has the same water potential as the cells
(b) has a constant pH
(c) is cold.
Suggest a reason for each of (a), (b) and (c).

Ultracentrifugation separates the components of cells that have been broken up. The separation is done using a centrifuge. This is an instrument that spins tubes of fluid around at high speeds. As they spin, solid particles in the fluid are thrown to the bottom of the tube, forming a pellet. We need to produce pellets with only one type of organelle. We can ensure which type of organelle is found in the pellet by controlling:

- the speed at which the centrifuge spins
- the time for which the centrifuge spins.

The flowchart in Figure 1.8 summarises the stages of cell fractionation and ultracentrifugation.

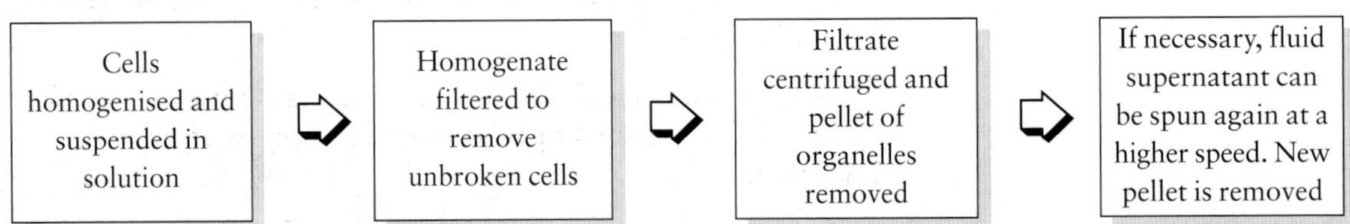

Figure 1.8
The stages of cell fractionation and ultracentrifugation.

Table 1.6 shows the times and speeds used to produce pure pellets containing different types of organelle. As you might expect, the larger components form pellets at slower speeds than smaller components.

Speed of centrifugation (*g*)	Time of centrifugation (minutes)	Organelles in pellet
500 to 1000	10	Nuclei and chloroplasts
10 000 to 20 000	20	Mitochondria and lysosomes
100 000	60	Rough ER and ribosomes

Table 1.6 Ultracentrifugation of cell homogenate to obtain samples of organelles (g is the acceleration due to gravity, which has a value of 9.8 m s^{-2})

Q 7 Which organelles would remain in the supernatant if an animal cell homogenate was centrifuged at 800g for 10 minutes? How would you obtain a pure sample of the smallest of these organelles?

Summary

- Cells are the smallest unit of structure in organisms.
- Cells can be seen using microscopes. A microscope enlarges the image of cells. This is called magnification. To show us small structures within cells, a microscope must also be able to distinguish between objects that are close together. This is called resolution.
- An optical microscope passes light through a specimen to be viewed. Its maximum magnification is about 1500 times and its limit of resolution is about 2 μm.
- An electron microscope uses beams of electrons. Its maximum magnification is about 500 000 times and its limit of resolution is about 1 nm. A transmission electron microscope passes beams of electrons through a specimen to be viewed. A scanning electron microscope bounces beams of electrons off the surface of the specimen to be viewed.
- Prokaryotic cells do not have a true nucleus. Only bacteria have prokaryotic cells.
- Eukaryotic cells have a nucleus that is separated from the cytoplasm by a nuclear envelope. Organisms other than bacteria have eukaryotic cells.
- Prokaryotic cells are surrounded by a cell wall and a plasma membrane. Some are surrounded by a capsule. Ribosomes are the only organelles found in their cytoplasm.
- Eukaryotic cells are surrounded by a plasma membrane. Some are surrounded by a cell wall (e.g. fungi and plants). Their cytoplasm contains a variety of organelles.
- Each organelle has a particular function. Mitochondria make ATP. Chloroplasts make carbohydrates in photosynthesising cells. Ribosomes make proteins. Endoplasmic reticulum forms channels through cytoplasm that help to distribute substances quickly around cells. Golgi apparatus, found in some cells, stores substances that cells have made before they are carried to the surface of the cell by small vesicles. Lysosomes are vesicles that contain protein-digesting enzymes.
- Cells have at least one chromosome that controls its activities. Prokaryotic cells have a single, circular chromosome that is in their cytoplasm. Eukaryotic cells have several linear chromosomes that are found in the nucleus.
- Cell organelles must be collected from cells before we can study their function. This is done by breaking cells into their components (cell fractionation) and then separating these components using a centrifuge (ultracentrifugation).

Examination questions

1 (a) The table shows some features of cells. Complete the table with ticks to show those features that are present in an epithelial cell lining the small intestine and those which may be present in a prokaryotic cell

Feature	Epithelial cell from small intestine	Prokaryotic cell
Golgi apparatus		
Mitochondrion		
Nuclear envelope		
Plasmid		
Ribosome		

(2 marks)

(b) (i) Explain why it is possible to see the detailed structure of a cell with an electron microscope but not with an optical microscope.

(2 marks)

(ii) Care must be taken when interpreting electronmicrographs. Some features seen using an electron microscope might not be present in the living cell. Suggest an explanation for this observation.

(1 mark)

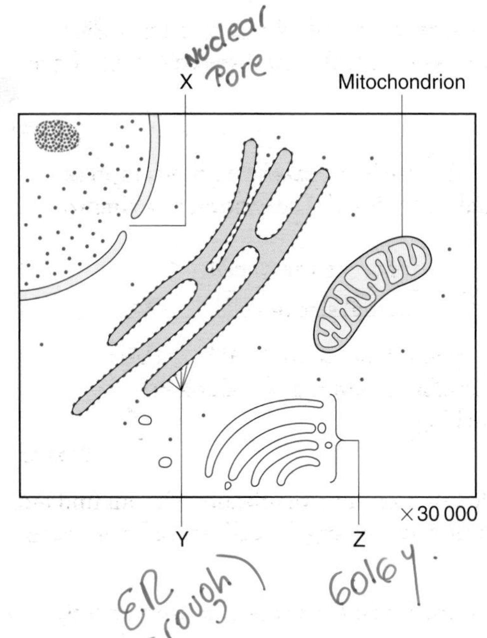

2 The drawing shows part of an animal cell.

(a) Name feature X.

(1 mark)

(b) Describe the function of organelle Y.

(1 mark)

(c) Describe one way in which the function of organelle Z is related to the function of organelle Y.

(1 mark)

Calculate the actual length of the mitochondrion in millimetres. Show your working.

(2 marks)

Assignment

Once you have read this chapter you will appreciate that although cells are very small, we have many tools and techniques available to examine them and find out more about their structure. But we have a problem. Almost everything we do to a cell alters it in some way. Because we must be sure that what we are looking at through a microscope is what we see in real life, biologists need to be able to interpret images correctly. In this assignment we shall look at one very specialised cell, a human red blood cell. We shall look at what we can find out using different types of microscope and show how we need to be careful in interpreting what we see.

In order to look at blood cells with an optical microscope, we can make a blood smear. The technique for doing this is shown in Figure 1.9. The photograph in Figure 1.10 shows the appearance of some of the red blood cells from this smear when seen with an optical microscope. Look at these cells carefully and then answer the following questions.

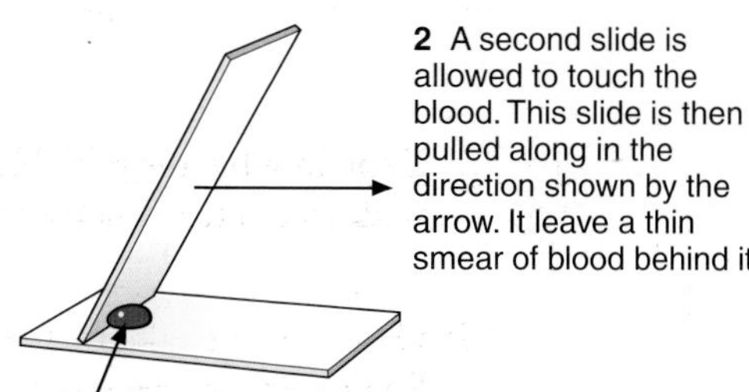

2 A second slide is allowed to touch the blood. This slide is then pulled along in the direction shown by the arrow. It leave a thin smear of blood behind it.
1 A small drop of blood is placed at one end of a microscope slide.

Figure 1.9
Making a thin smear of blood for examining with an optical microscope.

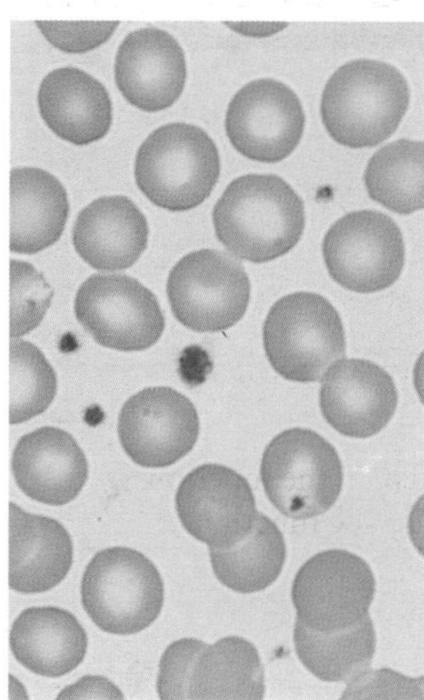

Figure 1.10
Red blood cells seen with an optical microscope.

1 The shape of a red blood cell is often described as a biconcave disc. What is the evidence from this photograph that these cells could have a biconcave shape?

(*1 mark*)

2 The magnification of a cell is the size it appears in a photograph or drawing, divided by its actual size. This can be written as a simple formula:

$$\text{magnification} = \frac{\text{apparent size of cell}}{\text{actual size of cell}}$$

The magnification of the photograph in Figure 1.10 is $\times$ 1000. Calculate the actual diameter of a human red blood cell in micrometres. Show your working.

(*2 marks*)

A red blood cell is surrounded by a plasma membrane. We can find out something about this membrane by treating the cells in different ways before examining them.

3 When the slide is waved backwards and forwards and dried rapidly before it is examined, the cells appear round and smooth, as in the photograph. When, however, the slide is left on the bench to dry slowly, the cells appear smaller and crinkled around their edges.

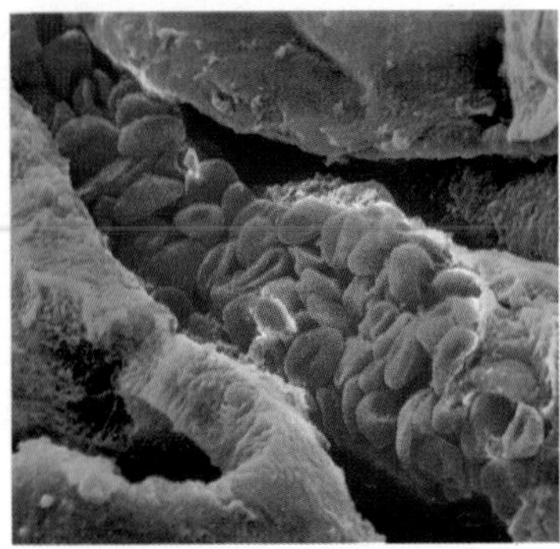

Figure 1.11
Red blood cells seen with a transmission electron microscope.

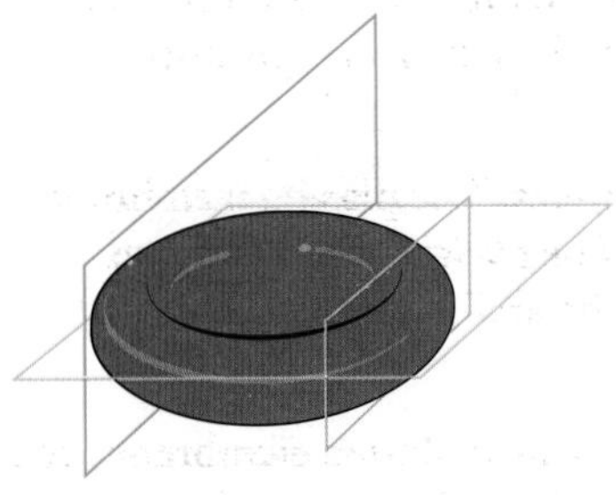

Figure 1.12
Sections have been cut through this red blood cell in different planes.

Suggest an explanation for the different appearance of the cells when they have dried out slowly.

(*2 marks*)

4 Detergents dissolve lipids. If a drop of detergent is added to the blood on the slide before it is examined, no cells can be seen at all. What does this suggest about the plasma membrane surrounding the red blood cell?

(*1 mark*)

Using an optical microscope of the sort found in a school or college laboratory, it is possible to find out a lot about red blood cells and the membranes that surround them. We can find out even more if we examine them with an electron microscope.

Look at Figure 1.11, which shows red blood cells as seen with a transmission electron microscope. In a transmission electron microscope, electrons pass through the specimen. It is not possible to look at a whole cell because it is much too thick to allow the electrons to pass through, so a thin section must be cut. Figure 1.12 shows a red blood cell that has been cut through in different planes.

5 (a) Make simple drawings to show the appearance of the cell if it were cut through each of the planes shown in Figure 1.12.

(*3 marks*)

(b) Suggest an explanation for the different shapes of the red blood cells shown in Figure 1.11.

(*1 mark*)

6 What evidence is there that the photograph in Figure 1.11 has been taken through an electron microscope?

(*1 mark*)

7 Rewrite the following table, matching each conclusion with the appropriate piece of evidence.

Conclusion	Evidence
The plasma membrane allows water molecules to pass through it	There are no mitochondria present
Many of the molecules in the cytoplasm are of the same type	The appearance of the cells when seen with a scanning electron microscope
Red blood cells do not use oxygen and cannot respire aerobically	The red blood cells in a smear that is left to dry slowly appear smaller and crinkled around the edges
Red blood cells are biconcave in shape	There is no nucleus present
Red blood cells do not contain DNA, so they cannot make proteins	The red blood cells in Figure 1.11 are a uniform dark colour

(*3 marks*)

2 Getting In and Out of Cells

Human cells need a constant supply of glucose to remain healthy. People who suffer from diabetes mellitus are unable to keep a constant supply of glucose in their cells. These people are called diabetics.

The British Diabetics Association estimates that 1.4 million people in the UK have been diagnosed with diabetes mellitus. The Association also estimates that a further 1 million people in the UK have diabetes mellitus but do not yet realise it.

Keeping a constant supply of glucose within cells is a complex process. Part of this process depends on a hormone called insulin. Many people are diabetic because they cannot produce insulin: they suffer from Type 1 diabetes. The boy in Figure 2.1 suffers from Type 1 diabetes. He has to inject himself regularly with insulin because he cannot produce it. Over 75% of diabetics, however, can produce insulin: they suffer from Type 2 diabetes.

Type 2 diabetes is caused by faults in plasma membranes. The plasma membranes of some sufferers can no longer 'recognise' insulin. The plasma membranes of other sufferers can 'recognise' insulin but have lost the ability to increase the movement of glucose through themselves.

The types of diabetes mellitus illustrate three ways in which plasma membranes act. In the rest of this chapter you will learn how cells use their plasma membranes to:

- 'recognise' chemical messengers, such as insulin
- control the movement of substances into the cell, such as glucose
- release substances that cells have made, such as insulin.

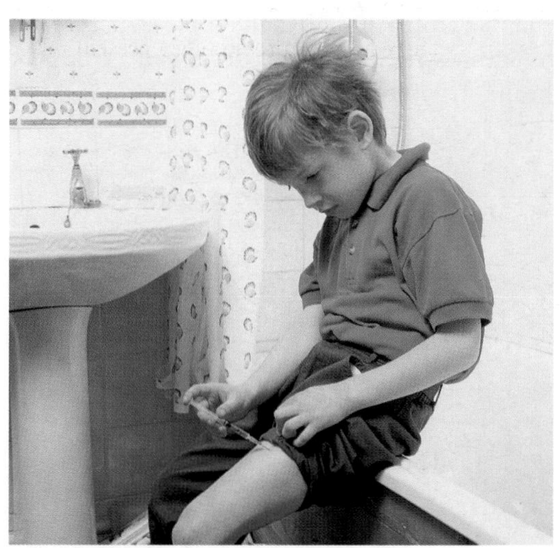

Cells and their surroundings

All cells are surrounded by a plasma membrane. A cell must get the raw materials it needs from its surroundings through its plasma membrane. A cell must also excrete the waste products it has made into the cell's surroundings through the plasma membrane.

Figure 2.1
This boy cannot produce a hormone called insulin. If he is to stay alive, he must inject himself with insulin each day.

Q 1 **Name one raw material that every cell in your body must have and one waste product that they all must excrete.**

Substances get through membranes in a number of ways. These are:

- diffusion
- facilitated diffusion
- osmosis
- active transport
- endocytosis
- exocytosis.

Before we can understand these processes, we must know something about the structure of the membrane itself.

Structure of plasma membrane

Membrane occurs inside cells as well as around them. Look at Figure 2.2, which shows a simplified plant cell. Its membranes are shown in blue. Oxygen molecules are produced inside chloroplasts, e.g. at the point labelled A in Figure 2.2. Work out from the diagram how many layers of membrane a molecule of oxygen must pass through to get out of the cell from the point labelled A.

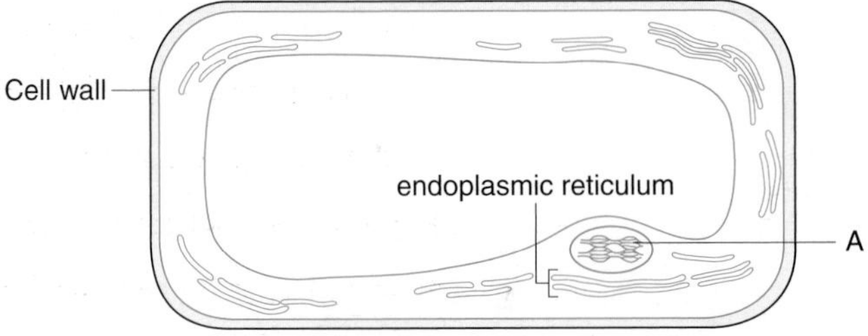

Figure 2.2
A simplified plant cell with the membrane shown in blue. Through how many layers of membrane must a molecule of oxygen produced at point A pass in order to get out of the cell?

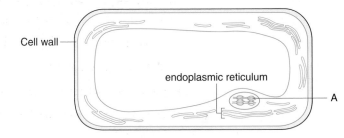

All the blue lines in Figure 2.2 represent membrane. Whether membrane is at the surface of the cell or in its cytoplasm, it has the same structure. Figure 2.3 shows what we think this structure is. It is described as the **fluid-mosaic model** of membrane structure.

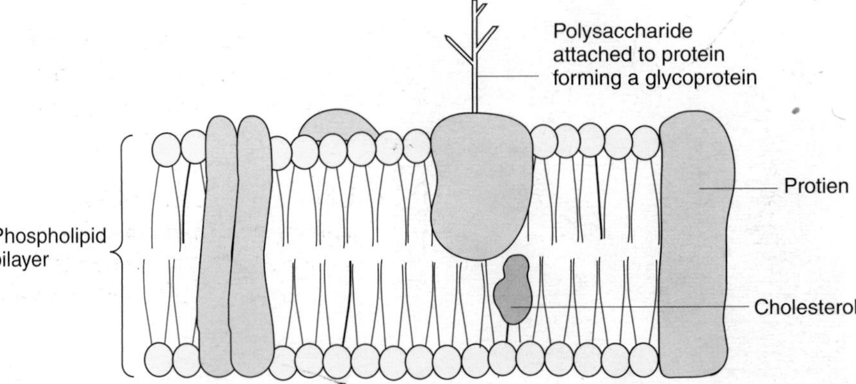

Figure 2.3
The fluid-mosaic model of membrane structure.

key term

Fluid-mosaic model

A membrane is a phospholipid bilayer studded with proteins, polysaccharides and other sorts of lipids. This patchwork of molecules is like a mosaic. Because the molecules move about within their respective layer, the membrane behaves like a fluid. This is why the model is called the fluid-mosaic model.

Q 2 Use the information in Figure 2.2 to suggest why we cannot see the structure of a membrane and need to produce a model of its probable structure.

A membrane is a **phospholipid bilayer**. This means that it has two layers of molecules called phospholipids. Each of these phospholipid molecules has two parts:

- a 'head' that will mix with water but not with fat (i.e. it is **hydrophilic**)
- two 'tails' that will mix with fat but not with water (i.e. they are **hydrophobic**).

In the phospholipid bilayer, the hydrophilic heads are always on the outside of the membrane. The hydrophobic tails are always on the inside of the membrane. Alone, this arrangement of phospholipids would form a barrier to water and to water-soluble substances. However, other molecules are scattered among the phospholipids. These include lipids (including cholesterol in the membranes of animals), proteins and polysaccharides. You will learn more about the structure of these substances in Chapter 3.

Q 3 Suggest why the hydrophilic heads are always on the outside of a phospholipid bilayer and the hydrophobic tails are always on the inside.

The proteins in membranes are of special interest to us. This is because they have a number of important functions. Proteins function as:

- **carriers** for water-soluble molecules (such as glucose)
- **channels** for ions (such as sodium and chloride ions)
- **pumps**, which use energy to move water-soluble molecules and ions
- **receptors**, which enable hormones and nerve transmitters to bind to specific cells
- **recognition sites**, which identify a cell as being of a particular type
- **enzymes**, which speed up chemical reactions at the edge of the membrane
- **adhesion sites**, which help some cells to stick together.

Polysaccharides are bound to some of the lipids and proteins studded in the phospholipid bilayer. Proteins that are bound to polysaccharides are called glycoproteins. It is these glycoproteins in the membrane that act as receptors and recognition sites.

Q 4 Cholesterol, found in animal cell membranes, reduces the sideways movement of phospholipid molecules within a membrane. Suggest why the presence of cholesterol might reduce the rate of movement of molecules through a membrane.

Diffusion

Because molecules and ions have inherent energy, they constantly move about. Figure 2.4 shows the effects of the movement of ink molecules and water molecules. In beaker A you can see a droplet of ink that has just been placed in water. Beaker A illustrates an important concept: that of the **concentration gradient**. 'Concentration' refers to the number of molecules or ions in a specified volume. 'Gradient' shows that one region contains more molecules or ions than a neighbouring region. In beaker A, there are more molecules of ink in the droplet than in the surrounding water. As a result, a concentration gradient occurs between the region containing the droplet and the surrounding region containing only water.

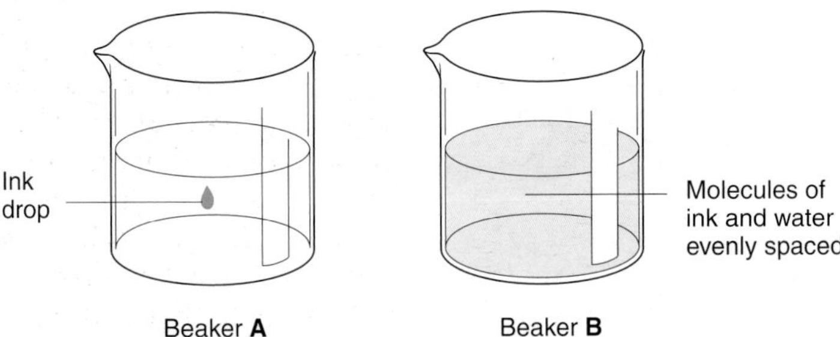

Figure 2.4
The effect of diffusion. A drop of ink was placed in the left-hand beaker of water. After a few hours, movement of ink molecules and water molecules caused the appearance of the right-hand beaker. By now, the molecules of ink and water are evenly spaced throughout the ink solution.

key term

Concentration gradient

A concentration gradient occurs when one region contains more of a molecule or ion than a neighbouring region. Molecules and ions move down a concentration gradient. When the concentration difference is large, the net movement of molecules or ions down the gradient is faster than when the concentration difference is small.

Beaker B in Figure 2.4 shows what happened when beaker A was left untouched. After a few hours, movement of the ink molecules has spread them evenly throughout the water of the beaker. During this time, ink molecules moved in all directions and they continue to do so. However, because they are now evenly spread out, more must have moved from where they were concentrated in the droplet to where they were less concentrated in the water. In other words, their net movement has been down the concentration gradient. This is the process of diffusion.

key term

Diffusion

Diffusion is the net movement of molecules or ions from a region of higher concentration to a region of lower concentration. (This can also be described as the net movement of molecules or ions down a concentration gradient.)

Extension box 1

More about gradients

In the ink solution described in the text, the dissolved ink is called the **solute** and the water in which the ink is dissolved is called the **solvent.** The cytoplasm of cells is mainly water, so water is always the solvent in cell solutions. However, there are many different solutes in cells.

If more than one solute is present in a solution, each solute still diffuses along its own concentration gradient. For example, in a human liver cell it is likely that glucose molecules will diffuse into the cell along a glucose concentration gradient. At the same time it is likely that urea molecules will diffuse out of the cell along a urea concentration gradient. The two concentration gradients do not interfere with each other.

The process of diffusion described so far applies to uncharged molecules. When a solute is charged, as ions are, diffusion occurs along an **electric gradient**. An electric gradient is caused by two adjacent regions having different electrical charges. If the electric gradient is strong enough, diffusion of ions can occur even against their own concentration gradient. Processes such as the flow of impulses in your nervous system depend on the balance of diffusion along both electric and concentration gradients.

Differences in pressure result in pressure gradients. These gradients can also influence the speed and direction in which solute molecules diffuse. You will see in Chapter 6 how the movement of molecules and ions between blood capillaries and the cells they supply is affected by pressure gradients.

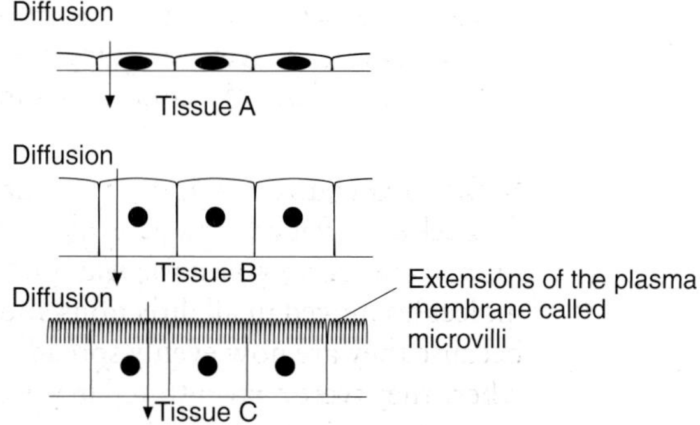

Figure 2.5
Cells of the same type form tissues. Diffusion across the thin cells in tissue A will be faster than across the thicker cells in tissue B. Although they are the same shape as cells in tissue B, the extension of the membrane of cells in tissue C increases their surface area. What effect will this have on the rate of diffusion across C compared with B?

The rate of diffusion is affected by a number of factors. One of these, differences in concentration, has already been described above. In biology, we are usually interested in the rate of diffusion across plasma membranes or across epithelia (see Chapter 5). These are often called **exchange surfaces.** Two further factors influence the rate of diffusion across exchange surfaces:

- the surface area across which diffusion occurs
- the thickness of the surface.

Figure 2.5 shows three groups of cells that line various parts of your body. The cells in A are thinner than those in B. As a result, diffusion will normally be faster across A than across B. The cells in C are similar in shape to those in B. However, the cells in C have microscopic extensions of their plasma membrane. As a result of this increase in surface area, diffusion will normally be faster across C than across B.

Q 5 Under what conditions might the diffusion rate across B in Figure 2.5 be faster than across A? Explain your answer.

We now have three factors affecting the rate of diffusion. Diffusion through an exchange surface is:

- faster when the difference in concentration is high (i.e. the concentration gradient is steep)
- faster when the exchange surface has a large area
- slower when the exchange surface is thick.

The relationship between these three factors is described by Fick's law. Although this law is shown as an equation, you will not be expected to use it to perform calculations.

key term

Fick's law The rate of diffusion through an exchange surface is proportional to:

$$\frac{\text{surface area} \times \text{difference in concentration}}{\text{thickness of surface}}$$

Q 6 Use Fick's law to explain the rates of diffusion across the exchange surfaces in Figure 2.5.

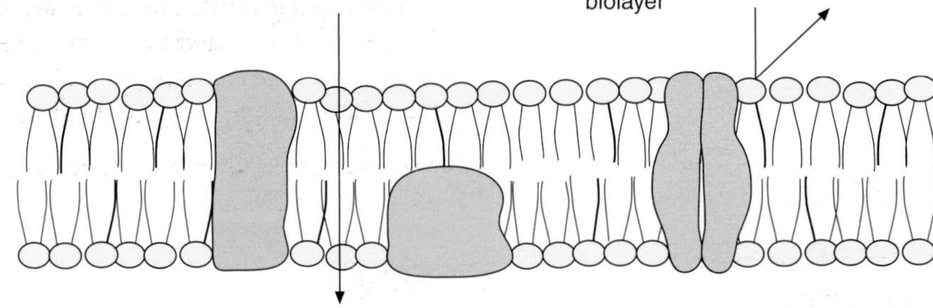
Oxygen, carbon dioxide and small uncharged molecules diffuse through phospholipid bilayer
Glucose, large water-soluble molecules and charged ions cannot pass through phospholipid biolayer

Figure 2.6
Oxygen, carbon dioxide and other small uncharged molecules can diffuse freely across the phospholipid bilayer of a plasma membrane. Glucose, other large water-soluble molecules and charged ions cannot.

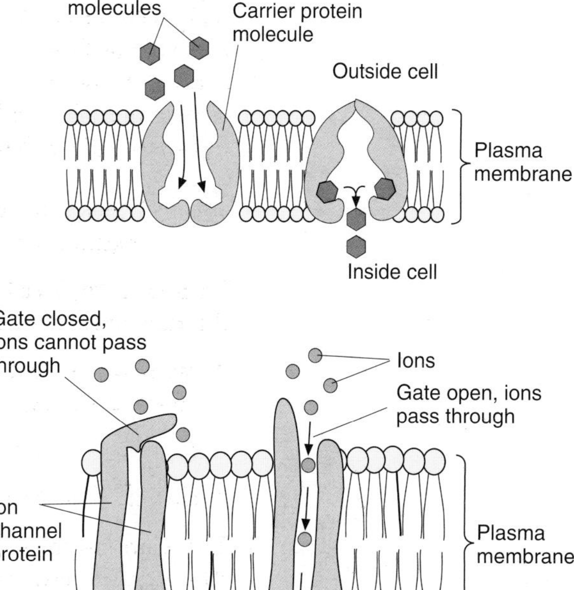

Figure 2.7
The two types of protein involved in facilitated diffusion. Carrier proteins bind to the diffusing molecule and take it through the membrane by changing shape. Ion channels help the diffusion of charged ions. Note that some ion channels have gates that open and close.

Q 7 **Suggest why (a) lipid-soluble molecules and (b) small molecules can diffuse through plasma membranes.**

Facilitated diffusion

Oxygen and carbon dioxide are small, uncharged molecules. They diffuse freely across the phospholipid bilayer of a cell's plasma membrane (Figure 2.6). We could say that the membrane is no barrier to their diffusion.

In contrast, large molecules, such as glucose, almost never diffuse across the bilayer. Neither do charged ions. Instead, these substances must cross the membrane through proteins that span both sides of the bilayer. Figure 2.7 shows the two types of protein that are involved in facilitated diffusion:

- **carrier proteins** bind to a specific type of diffusing molecule. When they have done so, they change shape, releasing the diffusing molecule at the other side of the membrane.
- **ion channels** are formed by proteins with a central pore that is lined with charged groups. The diffusion of charged particles, such as Ca^{2+}, Na^+, K^+, HCO_3^- and Cl^- ions, is helped by these ion channels. Some ion channels are gated (Figure 2.7) and so can open or close. These gates allow cells to regulate the flow of ions from one cell to another.

Note that, like simple diffusion, neither of these processes involves the use of energy by the cell.

Osmosis

We saw earlier in this chapter that a plasma membrane lets molecules and ions pass through itself. However, many types of large molecule cannot diffuse through a plasma membrane. Because the membrane only allows the passage of some molecules, it is called **partially permeable.**

key term

Partially permeable membrane

A partially permeable membrane will only let small molecules pass through it.

The cytoplasm of all cells contains large numbers of protein molecules that cannot diffuse out of the cells. These molecules become concentrated in the cytoplasm of the cell. As a result, a protein concentration gradient exists across cell surface membranes. This gradient affects the diffusion of water molecules across membranes. Figure 2.8 shows how this happens.

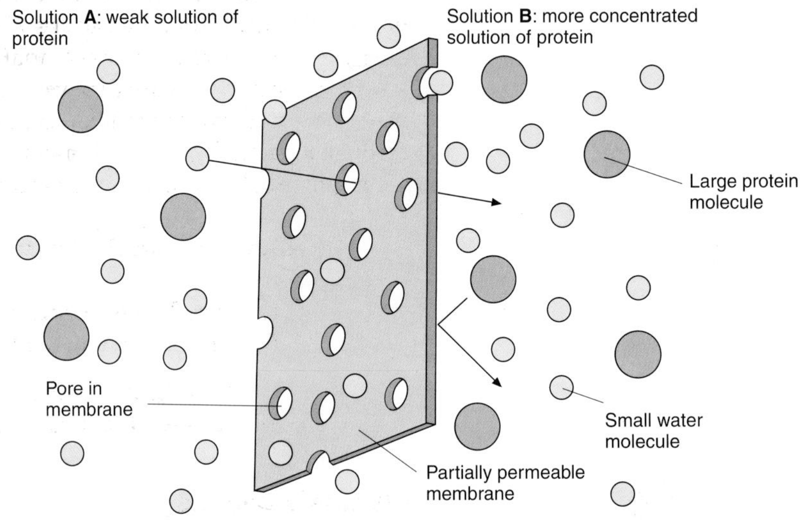

Figure 2.8
A model to show osmosis. Solution A is a weak solution of protein. Its water potential has a negative value, −3 kPa. This solution has a low concentration of protein molecules but a high concentration of water molecules. Solution B is a more concentrated protein solution. Its water potential has a more negative value than solution A, −7 kPa. Solution B has a higher concentration of protein molecules but a lower concentration of water molecules than solution A. Solution A is separated from solution B by a partially permeable membrane. As a result of osmosis, there will be a net movement of water from solution A to solution B through the partially permeable membrane.

On the left of Figure 2.8 there is a weak protein solution. On the right of Figure 2.8 is a stronger protein solution. The two are separated by a partially permeable membrane. Because of their inherent energy, water molecules will constantly move about. However, there will be a net movement of water molecules from the weaker solution to the stronger solution. In other words, water molecules will diffuse along the water concentration gradient. The protein molecules will also move about. They cannot diffuse along their concentration gradient because they cannot pass through the partially permeable membrane. This movement of water molecules happens with other solutes as well as protein and is called osmosis. In osmosis, there is a net movement of water molecules through a partially permeable membrane from where they are concentrated (i.e. a weak solution) to where they are more concentrated (i.e. a concentrated solution).

key term

Osmosis

Osmosis relates to the diffusion of water molecules. It is the net movement of water through a partially permeable membrane from a solution of less negative water potential to a solution of more negative water potential. (The water potential of pure water is zero. All solutions have a water potential with a negative value.)

Water potential

We can describe the movement of water molecules in Figure 2.8 in another way. This involves the use of the term **water potential**. Water potential is the ability of a solution to absorb water molecules by osmosis. It is measured in units of pressure called kilopascals (kPa). Pure water cannot absorb any more water by osmosis: it has a water potential of zero (0 kPa). A solution can absorb more water by osmosis. A solution always has a negative water potential (e.g. −6 kPa). The stronger the solution, the more negative its water potential. Using the concept of water potential, we can redefine osmosis as the net movement of water through a partially permeable membrane from a solution of less negative water potential to a solution of more negative water potential.

Q 8 **In each case, state the direction in which water will move by osmosis.**

(a) **solution P with a water potential of −1 kPa and solution Q with a water potential of −4 kPa**

(b) **solution R with a water potential of −1 kPa and solution S with a water potential of 0 kPa.**

Osmosis in cells

Osmosis has a significant effect on cells. Figure 2.9 shows what happens to human red blood cells when they are placed in solutions with different water potentials. In a solution with a more negative water potential than themselves, they lose water by osmosis and shrink. In a solution with a less negative water potential than themselves, they gain water by osmosis and eventually burst. Figure 2.10 shows that a similar process happens in plant cells when they are placed in solutions with different water potentials. Unlike red blood cells, plant cells have a cellulose wall. Because this wall is strong, it pushes on expanding cells. This stops them bursting in a solution of less negative water potential than their own cytoplasm.

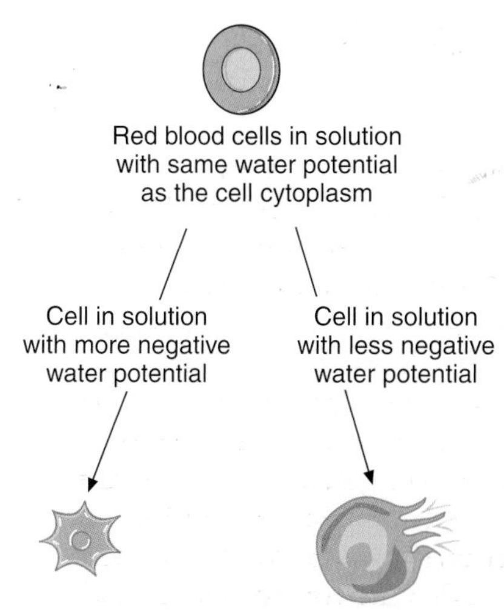
Red blood cells in solution with same water potential as the cell cytoplasm
Cell in solution with more negative water potential
Cell in solution with less negative water potential
Cell smaller and appears 'crinkled'
Cell swells and bursts

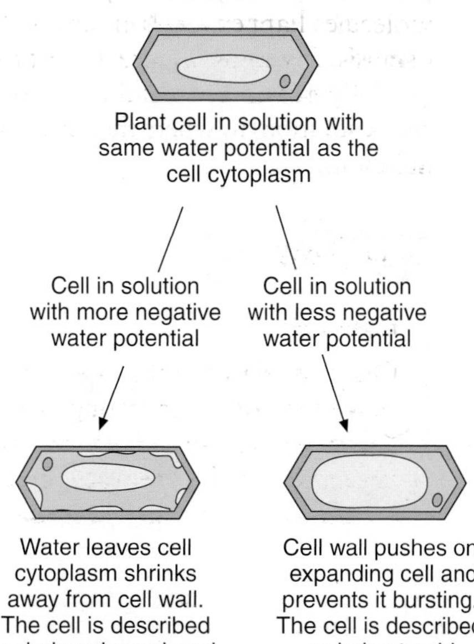
Plant cell in solution with same water potential as the cell cytoplasm
Cell in solution with more negative water potential
Cell in solution with less negative water potential
Water leaves cell cytoplasm shrinks away from cell wall. The cell is described as being plasmolysed
Cell wall pushes on expanding cell and prevents it bursting. The cell is described as being turgid

Figure 2.9 (above)
The effects of osmosis on human red blood cells. All animal cells are affected in this way.

Figure 2.10 (right)
The effects of osmosis on plant cells. Because the wall is able to press onto the expanding cell it stops these cells bursting in solutions with a less negative water potential than their own cytoplasm.

key term

Active transport

Active transport is the net movement of solute against a concentration gradient. The process always involves the use of energy by cells.

Active transport

Diffusion, facilitated diffusion and osmosis depend on the energy inherent to molecules and ions. Cells do not need to use energy when these processes occur. Because of this, these processes are described as **passive transport.** Sometimes cells move substances against their concentration gradient. In this case, cells need to use energy to make substances cross their membranes. Because of this, this process is called **active transport.**

Figure 2.11 shows that active transport involves the use of carrier proteins in the phospholipid bilayer of membrane. The substance to be transported binds to a carrier protein. This protein changes shape, releasing the substance to be transported on the opposite side of the membrane. So far, this is similar to facilitated diffusion. However, in active transport the carrier protein needs to be activated by an input of energy from the cell.

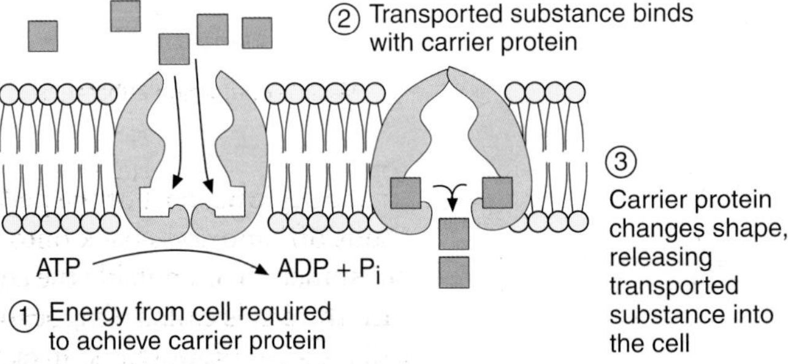
② Transported substance binds with carrier protein
③ Carrier protein changes shape, releasing transported substance into the cell
ATP
ADP + P$_i$
① Energy from cell required to achieve carrier protein

Figure 2.11
Active transport involves carrier proteins in the phospholipid bilayer of a membrane. The carrier molecules can only change shape following an input of energy from the cell.

Extension box 2

Energy for active transport

The energy used in active transport is released when a molecule of adenosine triphosphate (ATP) is broken down in cells. You will learn more about this reaction if you study Module 5 of the AQA specification. For the time being, all we need to know is that adenosine triphosphate is broken down to form adenosine diphosphate (ADP) and an inorganic phosphate group (Pi). The reaction involves an enzyme called ATPase. We can represent the reaction in a simple equation.

$$\text{ATP} \xrightarrow{\text{ATPase}} \text{ADP} + \text{Pi} + \text{energy}$$

Cells continuously produce ATP in a process called respiration. For this reason, the rate of active transport is affected by the rate of cell respiration. It is also slowed by poisons such as cyanide that slow respiration.

Endocytosis and exocytosis

Passive and active transport move only one to a few ions or molecules at a time. Endocytosis and exocytosis move substances across membranes in bulk. To do this, they use small membrane-bound sacs called vesicles. These vesicles are formed when they 'bud off' another membrane. Figure 2.12 shows both endocytosis and exocytosis.

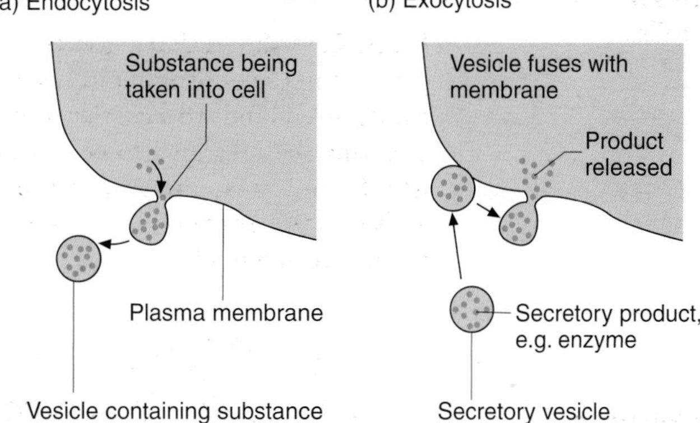

Figure 2.12
An overview of (a) endocytosis and (b) exocytosis.

In **endocytosis**, part of the plasma membrane sinks into the cell. It then 'buds off' and seals back onto itself. This produces a vesicle that contains substances from outside the cell. Endocytosis that brings solid material into the cell is called **phagocytosis**. Endocytosis that brings fluid into the cell is called **pinocytosis**. In **exocytosis**, a vesicle is formed in the cytoplasm. For example, a vesicle might 'bud off' from the edge of Golgi apparatus in the cytoplasm. Once formed, the vesicle moves to the plasma membrane. It then fuses with the plasma membrane so that its contents are pushed outside the cell. Hormones such as insulin (see start of chapter) are secreted from cells in this way.

Summary

- The membrane surrounding a cell is called the plasma membrane.
- The fluid-mosaic model of membrane structure shows that membranes have a phospholipid bilayer. This bilayer is studded with proteins and polysaccharides.
- Proteins in the plasma membrane perform a variety of functions. They act as cell receptors, as recognition sites and as adhesion sites. Proteins also assist the movement of molecules or ions through membranes.
- Diffusion is the net movement of molecules or ions down a concentration gradient. According to Fick's law, diffusion across exchange surfaces is fastest when the exchange surface has a large area, the exchange surface is thin and the concentration gradient is steep.
- Diffusion is sometimes helped by specific proteins that form part of the structure of membrane. This process is called facilitated diffusion.
- Water potential is the ability of a solution to absorb water molecules by osmosis. It is measured in units of pressure called kilopascals (kPa). Pure water has a water potential of zero (0 kPa). A solution always has a negative water potential (e.g. −6 kPa).
- Osmosis is the diffusion of water. It is the net movement of water through a partially permeable membrane from a solution of less negative water potential to a solution of more negative water potential.
- Diffusion, facilitated diffusion and osmosis depend on the energy inherent in molecules and ions. Because the cell does not use energy in these processes, they are referred to as passive transport.
- Active transport is a process by which cells move molecules or ions against their concentration gradient. It involves the use of energy by cells. Like facilitated diffusion, this process also involves carrier proteins in membranes.
- Endocytosis is the bulk movement of substances into a cell. In this case, substances become trapped in a vesicle that 'buds off' from the cell surface membrane. In phagocytosis the trapped substances are in solid form and in pinocytosis the trapped substances are in fluid form.
- Exocytosis is the bulk movement of substances out of a cell. These substances become trapped inside a vesicle formed in the cytoplasm. They are secreted when the vesicle moves to the plasma membrane and fuses with it.

Examination questions

1 (a) The diagram shows three adjacent plant cells.

A ψ_s = —10MPa ψ_p = —3MPa ψ = ...

B ψ = —12MPa

C ψ = —4MPa

(i) Calculate the water potential of cell A.

(1 mark)

(ii) Put arrows on the diagrams to show the movement of water between these cells.

(1 mark)

(iii) Explain why the water potential of a sucrose solution has a negative value.

(2 marks)

(b) It is possible to make an artificial membrane which has only a lipid bilayer. The diagram compares the permeability of such an artificial membrane with a biological cell membrane.

High permeability

Artificial membrane: H_2O, Glycerol, Na^+

Biological membrane: H_2O, Glycerol, Na^+

Low permeability

(i) Explain why the permeability to glycerol is the same in both membranes.

(1 mark)

(ii) Explain why the permeability to sodium ions is different in the two membranes.

(2 marks)

2 The drawings show the surface of epithelial cells lining the small intestine. The first is from a healthy mammal. The second is from a mammal whose intestine has been invaded by disease-causing *Escherichia coli* bacteria.

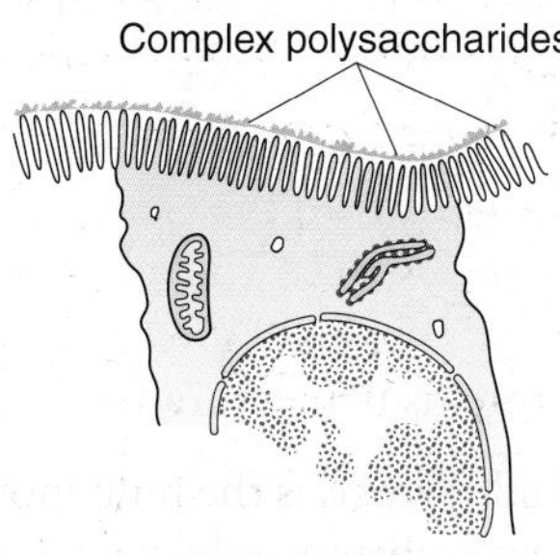

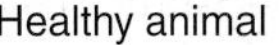
Healthy animal

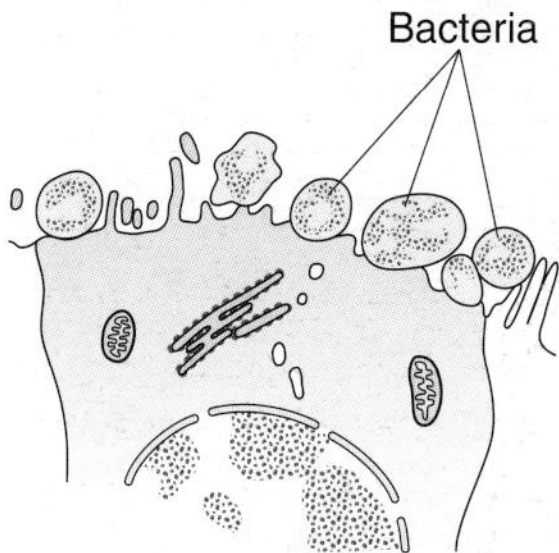

Animal invaded by *E. coli*

(a) (i) Give one piece of evidence from the drawings that shows that these cells were viewed with an electron microscope.

(1 mark)

(ii) Describe the effect of the *Escherichia coli* bacteria on the surface of the epithelial cells.

(1 mark)

(b) Use the diagram to describe and explain the effect the *Escherichia coli* bacteria would have on absorption of glucose by these cells.

Description *(1 mark)*

Explanation *(2 marks)*

Assignment

One of the skills that you need as a biologist is the ability to apply your knowledge to new situations. You should be able to use your understanding of basic principles to explain something that, at first sight, seems unfamiliar.

This assignment consists of a passage about the treatment of kidney disease. Read it carefully then answer the questions that follow. You will need to use information in this chapter about the way in which different substances get into and out of cells. You will also require some of your knowledge of cell structure (Chapter 1).

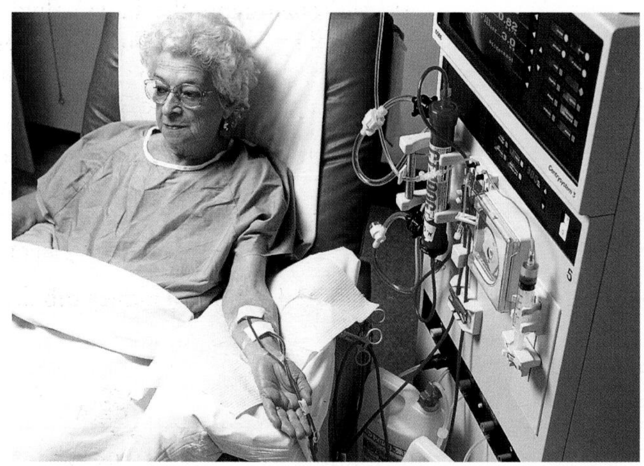

Figure 2.13
This patient has severe kidney disease and her blood is being treated by passing it through an artificial kidney.

An adult human kidney weighs approximately 100 g and is about the size of a clenched fist. It contains a million or so tiny tubes called kidney tubules or nephrons. In the first part of one of these nephrons, much of the water and soluble substances contained in the blood are filtered out and pass into the cavity of the nephron. As this filtrate passes down the tubule, all of the glucose, which is useful to the body, is reabsorbed and goes back into the blood. Waste products such as urea remain in the filtrate. A lot of the water present in the filtrate is also reabsorbed, leaving a concentrated solution of waste substances, urine.

Kidneys are vital. If they stop working, and the person is left untreated, a build up of waste products in the body rapidly results in death. However, patients whose kidneys no longer function effectively can be treated with a kidney machine or artificial kidney. Two or three times a week, a fine needle is inserted into a blood vessel in the patient's arm or leg and blood is pumped through the artificial kidney and back to the patient. The most important part of this artificial kidney is the dialyser. This contains a large number of small diameter tubes made from partially permeable membrane. Dialysis fluid is pumped through these tubes. The fluid contains specific concentrations of sodium, potassium and chloride ions and glucose, but it does not contain any urea. Waste products and excess ions pass from the blood through the membrane and into the dialysis fluid. Waste dialysis fluid is removed.

There is another, more recently developed, method of dialysis. This uses the patient's own body as an artificial kidney. A small cut is made in the body wall just above the navel and a tube is inserted into the body cavity. Dialysis fluid is poured into the abdomen through this tube. Once this has been done, a cap is placed over the end of the tube and the patient can continue with whatever he or she was doing and move around normally. The dialysis fluid in the cavity of the abdomen is separated from the blood supply by a partially permeable membrane called the peritoneum. This surrounds the organs in the abdominal cavity. Waste products pass from the blood, through the peritoneum into the dialysis fluid. Excess water is removed from the blood by putting additional glucose in the dialysis fluid. After several hours, the waste dialysis fluid is drained out of the abdominal cavity into a plastic bag and disposed of.

Now answer the questions below. The first two relate to the way in which a healthy kidney works (described in the first paragraph, lines 3–10). Marks have been added after each question. These should help you to decide on the amount of detail you are required to give in your answer.

1 Figure 2.14 shows a kidney tubule.

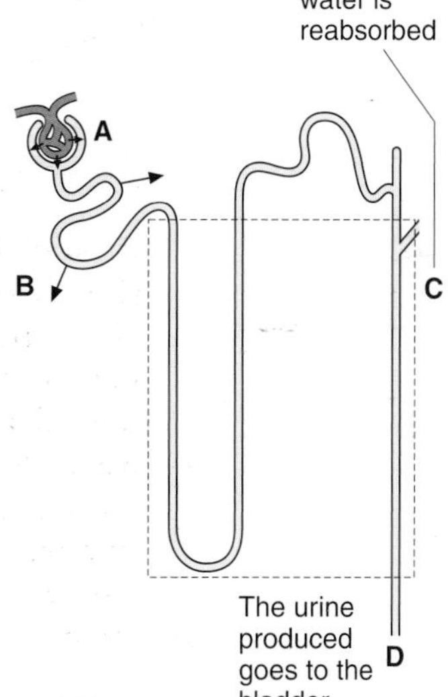

Figure 2.14
A nephron and the way in which it works.

Use information in the passage to write suitable labels explaining what happens at points A and B.

(2 marks)

2 (a) Under normal body conditions, all the glucose in the filtrate in the kidney is reabsorbed back into the blood (lines 6–7). Giving a reason for your answer, explain whether you think active transport, diffusion or osmosis are involved in the reabsorption of this glucose.

(b) Glucose is reabsorbed in the first part of the nephron. Figure 2.15 shows one of the cells that form the wall of this part of a nephron.

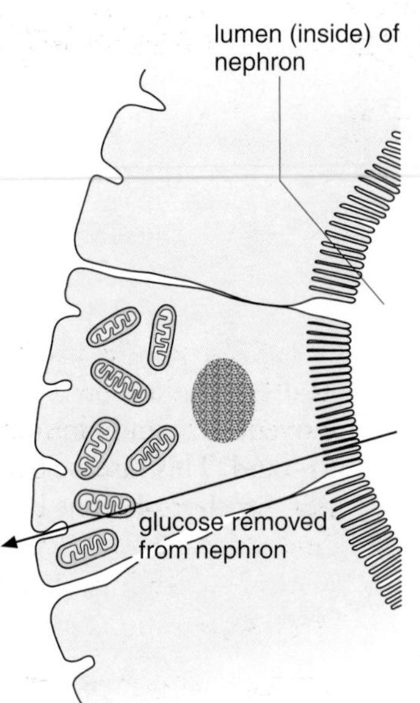

Figure 2.15
A cell from the wall of the first part of a nephron.

Explain how two of the features that can be seen in this cell are involved in the process by which glucose is reabsorbed.

(3 marks)

The artificial kidney described in Paragraph 2 depends on diffusion to remove waste products and excess ions from the blood. Your knowledge of this topic will help you to understand the way in which an artificial kidney works.

3 (a) Explain how diffusion enables the waste product, urea, to be removed from the blood in an artificial kidney.

(2 marks)

(b) Excess ions are removed but some remain in the blood. Explain why the dialyser only removes excess ions.

(2 marks)

4 Explain how the design of the dialyser prevents red blood cells and large molecules such as proteins being removed from the blood.

(2 marks)

5 Suggest the advantage of having a dialyser that contains a large number of small tubes rather than a smaller number of much larger tubes.

(2 marks)

Questions 6 and 7 relate to the method of dialysis described in Paragraph 3.

6 This method of dialysis is called continuous ambulatory peritoneal dialysis (CAPD). Explain why it is given this name.

(3 marks)

7 The passage refers to putting additional glucose in the dialysis fluid in order to remove excess water from the blood (lines 35–36). Explain why adding extra glucose will allow water to be removed from the blood.

(2 marks)

3 Biological Molecules

In 1986, a new disease began to affect cattle. Animals that were apparently healthy would occasionally stumble as they moved around. Gradually their movements would become more and more uncoordinated until they were unable to stand. This disease became known as bovine spongiform encephalopathy or BSE for short: bovine because it affects cattle, spongiform because the brains of animals that die of BSE have a sponge-like appearance when examined with a microscope and encephalopathy because it is a condition that affects the brain (Figure 3.1).

Figure 3.1
BSE affected large numbers of cattle in the UK. Fears that it could lead to humans being infected led to a world-wide ban on the sale of British beef.

The agent responsible for BSE is very different from the microorganisms that cause many other infectious diseases. It is not a virus or a bacterium, but a protein molecule called a prion. Prions are found in the brain cells of healthy animals but we do not know what they do. A cow that develops BSE becomes infected with a foreign prion, probably by eating food contaminated with material obtained from sheep infected with a similar disease known as scrapie. These foreign prions are very like the normal prions in the cow's brain cells but their molecules differ in shape. Instead of being coiled into a helix, they are folded into sheets. The abnormal foreign prions cause the cow prions to change shape into these folded sheets. The sequence of events becomes a chain reaction, converting more and more of the normal prions into abnormal ones. The abnormal prions build up into fibres, which upset the function of the brain cells and eventually lead to their death.

Protein molecules such as prions are only one of many sorts of molecule found in the cells of living organisms. Cells contain many small molecules such as water, which makes up approximately 80% of the mass of a typical cell. In addition, they contain inorganic ions, such as those of calcium and sodium, that are essential for the cell to function. However, they also have a number of larger molecules that are built up from smaller chemical building blocks. This chapter will concentrate on these larger molecules.

Large molecules and small molecules

Molecules such as those of the carbohydrates, proteins and lipids found in living organisms are called organic molecules because they contain carbon. Carbon atoms are unusual in that they can form chemical bonds with other carbon atoms as well as with the atoms of different elements. Because of this many of the organic molecules found in living organisms are very large in size and are known as macromolecules. Many of these large molecules are made up of smaller molecules, which act as building blocks. Smaller building blocks that are identical or very similar to each other are known as monomers and they join together to form a polymer.

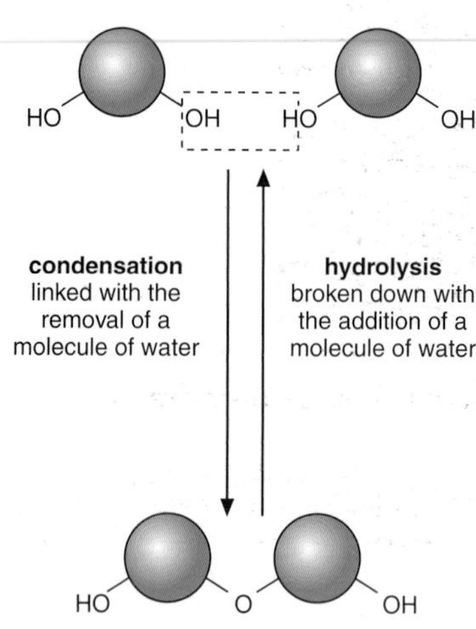

Figure 3.2
Monomers join together by condensation to form a polymer. This diagram shows two monomers joining together. When a large number of monomers are joined like this we get a polymer. A polymer can be broken down to its monomers by hydrolysis.

Q 1 **Starch, protein and fats are all large molecules. Starch is made up from many glucose molecules and proteins from a large number of amino acids. A fat consists of a molecule of glycerol and three fatty acid molecules. Which of the molecules mentioned is:**

(a) a monomer
(b) a polymer
(c) a macromolecule?

Figure 3.2 shows how two monomers can be joined together. This involves a chemical reaction known as condensation in which a molecule of water is formed. This water molecule comes from the hydrogen atom that is removed from one of these monomers and a hydroxyl (OH) group from the other. Because parts of the monomers have been removed, we refer to the parts that remain as residues once they have been joined. Joining a lot of monomers together in this way produces a polymer. Polymers may be broken down to the monomers from which they are formed by hydrolysis. This is the opposite reaction to condensation and involves the addition of water molecules.

Carbohydrates

A carbohydrate molecule contains carbon, hydrogen and oxygen. There are twice as many hydrogen atoms as oxygen atoms, the same proportion as in water. Carbohydrates are divided into three main types:

- monosaccharides are monomers. A monosaccharide is therefore a single sugar. Different monosaccharides contain different numbers of carbon atoms. Biologically important monosaccharides generally contain three (trioses), five (pentoses) or six (hexoses) carbon atoms.
- disaccharides contain two sugar residues
- polysaccharides are very large molecules containing many sugar residues.

key term

Carbohydrates have the general formula $C_xH_{2y}O_y$.

Figure 3.3
Carbohydrates play an important part in the biology of these orchids. Cellulose in the cell walls of the stems helps to provide support. As a result of photosynthesis, glucose is produced in the leaves. This is transported through the stem mainly as sucrose and stored as starch. The flowers produce a large amount of nectar, which is also rich in sucrose.

Table 3.1 summarises some information about these carbohydrates.

Type of carbohydrate	Examples	Made up from	Biological importance
Monosaccharide			
Triose	Triose		Intermediate product in the biochemical pathways of respiration and photosynthesis
Pentose	Ribose Deoxyribose		These sugars are found in the nucleic acids RNA and DNA
Hexose	Glucose Fructose		Important source of energy in respiration Found in many sweet-tasting fruits
Disaccharide	Sucrose	Glucose + fructose	The form in which carbohydrates are transported in plants
	Maltose	Glucose + glucose	Formed from the digestion of starch
	Lactose	Glucose + galactose	The carbohydrate found in milk
Polysaccharide	Starch	Glucose	The main storage carbohydrate in plants
	Glycogen	Glucose	The main storage carbohydrate in humans and other animals
	Cellulose	Glucose	An important component of plant cell walls

Table 3.1

Q 2 How many carbon atoms are there in a maltose molecule?

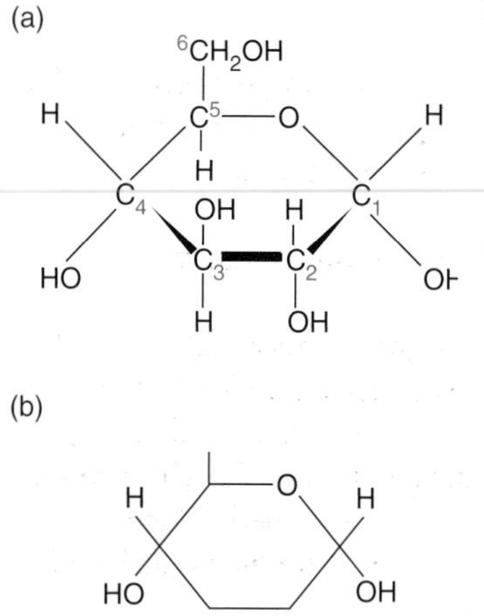

Glucose and other sugars

Glucose is a monosaccharide. It is also a hexose so a molecule of glucose contains six carbon atoms and has the molecular formula $C_6H_{12}O_6$. This formula simply tells us how many atoms are contained in the molecule; the structural formula shown in Figure 3.4 is more useful because it shows us how the atoms are arranged.

If you look at Table 3.1 again you will see that there are several different kinds of hexose sugar. All of them have the molecular formula $C_6H_{12}O_6$ but they differ from each other because their atoms are arranged in slightly different ways. Figure 3.5 shows the simplified structural formulae of four different hexose sugars. The slight differences in the way their atoms are arranged give them slightly different properties.

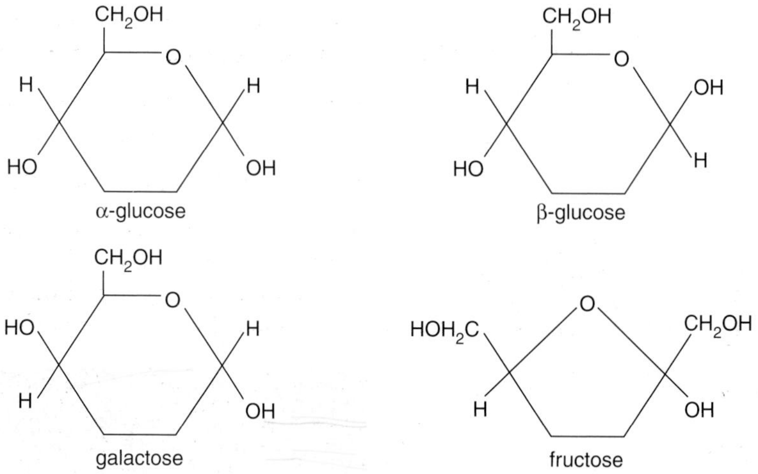

Figure 3.4
Diagram (a) shows the structural formula of a glucose molecule. The small numbers in red allow us to refer to particular carbon atoms. Diagram (b) is a simplified version of this and is all you need to learn.

Figure 3.5
The structural formulae of some hexose sugars. If you compare these with Figure 3.4, you will see that the molecule in Figure 3.4 is α-glucose.

Hexose sugars like α-glucose are the monomers from which many other carbohydrates are formed. Two α-glucose molecules may be joined by condensation to form a molecule of the disaccharide maltose. The bond forms between carbon 1 of one α-glucose molecule and carbon 4 of the other and is called a **glycosidic bond** (see Figure 3.6).

Figure 3.6
Two α-glucose molecules may join together by condensation to give a molecule of the disaccharide maltose.

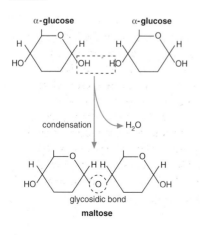

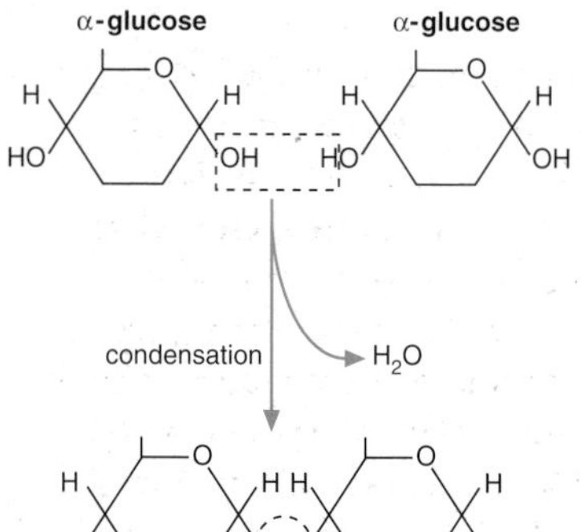

Q 3 Maltose is formed by the condensation of two α-glucose molecules. What is the molecular formula of maltose?

In a similar way, other disaccharides can be formed. Lactose, for example, the sugar found in milk, is formed from α-glucose and galactose while sucrose is formed from α-glucose and fructose.

When sugars such as α-glucose are boiled with Benedict's solution, an orange precipitate is formed as Cu(II) ions in Benedict's solution are reduced to Cu(I) ions. Because the arrangement of chemical groups in the sugar molecule enables it to reduce the Benedict's solution it is known as a **reducing sugar**. Glucose, fructose, maltose and lactose are all reducing sugars but sucrose is not. It is a **non-reducing sugar** and will only produce a positive test with Benedict's solution if it is first hydrolysed by boiling it with dilute acid. This splits the sucrose molecules into glucose and fructose, which are reducing sugars.

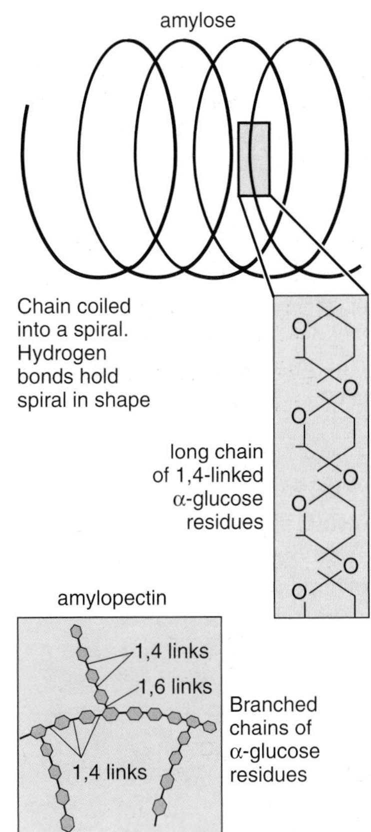

Figure 3.7
Starch consists of amylose and amylopectin.

Starch and other polysaccharides

Starch is a mixture of two compounds, amylose and amylopectin. Both of these molecules are polymers containing a large but variable number of α-glucose molecules linked to each other by condensation.

Q 4 Explain why the starch found in different plants may vary.

Amylose consists of long straight chains of α-glucose molecules linked by 1,4-glycosidic bonds (see Figure 3.7). These chains are coiled to form a spiral, each coil in the spiral containing about six glucose residues. The coils of the spiral are held in place with hydrogen bonds. **Amylopectin** is also a polymer of α-glucose but its molecules are branched. This is because some of the glucose residues are joined by 1,6-glycosidic bonds.

Starch has a number of properties that make it an efficient storage molecule (see Figure 3.8), for example:

- molecules of both amylose and amylopectin are compact. This means that a lot of starch can be stored in a relatively small space.
- it is easily broken down to glucose when needed for respiration
- starch is insoluble so it is more likely to remain in one place. In addition, as it is insoluble, it will not affect the water potential of the cells in which it is stored and bring about the movement of large amounts of water by osmosis.

Figure 3.8
Starch is a very important storage substance in plants. Starch stores such as those found in these barley grains form a vital energy source for most humans.

Extension box 1

The main source of energy in the human diet is starch. In the UK, starch makes up about 30% of what we eat. The story of starch digestion, however, is not a simple one. Although all starch can be hydrolysed by the amylase enzymes found in the gut, there are many factors that determine the rate at which it is broken down. The starch found in processed foods, such as many breakfast cereals, is known as rapidly digestible starch. As its name suggests, it is broken down readily. Starch found in unripe bananas and potatoes is not broken down so readily mainly because it is located in granules. This type of starch is known as resistant starch. It often travels the length of the small intestine and enters the large intestine or colon without being digested.

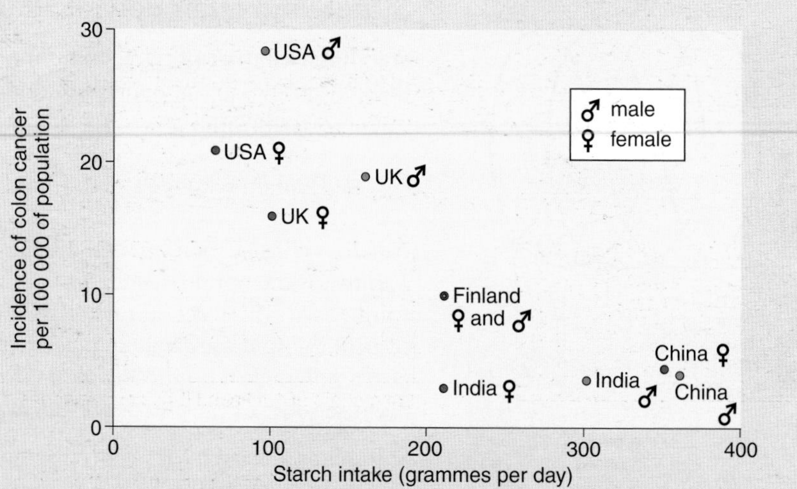

Figure 3.9
When the incidence of colon cancer is plotted against the daily intake of starch, the graph shows a negative correlation; the lower the starch intake, the higher the risk of colon cancer.

Figure 3.10
Eating a banana a day might be a good idea but it would have to be a green one that contains substantial amounts of resistant starch. In ripe bananas, those whose skins are yellow with black spots, most of the starch has been converted to sugars.

Studies such as the one that produced the results in Figure 3.9 have shown that there are clear links between the amount of starch in the diet and the incidence of colon cancer. Scientists have suggested that this link may be explained by the amount of resistant starch in the diet. As resistant starch enters the colon, it is broken down by the bacteria living there. These bacteria produce small fatty acid molecules such as butyric acid from digestion of the starch. Bacterial action may be helpful in two ways. The bacteria have a plentiful food source so they multiply rapidly. This helps to increase the rate of movement of faeces through the colon. Potential cancer-causing substances in the faeces therefore do not spend as long in contact with the cells that line the colon. Secondly, laboratory experiments have shown that butyric acid is very effective at preventing the growth of cancer cells.

Q 5 **For most countries, the figures for starch intake in males are higher than those for starch intake in females. Suggest why.**

Glycogen is another important polysaccharide. It is the main storage carbohydrate in humans. During a period of strenuous exercise, the body's glucose supplies are rapidly used up. Supplies can be replenished by breaking down glycogen stored in the liver and in the muscles. The molecules of glycogen are very similar to those of amylopectin but they are even more branched.

Cellulose is another polysaccharide. Its function is structural and it forms a very important component of plant cell walls. The monomer that forms cellulose is β-glucose. If you look back at Figure 3.5 you will see that β-glucose differs from α-glucose in that the H and OH groups on carbon 1 are the other way up. The β-glucose molecules in cellulose are also linked by condensation (Figure 3.11). In order to get the OH groups in carbon atoms 1 and 4 in the right position for a 1,4-glycosidic bond to be formed, one of the β-glucose molecules has to be 'flipped over'. This results in cellulose consisting of long straight chains with alternate β-glucose residues flipped over.

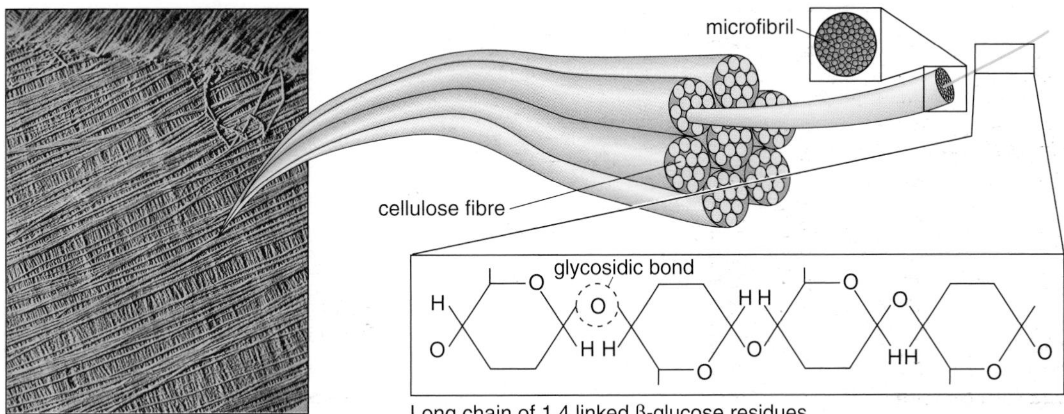

Figure 3.11
Cellulose and plant cell walls.

Cellulose molecules lie side by side and are linked to each other by hydrogen bonds to form bundles called **microfibrils**. The microfibrils are in turn held together in fibres. A cell wall is built up of many of these fibres running in different directions. Cellulose fibres have a high **tensile strength**. This means that they can withstand a very large pulling force without breaking and this helps to give the plant cell wall its strength.

Proteins

key term

Proteins are very large molecules made up of long chains of amino acids.

Each of the substances we considered in the previous section was built up from a single type of monomer. Not surprisingly, there are relatively few different sorts of polysaccharide because there are relatively few types of carbohydrate monomer. Proteins are very different. The basic building blocks of proteins are amino acids. Because there are 20 different amino acids found in proteins and because they can be put together in any order, there is a huge number of different proteins and they have many functions in a living organism (see Figure 3.12).

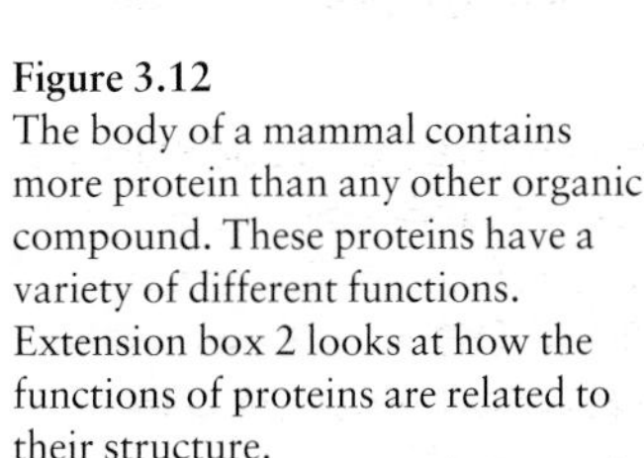

Figure 3.12
The body of a mammal contains more protein than any other organic compound. These proteins have a variety of different functions. Extension box 2 looks at how the functions of proteins are related to their structure.

Q 6 **A bacterial cell contains 1050 different types of protein but only one sort of polysaccharide. Explain why.**

Amino acids

The 20 different amino acids that occur naturally and make up proteins have the same general structure. There is a central carbon atom, called the α-carbon, to which four groups of atoms are attached. These are an amino group ($-NH_2$), a carboxylic acid group ($-COOH$), a hydrogen atom ($-H$) and another group often referred to as the R group. The first three groups are always the same. The R group differs from amino acid to amino acid. Figure 3.13 shows the general structure of an amino acid. It also shows the molecular structures of three particular amino acids found in human proteins.

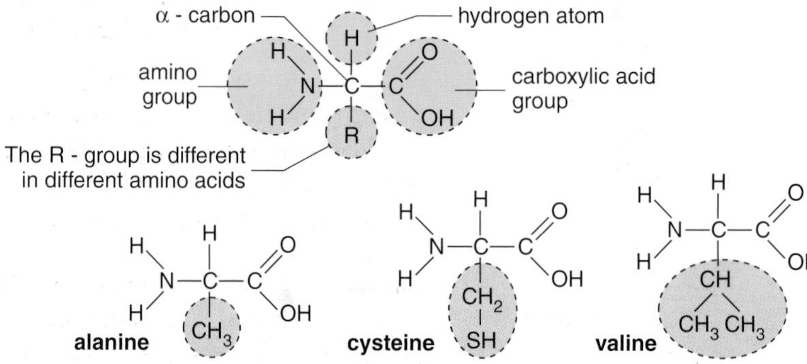

Figure 3.13
The structure of amino acids.

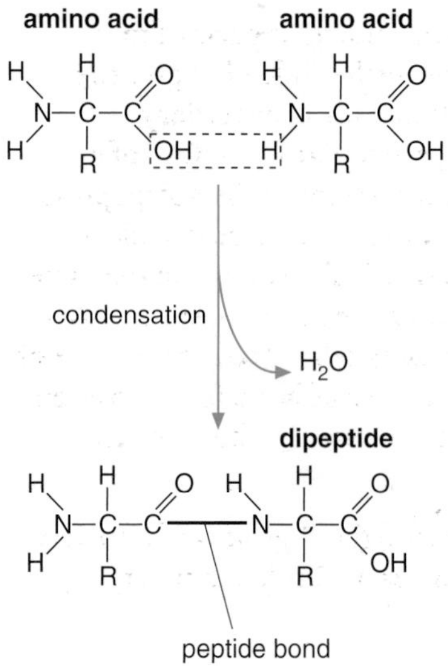

Figure 3.14
Joining amino acids.

Amino acids can be joined together by condensation. A hydrogen atom is removed from the amino group of one amino acid and this combines with an $-OH$ group removed from the carboxylic acid of the other, forming a molecule of water. The bond formed between the two amino acid residues is known as a **peptide bond**. Joining two amino acids together produces a **dipeptide**. When a small number of amino acids are joined in this way, we have a **peptide**. Longer chains are called **polypeptides**. As with other biologically important polymers, polypeptides can be broken down to their constituent amino acids by hydrolysis. Figure 3.14 shows how a dipeptide can be formed from two amino acids.

Q 7 **A peptide contains nine amino acids. How many peptide bonds will there be in a molecule of this peptide?**

Q 8 **Use Figure 3.14 to explain the meanings of the terms *N-terminus* and *C-terminus* for the ribonuclease molecule shown in Figure 3.15.**

Polypeptides and proteins

We can join more amino acids together in this way to give a polypeptide chain. A **protein** consists of one or more of these polypeptide chains folded into a complex three-dimensional shape. Different proteins have different shapes. These shapes are determined by the order in which the amino acids are arranged in the polypeptide chains. This is the **primary structure** of the protein. Genes carry the genetic code that enables cells to make polypeptides and ensures that the sequence of amino acids is the same in all molecules of a particular polypeptide. Changing a single one of these amino acids may be enough to lead to a change in the shape of the protein and prevent it from carrying out its normal function. Figure 3.15 shows the primary structure of an enzyme called ribonuclease,

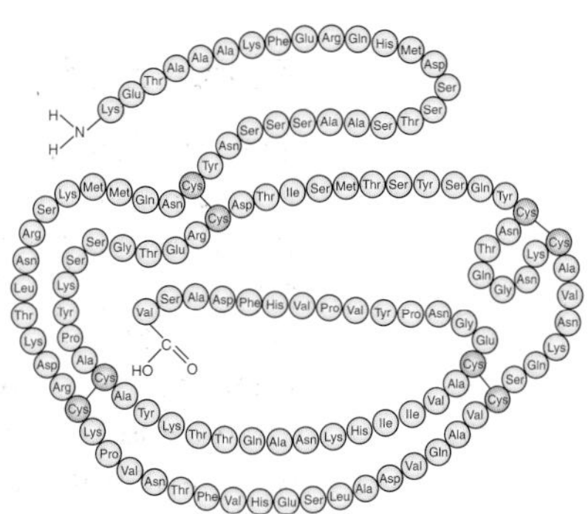

Figure 3.15
The primary structure of a protein, the enzyme ribonuclease. The names of the amino acids have been abbreviated.

which breaks down, or hydrolyses, ribonucleic acid (RNA). It is made up of 124 amino acids linked together with peptide bonds.

The polypeptide chain may form a particular shape, which is held in place by hydrogen bonds between the amino acid residues. This produces the **secondary structure** of the protein (Figure 3.16). Sometimes the chain of amino acids, or part of it, may coil to form a spiral known as an α-helix. In other proteins, a β-pleated sheet is formed. This occurs when two parts of the chain lie parallel to each other and hydrogen bonds link one part of the chain to the other. Whether or not an α-helix or a β-pleated sheet is formed depends on the sequence of amino acids in the polypeptide. Some sequences are more likely to form an α-helix while others form β-pleated sheets.

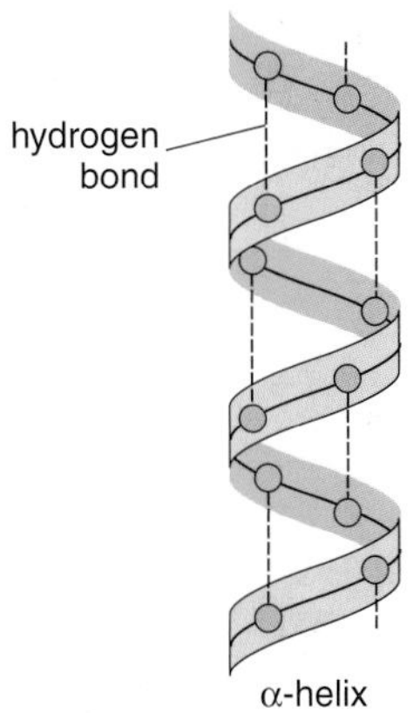

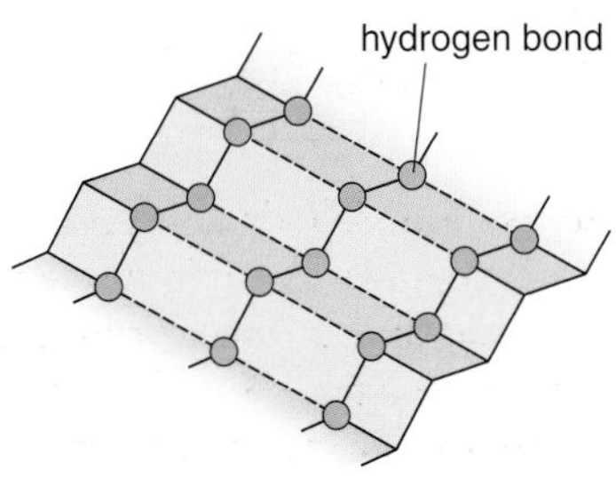

Figure 3.16
There are two main types of secondary structure in a protein. In an α helix, the polypeptide chain is coiled. In a β pleated, the polypeptide is folded.

The secondary structure involves twisting or folding of parts of the polypeptide chain. The whole chain may be further folded to give the protein molecule a complex globular shape known as its **tertiary structure**. As with the secondary structure, the tertiary structure of a protein is determined by the sequence of amino acids in the polypeptide chain. Since all molecules of a particular protein, such as ribonuclease, have the same sequence of amino acids, they will always fold in the same way to produce molecules with the same three-dimensional shape. This shape is extremely important and is very closely related to the function of the protein (see Extension box 2). Different types of bond form between different amino acids and help to maintain the shape of the protein. These bonds include:

- hydrogen bonds, which are formed between the R groups of a variety of different amino acids. These bonds are easily broken but very numerous.
- disulphide bonds formed between residues of the sulphur-containing amino acid cysteine. These are fairly strong bonds which are particularly important in structural proteins such as those found in skin and hair.

Q 9 **In which of the following are hydrogen bonds present**
(a) **maltose**
(b) **starch**
(c) **the primary structure of a protein**
(d) **the secondary structure of a protein?**

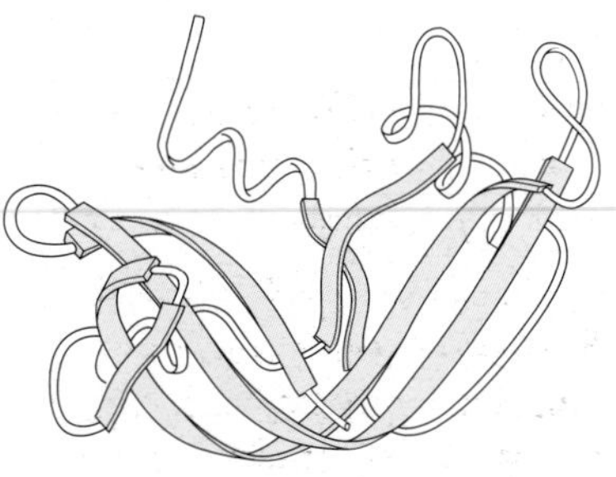

Heating a protein causes the bonds that maintain its tertiary structure to break. The protein is said to be **denatured.** As a result, the molecules lose their shape and can no longer carry out their function (see Chapter 4). Figure 3.17 shows the secondary and tertiary structures of the ribonuclease molecule, whose primary structure is illustrated in Figure 3.15.

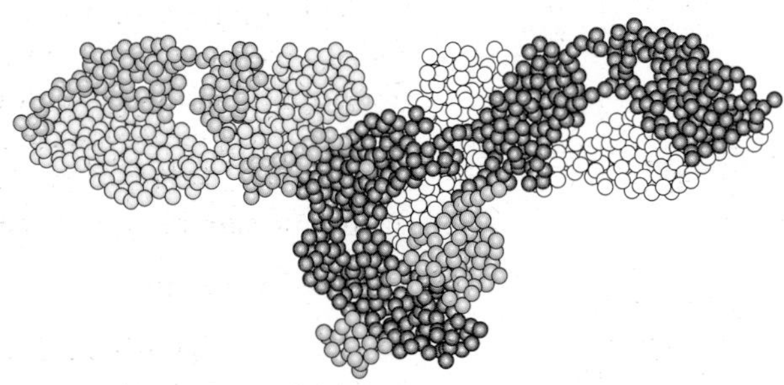

Figure 3.17 (above)
This diagram shows both the secondary and tertiary structure of a molecule of ribonuclease. The three spiral portions of the polypeptide chain represent the parts which are coiled into an α-helix. The flat pieces show where the polypeptide chain is folded into a β-pleated sheet.

Figure 3.18 (top right)
Antibody molecules are produced by the body in response to a foreign molecule or cell, such as that of an invading bacterium. Antibodies are proteins that have a quaternary structure. This model shows the two long polypeptide chains and the two shorter ones that make up an antibody molecule.

Some proteins are made up from more than one polypeptide chain. The way in which two or more polypeptide chains combine gives the protein its **quaternary structure** (see Figure 3.18). The chains are held together by the same sorts of chemical bond that maintain the tertiary structure. The enzyme ribonuclease does not have a quaternary structure because it consists of only one polypeptide chain. Each molecule of haemoglobin is made up of four polypeptide chains so it can be described as having a quaternary structure.

Haemoglobin

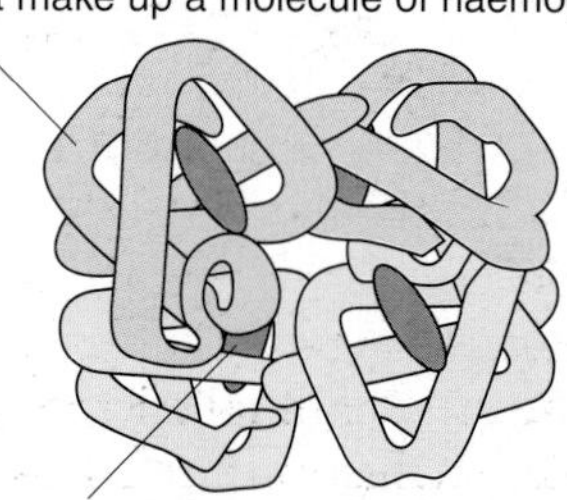
One of the four polypeptide chains that make up a molecule of haemoglobin
A haem group:
Each polypeptide chain is attached to a haem group. This group is important in transporting oxygen
Each polypeptide chain in a molecule of collagen is coiled into a helix
A collagen molecule consists of three polypeptide chains coiled round each other

Figure 3.19 (right)
Haemoglobin is the red oxygen-carrying pigment found in the blood of humans and many other animals. It is an example of a globular protein. Collagen, on the other hand, is a fibrous protein. It is found in many parts of the human body, including the skin, cartilage and the walls of larger blood vessels.

Protein molecules have one of two basic shapes: they may be curled up into a ball, in which case they are known as **globular proteins**, or they may be long and thin, and known as **fibrous proteins**. Globular proteins are usually soluble and play an important role in the metabolism of living organisms. Examples of globular proteins are the carrier proteins found in cell membranes, enzymes and haemoglobin (see Figure 3.19). Fibrous proteins, on the other hand, as their name suggests, form long chains. They are insoluble and have important structural functions.

Extension box 2

Locks, keys and proteins

Globular proteins have many functions in living organisms. Their molecules have a tertiary structure that gives them a distinctive shape and this shape is related to their function (see Figure 3.20).

You will come across many globular proteins during your A-level course. They have different functions but the basic principle concerning the way in which they work is always the same. Each protein has a unique sequence of amino acids that gives it a particular tertiary structure and a distinctive shape. This shape allows other molecules to fit specific sites on its surface, and the protein to carry out its function.

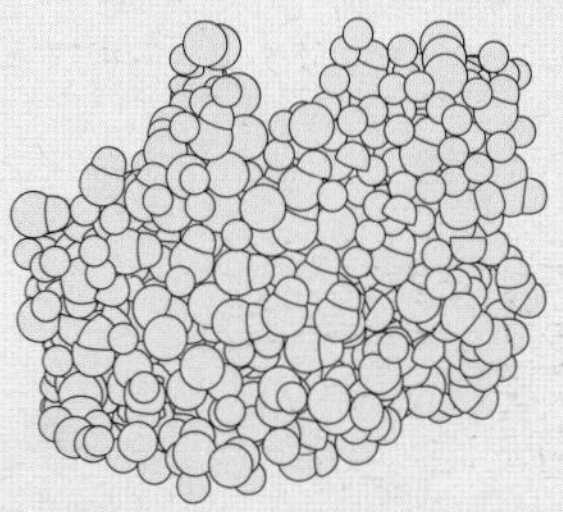

Figure 3.20
This model shows the three-dimensional shape of the enzyme ribonuclease. This is a protein whose function depends on the shape that results from its tertiary structure.

Proteins as enzymes

We have seen that ribonuclease is an enzyme that breaks down the nucleic acid RNA into smaller components. It is a globular protein so it has a distinctive tertiary structure. Like all enzymes, it functions because part of its molecule forms what is called an active site. RNA molecules will fit into this site where they are broken down into their components. The tertiary structure of a protein is held by chemical bonds between the amino acids. One type of bond is the disulphide bond, which is formed between amino acids that contain sulphur. When ribonuclease is incubated with mercaptoethanol, these bonds are broken and the enzyme loses its tertiary shape. It no longer works. This is shown in Figure 3.21. Enzymes are described in detail in Chapter 4.

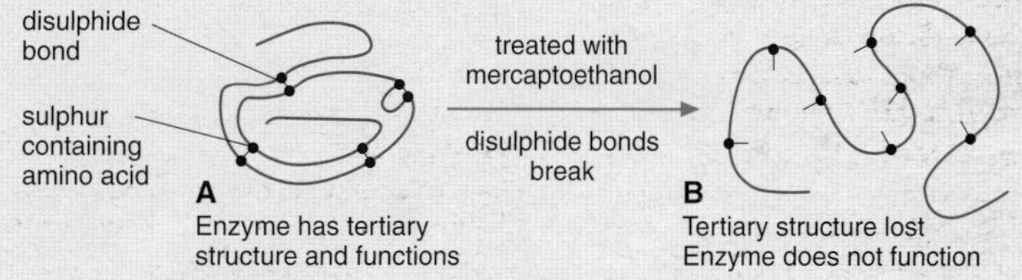

Figure 3.21
Diagram A represents a molecule of ribonuclease. The dots represent cysteine, an amino acid that contains sulphur. Disulphide bonds form between these amino acids and maintain the tertiary structure and the shape of the enzyme. In diagram B, the enzyme has been treated with mercaptoethanol. The bonds between the cysteine residues have broken. The ribonuclease molecule has unravelled and lost its tertiary structure. It will no longer function.

Proteins as antibodies

The production of antibodies is one of the main ways in which the white cells in the body respond to viruses and bacteria. Antibodies recognise these harmful microorganisms and attach to particular molecules on their surfaces. This triggers off mechanisms that lead to their destruction and the infection is controlled (see Figure 3.22). Antibodies, however, are very specific. One that is effective against a chickenpox virus, for example, will not recognise a measles virus. Once again, it is all down to shape (see Figure 3.23).

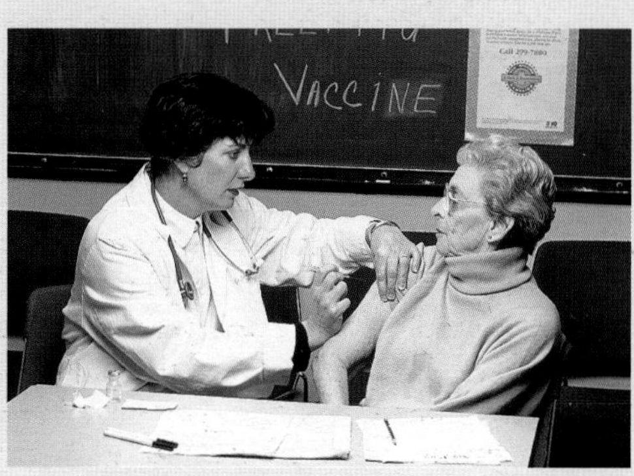

Figure 3.22
This person is receiving her yearly influenza injection, which involves giving the individual concerned a vaccine containing harmless, inactive influenza viruses. These stimulate some of her white cells to produce influenza antibodies if she is exposed to living influenza viruses. A different vaccine has to be used each year. This is because there are many different strains of influenza virus. An antibody that recognises one strain may not recognise another. Once again, it is a matter of the shape of a protein, this time an antibody, not matching and therefore fitting the shape of another found on the surface of the influenza virus.

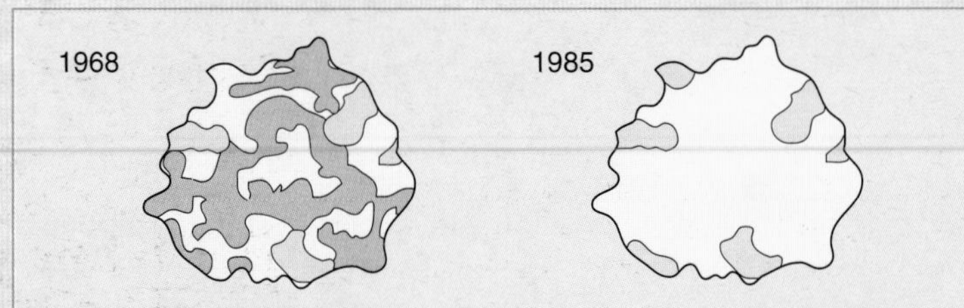

Figure 3.23
These diagrams show changes in the shapes of the protein coats of a particular strain of influenza virus between 1968 and 1985. The red-coloured areas are spikes on the protein coat of the virus.

Q 10 Explain why a vaccine that was effective against the strain of influenza virus found in 1968 would not be effective against the strain found in 1978.

Proteins in the nervous system

Nerve cells or neurones do not join directly to one another. There are small gaps between them called synapses (see Figure 3.24). An impulse is transmitted along a nerve until it reaches a synapse. It then causes molecules of a neurotransmitter substance to be released. The neurotransmitter molecules diffuse across the gap and fit into protein receptor molecules. This results in a nerve impulse being triggered in the second neurone.

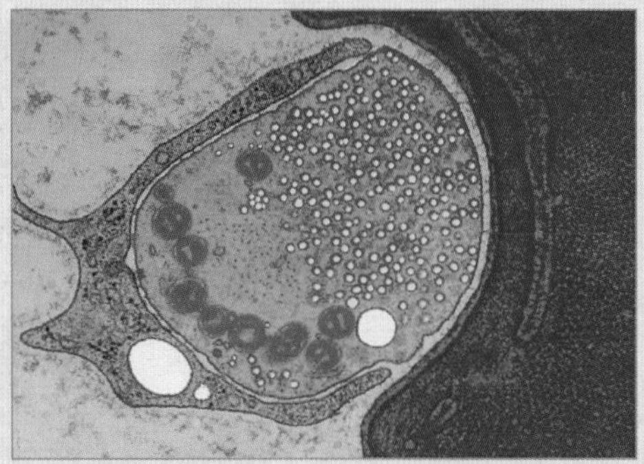

Figure 3.24
This electronmicrograph shows a synapse between two nerve cells. The round structures are vesicles containing neurotransmitter molecules.

There are many different neurotransmitters. An important neurotransmitter in synapses in the brain is a substance called serotonin (Figure 3.25). Serotonin has molecules that have very similar shapes to those of a number of drugs that affect the brain. The similar shapes of these drug molecules means that they can all fit into the proteins that form the serotonin receptors.

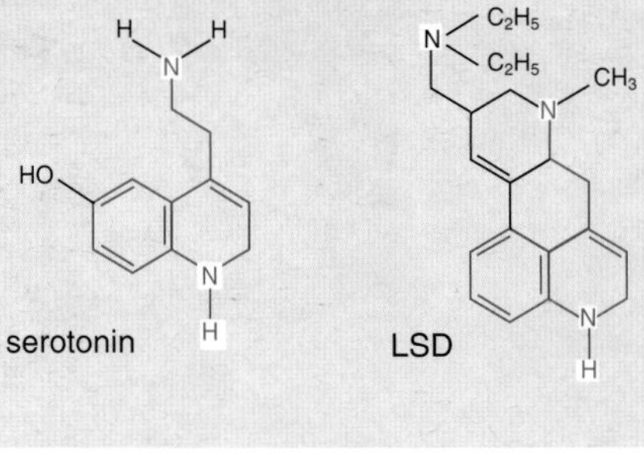

serotonin

LSD

Figure 3.25
The molecules of serotonin and LSD, drugs that affect the brain. Notice how they are very similar in shape.

Lipids

Triglycerides

Figure 3.26
The body shape of this seal is due to a thick layer of fat under its skin. This insulates it from the cold water in which it lives. Fats also act as an important store of energy.

> **key term**
>
> The name **lipid** is used to describe a range of substances. Some of the most important of these are the **triglycerides**, usually known as fats and oils.

A triglyceride consists of three fatty acid molecules joined to a molecule of glycerol. Each **fatty acid** consists of an acid COOH group joined to a long hydrocarbon tail consisting of carbon and hydrogen atoms. The length of this hydrocarbon tail varies but in many of the fatty acids found in triglycerides there are between 14 and 16 carbon atoms. Some fatty acids have one or more double bonds between the carbon atoms in the tail. These are **unsaturated** fatty acids, so-called because they are unsaturated with hydrogen atoms (see Figure 3.27). The presence of double bonds produce kinks in the hydrocarbon chains. This stops the hydrocarbon chains of neighbouring triglycerides lying too close together and makes the lipid more fluid, lowering its melting point. As a consequence triglycerides that contain unsaturated fatty acids are often liquid at room temperatures. They are known as oils. Fatty acids that do not contain double bonds in their hydrocarbon chains are described as **saturated.** They usually have a higher melting point than unsaturated fatty acids and give rise to fats that are usually solid at room temperatures. Oils are more commonly found in plants while fats are usually associated with animals.

Q 11 Explain what is meant by a polyunsaturated fatty acid.

Glycerol is a type of alcohol. Its chemical structure is shown in Figure 3.27. Each of the three carbon atoms in a molecule of glycerol is associated with an OH group that is able to combine with a fatty acid (Figure 3.28). This reaction is a condensation, so when a triglyceride is formed, three molecules of water are also produced.

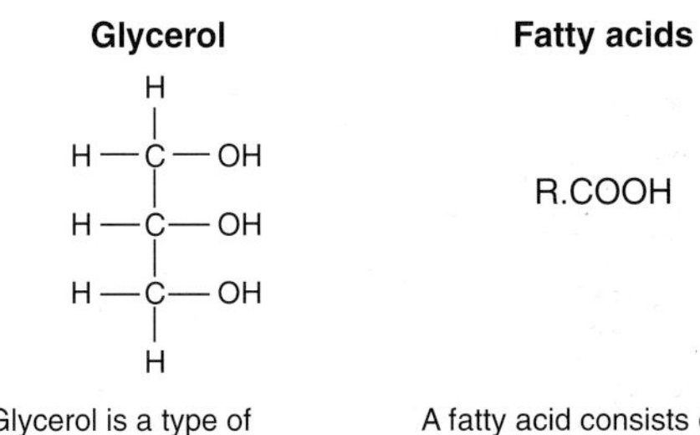

Glycerol is a type of alcohol. It has three OH groups each of which can combine with a fatty acid.

A fatty acid consists of an acid COOH group and a long hydrocarbon chain consisting of carbon and hydrogen atoms. This is represented by the letter R

Saturated fatty acid

Unsaturated fatty acid

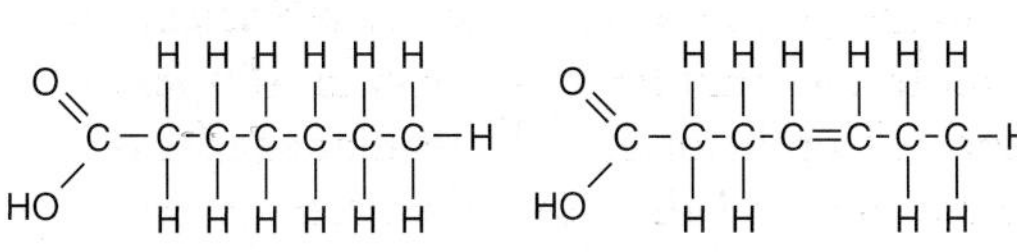

With the exception of the last one, each of the carbon atoms in the hydrocarbon chain of a saturated fatty acid is joined to two hydrogen atoms

In unsaturated fatty acids, there are double bonds between some of the carbon atoms. The hydrogen chain is not saturated with hydrogen atoms

Figure 3.27
Glycerol and fatty acids.

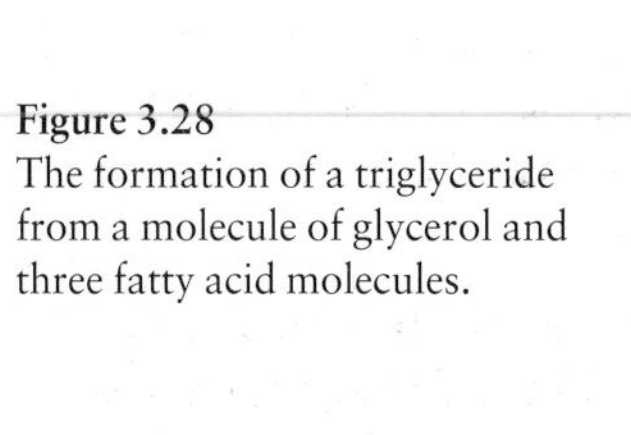

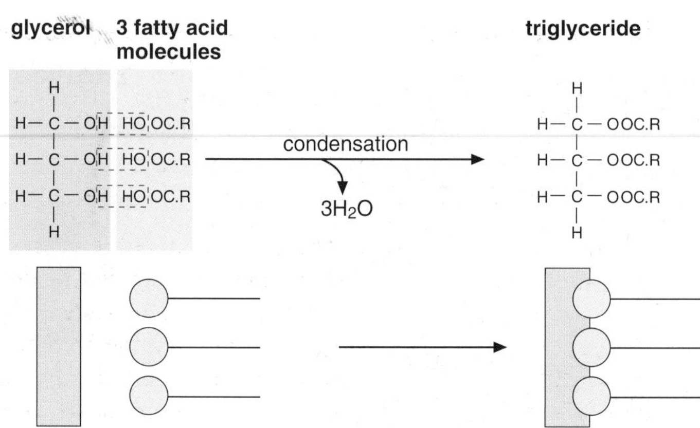

Figure 3.28
The formation of a triglyceride from a molecule of glycerol and three fatty acid molecules.

Phospholipids

Think of a phospholipid as having a 'head' consisting of the glycerol and phosphate and a 'tail' containing the long hydrocarbon chains of the two fatty acids (see Figure 3.29). The presence of the phosphate group means that the charge on the head of the molecule is unevenly distributed. It is said to be **polar** and is attracted to water. The head end of the molecule is described, therefore, as being **hydrophilic** or 'water-loving'.

key term

Phospholipids are very similar to triglycerides except that one of the fatty acids is replaced with a phosphate group.

The hydrocarbon tails do not have this uneven charge distribution. They are therefore **non-polar** and will not mix with water. The tail end of the molecule is described as being **hydrophobic** or 'water-hating'. This property means that if phospholipids are placed in water they will arrange themselves in a double layer with their hydrophobic tails pointing inwards and their hydrophilic heads pointing outwards. This double layer is called a **phospholipid bilayer** and it forms the basis of cell membranes (see Chapter 2).

Q 12 How does a phospholipid differ from a triglyceride?

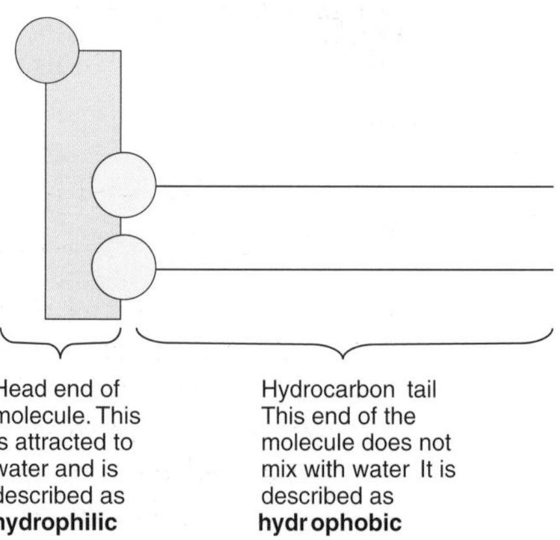

Figure 3.29
A phospholipid molecule. Because the head end of the molecule and the hydrocarbon tail have different properties, phospholipids arrange themselves in bilayers when placed in water.

Extension box 3

Figure 3.30 (above)
Oilseed rape has become increasingly popular as a crop in the UK over the past 25 years. Its bright yellow flowers are a familiar sight in spring. After the petals have withered and fallen from the flowers, the pods containing the seeds ripen and are harvested. Oil is extracted from the seeds.

Figure 3.31 (right)
A molecule of erucic acid.

Fats and oils are often found in seeds. A large amount of energy is released when a triglyceride is respired, so triglycerides act as energy stores for the young plant as the seed germinates. Plants whose seeds store triglycerides are often important agricultural crops (see Figure 3.30). Every year almost 80 million tonnes of fats and oils are produced world-wide from this source.

There are many different varieties of oilseed rape and the oils that the seeds contain vary slightly in chemical composition. These differences are due to the fatty acids in the triglycerides of the seeds. In the 1970s, it was found that the presence in the diet of one of the fatty acids in rape seed oil, erucic acid (Figure 3.31), was linked with the accumulation of fat in the heart muscle of young animals. It was thought that if this happened in humans, it could lead to serious heath problems, so plant breeders set about breeding new strains of oilseed rape that were low in erucic acid.

```
  H H H H H H H H   H H H H H H H H H H H H     O
  | | | | | | | |   | | | | | | | | | | | |   //
H-C-C-C-C-C-C-C-C=C-C-C-C-C-C-C-C-C-C-C-C-C-C
  | | | | | | |       | | | | | | | | | | |   \
  H H H H H H H       H H H H H H H H H H H     OH
```

Q 13 What type of fatty acid is erucic acid? Is it saturated, monounsaturated or polyunsaturated?

Varieties of oilseed rape low in erucic acid are now grown for use in the food industry. Oil from these plants is used in the production of margarine and cooking oil. Varieties high in erucic acid are still grown. The oil from these plants is used to produce erucimide, a substance that is used for coating plastics such as credit cards and plastic bags to prevent them becoming sticky.

Separating and identifying molecules using chromatography

Biochemists use chromatography to separate and identify the various substances present in a mixture. Paper chromatography uses a strip or square of absorbent paper (see Figure 3.33) but there are other sorts of chromatography that rely on different techniques. The substances to be separated are extracted and dissolved in a suitable solvent. A pencil line is ruled near the bottom of the paper. This forms the **origin**. A fine pipette or capillary tube is used to put drops of the solvent containing the mixture on the origin. A very small drop is placed on the paper, it is allowed to dry and another spot is put on top of the first. This process is repeated a number of times, producing a tiny concentrated spot.

Figure 3.32
In autumn, the leaves fall from many woody plants. A layer of cork forms at the bottom of the leaf stalk and this traps some of the sugars produced in leaves during photosynthesis. These sugars are converted to anthocyanins, which give leaves of plants such as this Virginia creeper their characteristic red colour. Leaf pigments are just one group of substances that can be investigated with chromatography.

The chromatography paper is now suspended in a jar containing a little solvent. The paper is arranged so that the bottom of it just dips into the solvent. The lid is placed on the jar so that the air inside becomes saturated with solvent vapour and the apparatus is left on one side. The solvent moves up the paper. The different substances in the mixture also move but they move at different rates and are separated out. When the solvent has nearly reached the top of the paper, the paper is removed and the position of the **solvent front** is marked with a pencil.

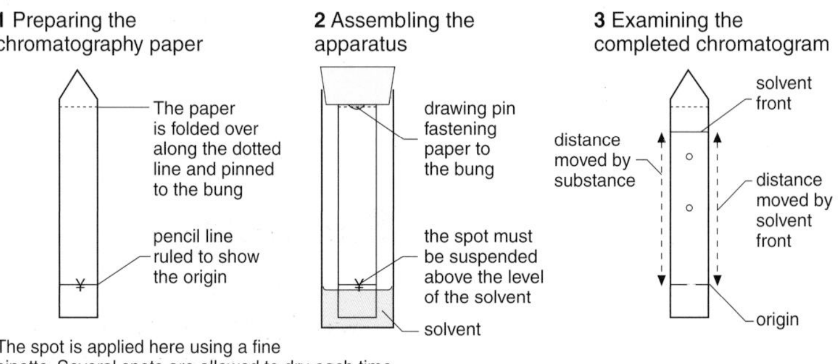

Figure 3.33
Carrying out paper chromatography.

If you are separating coloured substances such as the pigments in a leaf, you will see different coloured areas on the paper, each one representing a different pigment. Many substances are not coloured, however, and the chromatogram has to be developed before the positions of the individual substances can be seen. Developing is usually done by spraying the chromatogram or dipping it in an indicator that changes colour when it is mixed with the substance concerned.

The process described so far allows different substances to be separated but it does not identify them. To do this we calculate the **Rf value**. This is a measure of how far the substance has moved compared with the distance moved by the solvent. It can be calculated from a simple equation:

$$\text{Rf value} = \frac{\text{distance moved by substance}}{\text{distance moved by solvent front}}$$

Each compound in the mixture has a unique Rf value so we can use this to identify which spot represents which substance.

Q 14 Why does the Rf value of a substance always have a value less than 1?

Sometimes the spots containing the substances do not separate very much. When the chromatogram is removed some of the spots are very close together. In situations like this we can use two-way chromatography (Figure 3.34). A square piece of paper is used. When the solvent front is near the top of the paper, the chromatogram is removed and turned through 90° so that the spots are at the bottom of the paper. The paper is then suspended in a different solvent, which is again allowed to run up the paper.

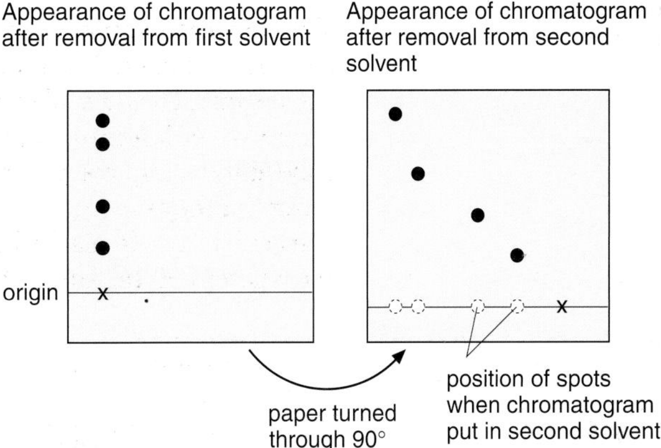

Figure 3.34
Two-way chromatography separates the spots much more.

Summary

- Biological molecules are based on a small number of chemical elements and frequently consist of monomers joined by condensation into polymers. These polymers may be broken down by hydrolysis to give the monomers from which they are built.
- Starch and glycogen are formed by the linking of α-glucose molecules with the formation of glycosidic bonds
- Cellulose is formed by the linking of β-glucose molecules
- Amino acids may be linked together with peptide bonds to produce dipeptides and polypeptides. Polypeptides form proteins which show different levels of structure.
- A molecule of glycerol and three molecules of fatty acids combine to produce a triglyceride
- Chromatography can be used to separate and identify molecules in a mixture.

Examination questions

1 The table shows the number of carbon atoms contained in some substances.

Substance	Number of carbon atoms
Amino acid	2–11
Glycerol	
Glucose	
Starch	Large and variable number

(a) Complete the table to show the number of carbon atoms in glycerol.

(1 mark)

(b) Explain why the number of carbon atoms

(i) may vary from 2 to 11 in an amino acid molecule;

(1 mark)

(ii) may be described as variable in a starch molecule.

(1 mark)

(c) Explain why the type of chemical reaction in which glucose is converted to starch may be described as condensation.

(2 marks)

2 Describe how the process of paper chromatography is carried out in order to separate substances in a mixture.

(3 marks)

(b) Some fatty acids were synthesised in which all the carbon atoms were radioactive. A mixture of three of these fatty acids was separated using paper chromatography. The level of radioactivity in each of the fatty acids was measured by passing the chromatography strip under a Geiger counter and the level was plotted as a graph. The results are shown in the diagram.

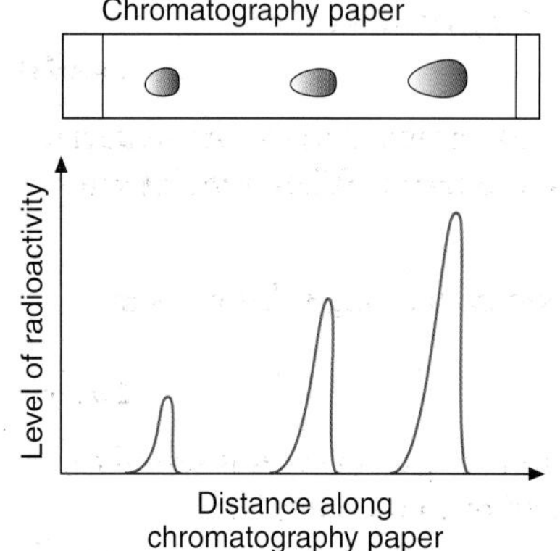

Figure 3.35

(i) The *total* level of radioactivity in each of the fatty acids was measured by estimating the total area under each curve rather than the height of each peak. Explain why.

(1 mark)

(ii) Suggest **one** reason why the three fatty acids showed different levels of radioactivity

(1 mark)

Assignment

Figure 3.36
The ability to feed their young on milk is a characteristic of all mammals.

This exercise is designed to introduce you to the skills involved in interpreting data. The first two questions will help you to understand the data presented in the table. You will then have to change or translate the information in the table into another form, a bar chart. The final part of the exercise involves interpretation of information.

All mammals share certain characteristics. They have hair on their body surface and have sweat glands that help in controlling body temperature. They also feed their young on milk produced from the mammary glands of the female (see Figure 3.36).

Milk must provide a young mammal with the nutrients it requires before it is old enough to eat food similar to that of an adult. In addition to protein, carbohydrate and lipid, milk also contains other substances such as vitamins and mineral ions. The relative amounts of these substances, however, differ from one species of mammal to another. Table 3.2 shows the composition of human milk and seal milk.

Species	Mass of substance (g per 100 g of milk)			
	Protein	Lactose	Lipid	Ash
Human	1.2–1.5	7.0	3.8	0.2
Harp seal	13.8	0	36.5	0.6

Table 3.2

1 Explain why the figures are given per 100 g of milk.

(2 marks)

2 All the water can be removed from a milk sample and the dry material that is left can then be heated strongly and burnt. What remains after burning is called ash.

(a) Explain why there will be no carbon-containing substances in the ash.

(2 marks)

(b) Use the information in the question to suggest why it is useful to measure the amount of ash obtained from the sample.

(1 mark)

3 Draw a bar chart to compare the mass of these substances in human and harp seal milk.

(4 marks)

Harp seals live in the Arctic. They give birth to their young on large areas of ice. These areas have few predators but there disadvantages for the young seal because there is very little shelter and it is extremely cold. When it is born, a typical young harp seal weighs 10.8 kg. After 9 days of feeding on its mother's milk, its body mass will have increased to 34.4 kg.

4 (a) Calculate the percentage increase in mass of the young harp seal over this 9 day period. Show your working.

(2 marks)

(b) Explain how the increase in mass of the young harp seal can be related to the composition of the milk on which it feeds.

(2 marks)

The composition of the milk that is produced depends to a certain extent on the food that is eaten. In a study of human breast milk, samples were collected from two groups of women. Those in one group were vegans and only ate food obtained from plants. Those in the other group, the control group, ate food obtained from both animals and plants. Table 3.3 shows the concentrations of different fatty acids in the milk samples.

Fatty acid	Number of double bonds in hydrocarbon chain	Number of carbon atoms in hydrocarbon chain	Concentration of fatty acid in mg per g per gram of fat	
			Vegan group	Control group
Lauric	0	12	39	33
Myristic	0	14	68	80
Palmitic	0	16	166	276
Stearic	0	18	52	108
Palmitoleic	1	16	12	36
Oleic	1	18	313	353
Linoleic	2	18	317	69
Linolenic	3	18	15	8

Table 3.3

5 (a) Explain why the first four fatty acids in this table may be described as saturated.

(1 mark)

(b) Calculate the total concentration of saturated and unsaturated fatty acids in the vegan group and in the control group. Write your answers in a suitable table.

(2 marks)

(c) Describe and explain the main differences between the occurrence of saturated and unsaturated fatty acids in milk produced by the vegan group and the control group.

(3 marks)

4 Enzymes

The greenish-yellow glow of a firefly on a summer night in Malaysia is produced by one of the most interesting enzymes ever discovered. It is called luciferase and it catalyses the breakdown of a protein called luciferin. During this reaction, most of the energy is released as light rather than heat, causing the glow we can see in the dark. Colonies of bacteria that contain a similar light-releasing enzyme inhabit four species of 'flashlight fish', deep-sea fish with glowing pockets directly beneath their eyes (Figure 4.1). Each fish provides nourishment for the millions of bacteria in its eye pockets; in return, the luminous bacteria help the fish prey and mate in the inky reaches of the deep ocean.

Luciferase acts on luciferin only if energy is present in the form of adenosine triphosphate (ATP):

$$\text{luciferin} \xrightarrow[\text{luciferase}]{\text{ATP}} \text{product} + \text{light} + \text{heat}$$

Researchers have been using this 'glowing enzyme' to detect how much ATP is present in certain systems. For example, luciferase and luciferin can be added to blood stored in blood banks. If the red blood cells are stored too long they begin to degenerate and leak ATP molecules into surrounding fluid. These molecules will in turn drive the breakdown of luciferin by luciferase and the blood will glow in the dark. Bacteria also release ATP and the same method is used to determine if they are present in urine so that infections may be detected and treated.

Luciferase and similar enzymes are truly illuminating tools for biologists, as well as providing headlights for fish and tail-lights for fireflies.

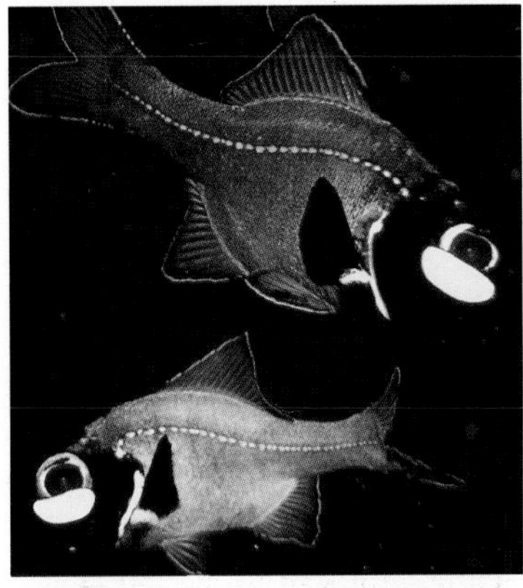

Figure 4.1
Flashlight fish.

Enzymes are amazing molecules. They are produced by living cells, with each cell containing several hundred enzymes. They are extremely specific, generally reacting with only one substrate. They can speed up the rate of chemical reactions by as much as a million times and at the end of the reaction they remain unchanged. Within a cell, in the absence of enzymes reactions would take place at too slow a rate to sustain life. To increase the rate of reactions, high temperatures would be necessary and this would be lethal to a cell.

Q 1 **By what process are enzymes made within the cell?**

key term

Enzymes are globular proteins that catalyse chemical reactions in living organisms.

Figure 4.2
The position of amino acids in the primary structure and how they come together to form the active site.

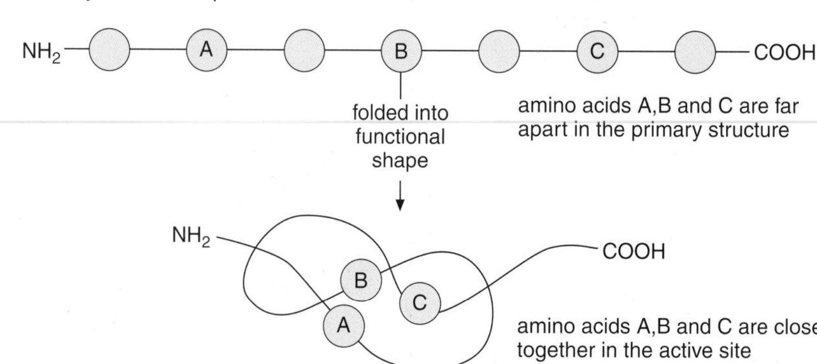

All enzymes are proteins

To understand how enzymes work it is necessary to understand the structure of proteins.

A protein is made up of a unique sequence of amino acids, known as its primary structure. However, the protein molecule does not lie flat: in a watery environment, such as inside living cells, the primary structure of the molecule will fold of its own accord into a unique and precise three-dimensional shape, held together by a number of different bonds (see Chapter 3).

The shape of the enzyme has to be just right: if the enzyme molecule is deformed, even slightly, it will not function.

Each different enzyme has its own highly specific shape, with a 'pocket' at a particular position. The pocket is known as the **active site**, and it is here that a substrate binds.

Only a few amino acids, normally between 3 and 12, in the chain of an enzyme are actually involved in the enzyme-controlled process. The specific amino acids that form the active site come together from different positions in the primary structure of the protein (see Figure 4.2). It is here that binding of the substrate occurs. The remaining amino acids make up the bulk of the enzyme, which maintains the correct globular shape of the molecule.

key term

Activation energy is the energy needed to bring molecules together so that they will react with each other.

Activation energy

Consider a molecule of sucrose. It consists of a molecule of glucose and a molecule of fructose joined by a glycosidic bond. Sucrose can be hydrolysed into glucose and fructose by the addition of water to break the glycosidic bond (Figure 4.3).

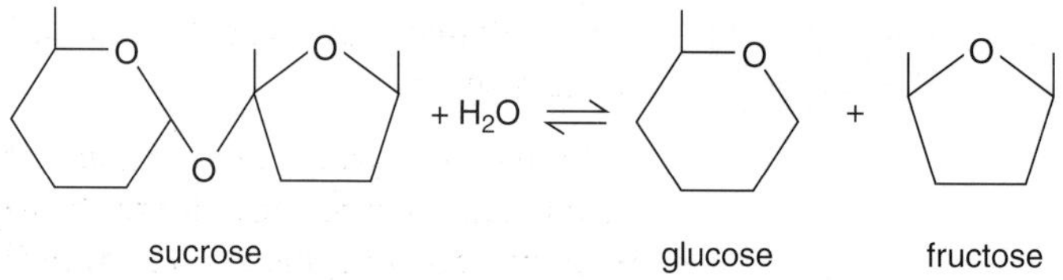

sucrose $+ H_2O \rightleftharpoons$ glucose + fructose

Figure 4.3
The reaction between sucrose and water will occur slowly due to random collisions.

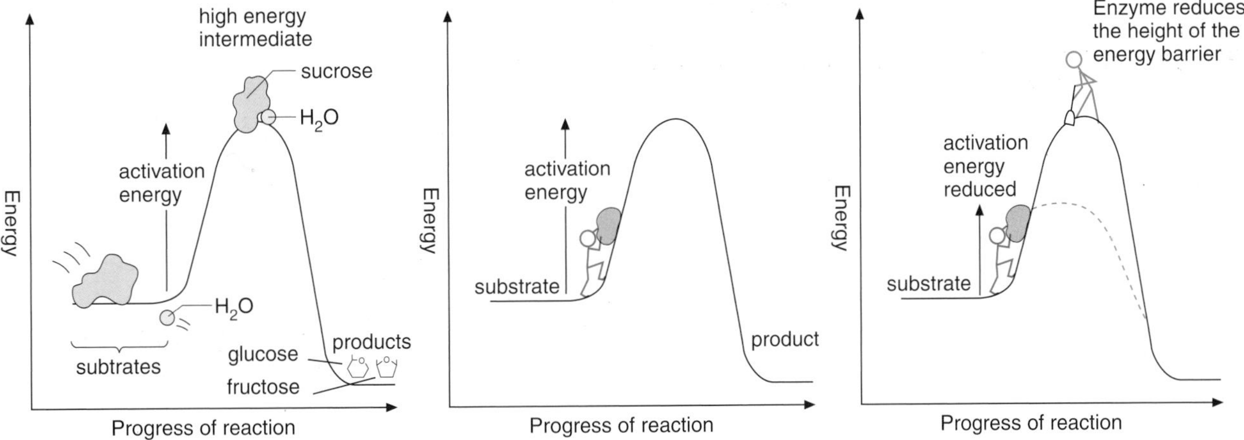

Figure 4.4
Energy diagram. Sucrose and water must collide with sufficient energy to form the unstable high-energy intermediate if they are to react form glucose and fructose.

Figure 4.5
Energy diagram illustrating that energy needs to be added to a reaction to allow it to proceed.

Figure 4.6
Energy diagram illustrating that enzymes reduce the activation energy.

Without an enzyme, molecules react by randomly colliding, but energy has to be applied. For a reaction to take place molecules, known as **substrates**, have to collide with enough energy to break and form bonds, creating **products**. In this case, the sucrose and water collide to form glucose and fructose. This reaction could be carried out in a test-tube but it would occur very slowly.

The energy required to make substrates react is called the **activation energy**. This can be illustrated in an energy diagram (see Figure 4.4). At the start of the reaction, sucrose and water have a certain amount of energy. They collide with one another and form an unstable high-energy intermediate which quickly changes into glucose and fructose.

In rearranging the bonds of sucrose and water some energy is released, so the products have less energy than the initial molecules. The minimum amount of energy needed to start the reaction, leading to the formation of an unstable intermediate, is the activation energy.

Every chemical reaction thus has an energy barrier that has to be overcome before a reaction can occur. A comparison that is often used is that of a boulder resting on top of a hill (Figure 4.5). Although it will naturally roll down the hill the boulder is prevented from doing so by a small mound of earth.

There are two ways to get the boulder to roll down the hill. You can supply enough energy to push it to the top of the mound, where it can then roll down by itself. This is equivalent to supplying heat to start a reaction.

Alternatively, you could dig away at the mound, reducing the energy needed to push the boulder to the top of the mound. This is equivalent to supplying an enzyme to a reaction (Figure 4.6).

substrate enzyme enzyme—substrate complex

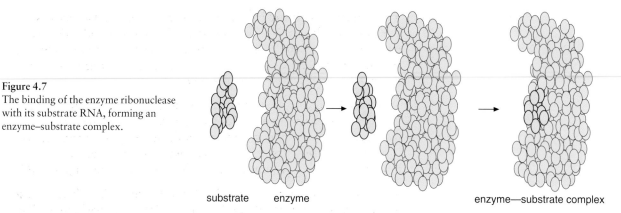

Figure 4.7
The binding of the enzyme ribonuclease with its substrate RNA, forming an enzyme–substrate complex.

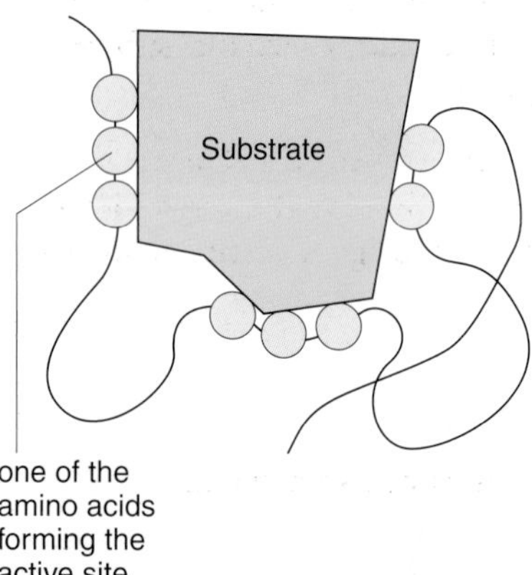

Figure 4.8
Diagram illustrating the position of the amino acids forming the active site.

Instead of actually supplying energy, the enzyme reduces the height of the energy barrier and therefore reduces the activation energy necessary for a reaction to take place. This may not seem very impressive but the involvement of an enzyme increases the rate of reaction by a billion times! Also, by lowering the activation energy, the reaction can take place at the temperatures and pressures which exist inside living cells.

Binding for activity

Substrate molecules bind to the active site of an enzyme. See for example the binding of the enzyme ribonuclease with its substrate RNA, forming an enzyme–substrate complex (Figure 4.7).

The substrate molecules are usually very much smaller than the enzymes. When substrates bind to the enzyme they slot neatly into the active site and are close to the amino acids which form it (Figure 4.8).

At first it was thought that an enzyme's active site was merely a negative impression of its substrate. This idea was called **'lock and key'** because the substrate seems to fit into the active site as a key fits into a lock (Figure 4.9).

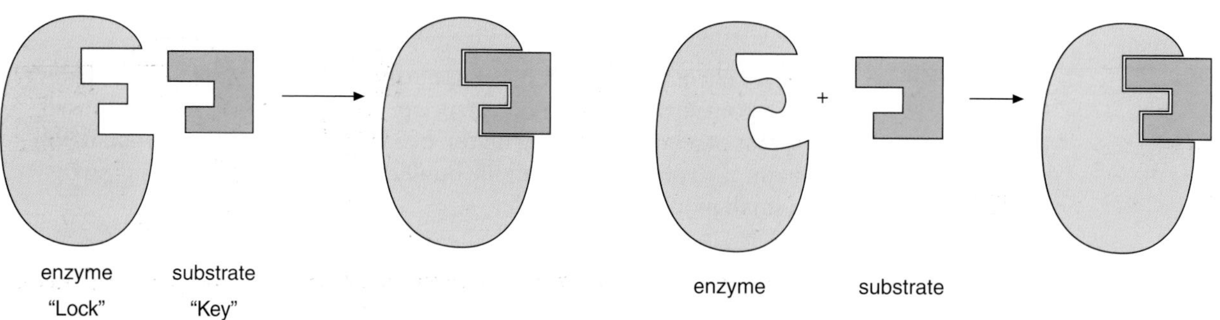

Figure 4.9
The lock and key hypothesis.

Figure 4.10
The induced fit theory. The change in the active site brings the amino acids into their correct positions in the active site so a reaction can occur.

The enzyme was pictured as a rigid structure, which had an active site that was complementary to the substrate. However, this idea did not explain how other molecules could alter an enzyme's activity by binding to it at some other site that was not the active site. If the enzyme were flexible then such an effect could be explained.

When a substrate combines with an enzyme, it induces changes in the enzyme's shape. The amino acids which make up the active site are moulded into a precise form, which enables the enzyme to perform its catalytic function effectively. For this reason biologists refer to this modified theory as the **induced fit** of substrate to enzyme (Figure 4.10).

Thus the shape of the enzyme molecule is affected by the substrate, just as the shape of a glove is affected by the hand that is placed into it. The flexible enzyme molecule wraps itself around the substrate. The enzyme molecule in turn distorts the substrate in the resulting enzyme–substrate complex, causing a reaction to occur rapidly.

Once the reaction has occurred and products formed, which are a different shape to the substrate, they no longer bind to the active site and diffuse away. The flexible enzyme returns to its original shape, ready to bind to the next molecule of substrate.

Enzyme reactions

The enzyme combines reversibly with the substrate to form an enzyme–substrate complex:

enzyme + substrate → enzyme–substrate complex

$$E + S \rightarrow ES$$

The enzyme–substrate complex then breaks down to give the product and releases the enzyme in an unchanged form:

enzyme–substrate complex → product + enzyme

$$ES \rightarrow P + E$$

We can represent this using a simple diagram (Figure 4.11).

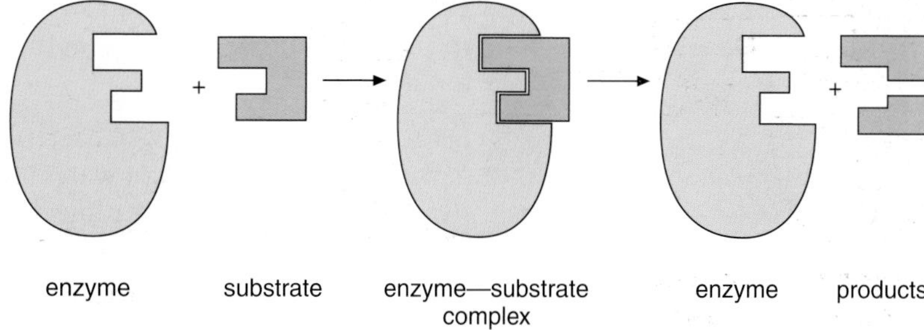

Figure 4.11 Illustration of an enzyme-controlled reaction.

Enzyme controlled reactions are affected by a number of factors. How they affect the reactions can be understood using the principle described above.

Effect of temperature on enzyme activity

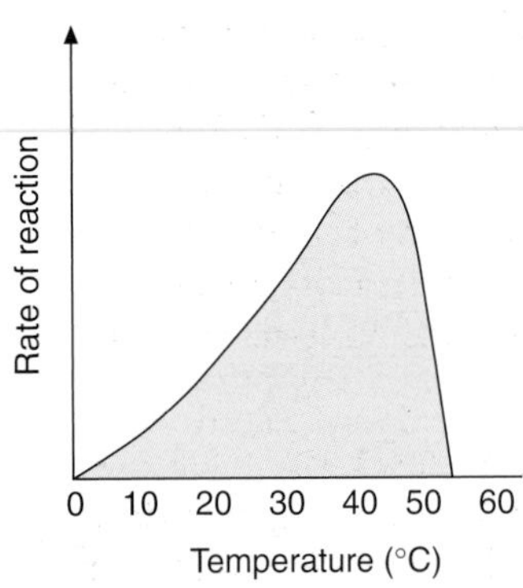

Figure 4.12
The effect of temperature on an enzyme-controlled reaction.

The graph in Figure 4.12 shows the rate of a typical enzyme-catalysed reaction and how it varies with temperature. The rate of an enzyme reaction is measured by the amount of substrate changed or the amount of product formed during a period of time.

At low temperature, say between 5°C and 30°C, increasing the temperature provides more heat energy. This increases the kinetic energy and makes both the enzyme molecules and the substrate molecules move faster. There will be an increase in the number of collisions between the substrate and the active site of the enzyme, resulting in more enzyme-substrate complexes and in turn the formation of more products.

As the temperature continues to increase, to above 40°C, the enzyme and substrate molecules move even faster. However, the structure of the enzyme molecule vibrates so energetically that some bonds holding the tertiary structure break. This is especially true of hydrogen bonds. The enzyme begins to lose its globular shape, which affects the active site such that the substrate will no longer fit into it. The enzyme is said to be denatured and will not regain its correct shape even if the temperature is lowered.

The temperature at which an enzyme catalyses a reaction at the maximum rate is called the optimum temperature. This can vary considerably for different enzymes, from 2°C to 78°C.

If the temperature is reduced to near or below freezing point, enzymes are inactivated, not denatured. They will regain their function when higher temperatures are restored.

Q

2 **Why is it an advantage for humans to have a constant body temperature of 37°C.**

3 **Egg albumen, when heated goes solid and white. What do you think has happened to the proteins in the egg white and why does the white not become runny again when the egg is cooled?**

Extension box 1

Until recently, heat-loving, **thermophilic**, organisms e.g. the bacterium *Thermus aquaticus*, which grows optimally at 70°C in hot volcanic springs, were thought to exist at the highest temperatures possible for life.

In the last 15 years however, bacteria that grow in temperatures beyond the boiling point of water (100°C) have been found. *Pyrococcus furiosus* (the 'furious fireball') and *Thermotoga maritima* known as **hyperthermophiles** are not able to grow below 70°C.

Many biological molecules, which include enzymes, DNA and RNA are denatured rapidly by heat; the theoretical limit for stability was thought to be close to 90°C, the temperature above which the DNA helix cannot form. However, nature defies such predictions; the record for thermo-tolerance is held by *Pyrodictium abyssi*, which can grow at

113°C. DNA from hyperthermophiles is no different from that of other organisms. However, in all cells, DNA is usually wrapped in proteins known as histones, which protect it from breakage but allow contact with the enzymes that replicate the DNA. The stability of DNA in hyperthermophiles is probably due to an unusually heat resistant (thermostable) form of the histones. Proteins consist of chains of amino acids (polypeptides) folded in a very precise manner. Heat damages proteins by breaking hydrogen bonds that maintain the tertiary shape of the protein.

Proteins and enzymes in hyperthermophiles are remarkably thermostable. For example, a starch-degrading enzyme, α-glucosidase, from *Pyrococcus* is optimally active at 115°C. However, enzymes from organisms that live at normal temperatures and the same type of enzymes from hyperthermophiles appear structurally similar. When the structures of enzymes are compared, it is clear that their overall structures are remarkably similar, so the difference that allows the protein from the hyperthermophiles to survive must lie within the fine detail.

Several factors may account for the thermostability of proteins in hyperthermophiles.

- More bonds may be used to maintain the protein shape.
- The loops of polypeptides extending from the protein surface may be absent or reduced.
- There may be fewer of the amino acids that are particularly unstable at high temperature.

Figure 4.13
A hot spring in Yellowstone National Park, USA.

More than 80% of the biosphere, including many areas of the oceans, never exceeds 5°C, yet life abounds there. Even colder habitats exist in polar regions, for example the surface of ice floes where high concentrations of salt depress the freezing point of the water. Here grow the cold-loving organisms, **psychrophiles**. The lowest temperature at which life is known to exist is −10°C. Sub zero temperatures cause water to freeze in the cell, denying the cell free water molecules for metabolism. Ice crystals which form at these temperatures can rupture cell membranes. Most psychrophiles grow very slowly and might be expected to have specially adapted cell components. Indeed, proteins from *Bacillus TA41* isolated from Antarctic seawater are adapted to be active at low temperatures. Less is known about psychrophiles than about other extremophiles because they are so difficult and slow to grow. Also their biotechnological potential is less obvious.

Uses in biotechnology for the enzymes from extremophiles are emerging rapidly. Most enzymes are not robust molecules and industry often employs reactions that proceed at relatively high temperatures.

- The polymerase chain reaction (PCR), which enables many copies of specific, DNA sequences to be made in a test tube.
- The food processing industry uses a great many enzymes, such as proteases, lipases and cellulases.
- Many domestic detergents operate best in warm, alkaline conditions. Biological washing powders already contain enzymes from thermophiles.

Effect of pH on enzyme activity

pH is a measure of the concentration of hydrogen ions in a solution. The higher the hydrogen ion concentration, the lower the pH. Most enzymes function efficiently over a narrow pH range. A change in pH above or below this range reduces the rate of enzyme activity considerably (Figure 4.14).

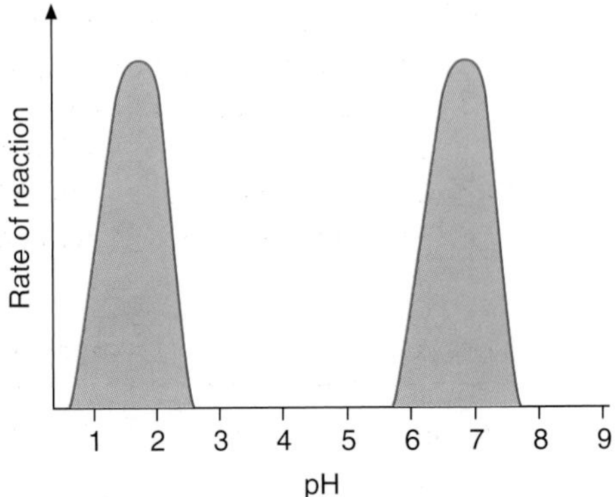

Figure 4.14
The effect of pH on the rate of an enzyme-controlled reaction.

pH has two main effects on the action of an enzyme:

- changes in pH lead to the breaking of the ionic bonds that hold the tertiary structure of the enzyme in place. The enzyme begins to lose its functional shape, particularly the shape of the active site, such that the substrate will no longer fit into it. The enzyme is said to be denatured.
- changes in pH affect the charges on the amino acids within the active site such that the enzyme will not be able to form an enzyme–substrate complex.

The pH at which an enzyme catalyses a reaction at the maximum rate is called the optimum pH. This can vary considerably from pH 2 for pepsin to pH 9 for pancreatic lipase.

Q 4 **Amylase is an enzyme found in the buccal cavity at pH 8. What happens to this enzyme when it finds its way into the stomach?**

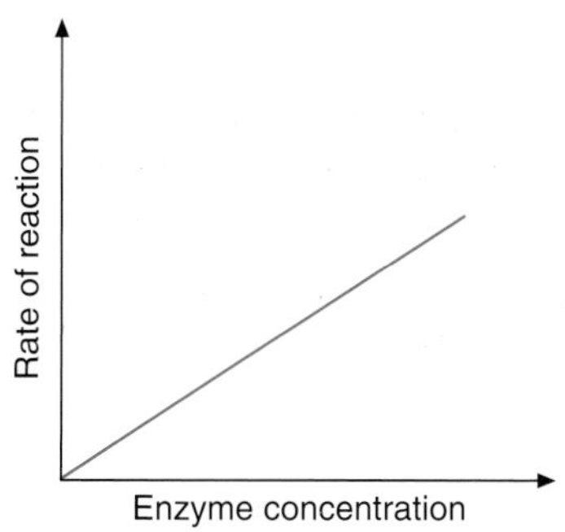

Figure 4.15
The effect of enzyme concentration on the rate of an enzyme-controlled reaction.

Effect of enzyme concentration

Provided that the substrate concentration is maintained at a high level, and other conditions such as pH and temperature are kept constant, the rate of reaction is directly proportional to the enzyme concentration (Figure 4.15). The reason for this is that if more enzymes are present then there are more active sites available for the substrate to slot into.

Effect of substrate concentration

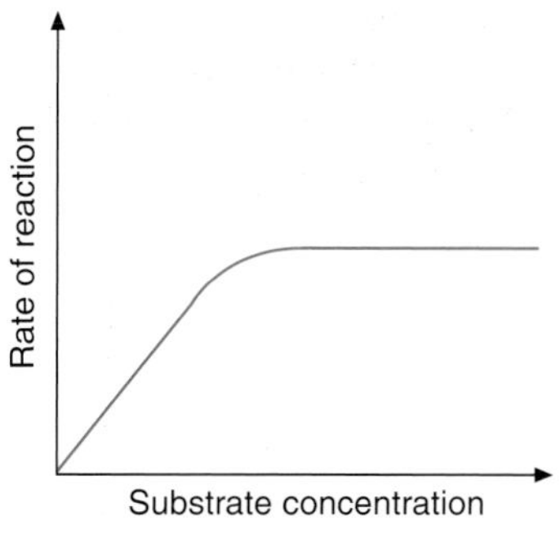

Figure 4.16
The effect of substrate concentration on the rate of an enzyme-controlled reaction.

As the substrate concentration increases, for a given enzyme concentration, then the rate of reaction also increases. As there are more substrate molecules present, an enzyme's active site can bind with more new substrates in a given time (Figure 4.16).

However, if the substrate concentration continues to increase, with a constant enzyme concentration, there comes a point where every enzyme's active site is forming enzyme–substrate complexes at its maximum rate. If more substrate is added, the enzyme simply cannot bind with the substrate any faster; substrate molecules are effectively queuing up for an active site to become vacant. Thus any extra substrate has to wait until the enzyme–substrate complex has dissociated into products and left the active site before it can complex with the enzyme. The enzyme is said to be working at its maximum possible turnover rate and the rate of reaction reaches a plateau. The maximum possible turnover rate is usually defined as the number of substrate molecules turned into product in one minute by one molecule of enzyme. Values range from less than 100 to many millions (see Table 4.1).

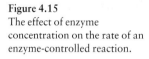

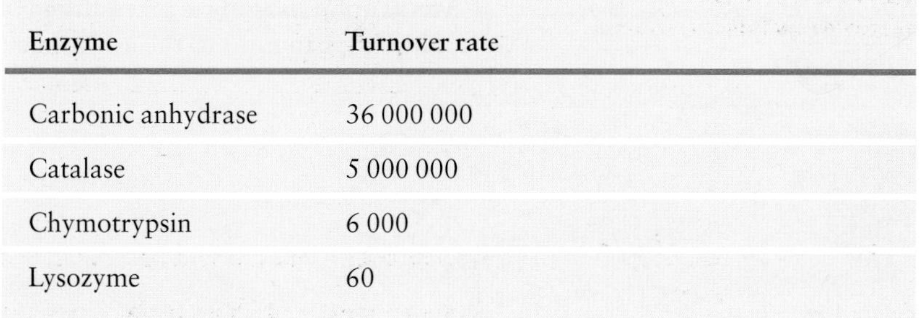

Enzyme	Turnover rate
Carbonic anhydrase	36 000 000
Catalase	5 000 000
Chymotrypsin	6 000
Lysozyme	60

Table 4.1

At high substrate levels, both enzyme concentration and the time it takes for dissociation of the enzyme–substrate complex molecule limit the rate of reaction.

Enzyme inhibition

Competitive inhibitors

key term

Enzyme inhibitors are molecules that can reduce the rate of an enzyme-controlled reaction.

These molecules have a shape similar to that of the enzyme's normal substrate so that they can fit into the active site and form an enzyme–inhibitor complex (Figure 4.17).

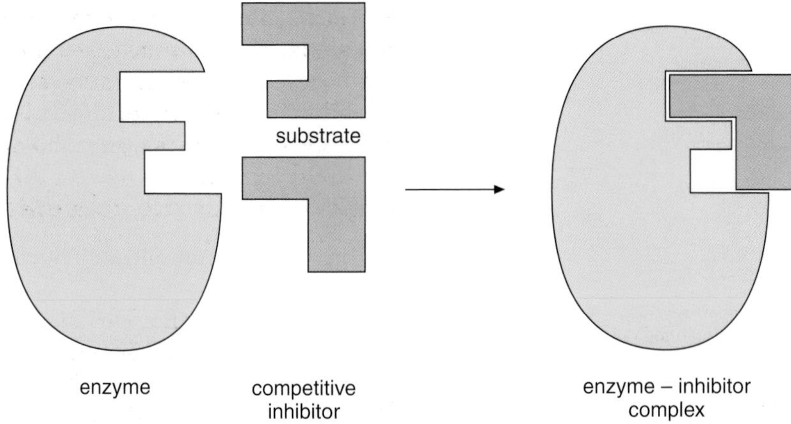

Figure 4.17
The competition between a competitive inhibitor and a substrate for the active site on an enzyme

If the competitive inhibitor molecule is in the active site, no other molecule can enter the site. While it remains there it prevents access for any molecules of the true substrate.

The inhibitor and substrate compete for the active site. The molecule that is most likely to form a complex with the enzyme will be the one present in the highest concentration. If the concentration of substrate increases, the level of inhibition is reduced. So at high substrate concentrations the inhibitor has little or no effect because the relatively small number of inhibitor molecules is overwhelmed by the number of substrate molecules present (Figure 4.18).

Methanol poisoning is an example of competitive inhibition. Methanol (CH_3OH) can bind to the active site of the enzyme dehydrogenase, whose

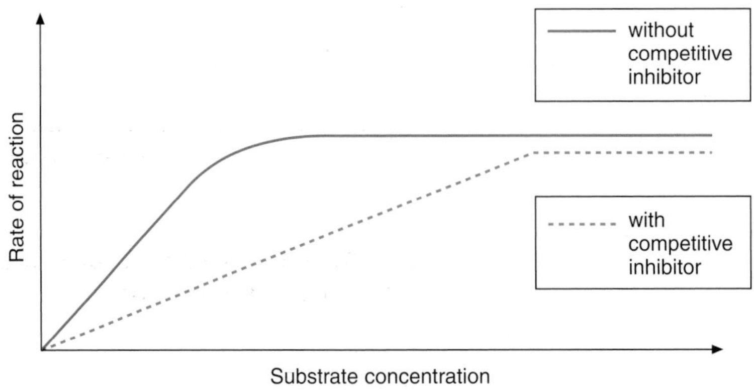

Figure 4.18
The effect of substrate concentration on the rate of a reaction with and without a competitive inhibitor.

true substrate is ethanol (CH_3CH_2OH). A person who has accidentally swallowed methanol is treated by being given large doses of ethanol, which competes with methanol for the active site.

Non-competitive inhibitors

This type of inhibitor has no real structural similarity to the substrate and forms an enzyme–inhibitor complex in a 'pocket' on the enzyme other than its active site (Figure 4.19). It has the effect of altering the globular structure of the enzyme and the shape of the active site, so that even though the true substrate is present, it is unable to bind with the enzyme. In this case, if the concentration of the substrate is increased, the degree of inhibition is not affected because the increased number of substrate molecules does not affect the inhibitor's ability to bind with the enzyme. Even at high concentrations of substrate therefore a non-competitive inhibitor will always reduce the rate of reaction (Figure 4.20).

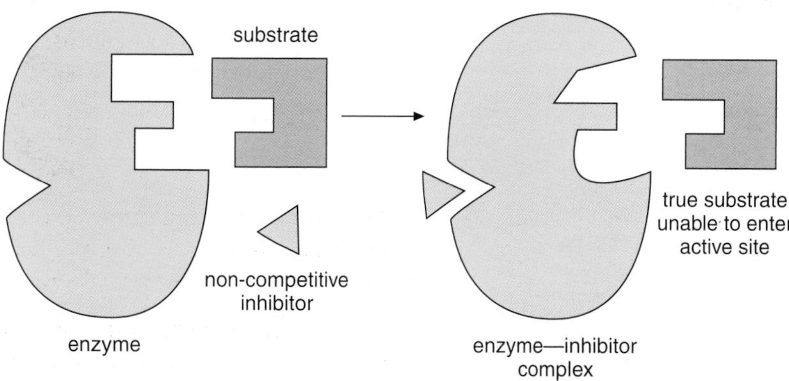

Figure 4.19
The effect of a non-competitive inhibitor on an enzyme in a metabolic pathway.

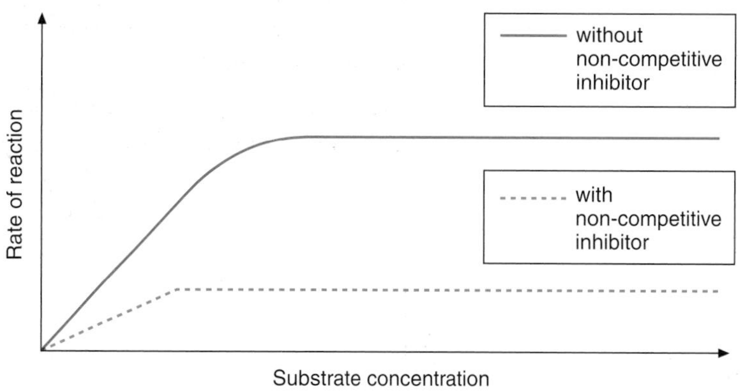

Figure 4.20
The effect of substrate concentration on the rate of a reaction with and without a competitive inhibitor.

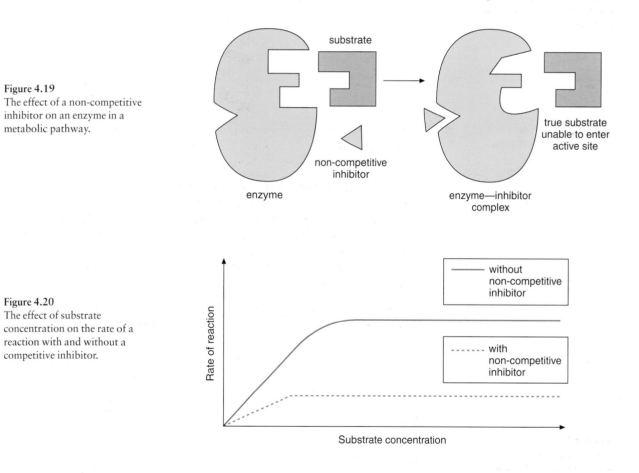

Q 5 **Write down the main differences between competitive and non-competitive inhibitors.**

End-product inhibition

Metabolic reactions within a cell are normally multi-stepped reactions, with each step being controlled by a single enzyme (Figure 4.21). The end products of this process may accumulate within the cell and it may be important for the reaction to stop when sufficient product has been made. This is achieved by non-competitive inhibition of an enzyme earlier in the reaction sequence by the end product.

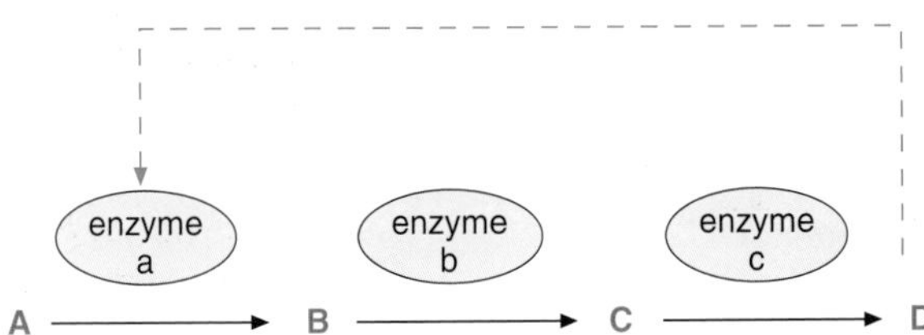

Figure 4.21
The effect of an end-product inhibitor on an enzyme earlier in a metabolic pathway.

This is an example of a negative feedback mechanism serving to control an aspect of metabolic activity.

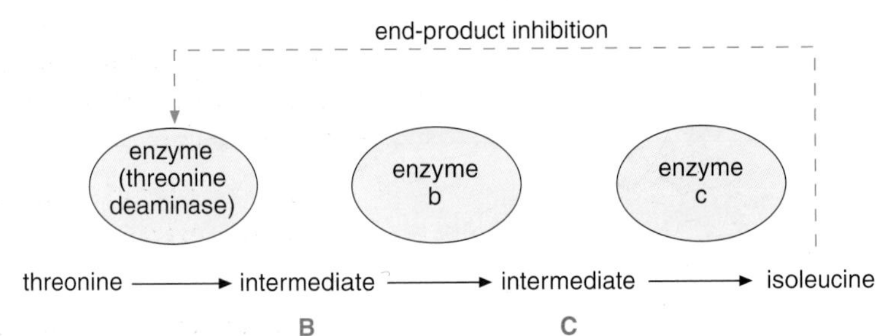

Figure 4.22
End-product inhibition. When sufficient isoleucine has been made, the isoleucine acts as a non-competitive inhibitor of threonine deaminase, changing the shape of the active site, thus preventing more threonine from reacting.

Summary

- Enzymes consist of globular proteins
- Enzymes act as catalysts by lowering activation energy through the formation of enzyme–substrate complexes
- There are two models of enzyme action: the lock and key and induced fit models
- Temperature, pH, concentration of enzyme and concentration of substrate affect the rate of enzyme-controlled reactions
- Competitive and non-competitive inhibitors decrease the rate of enzyme-controlled reaction.

Examination questions

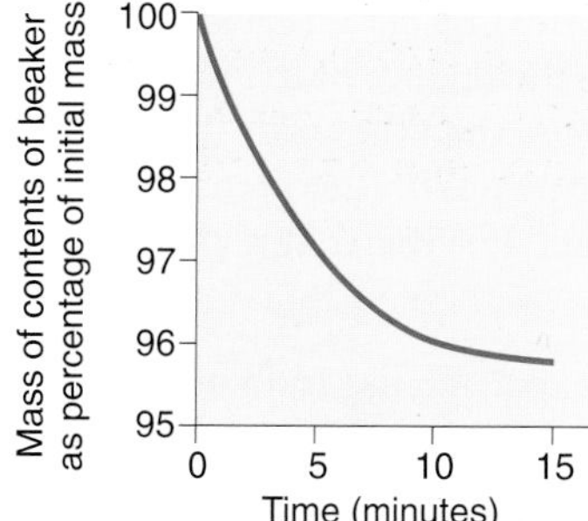

Figure 4.23

1 Enzymes are catalysts that catalyse specific reactions by lowering their activation energy. The lock and key and the induced fit models have been used to explain the way in which enzymes work.

(a) Explain what is meant by *activation* energy.

(*1 mark*)

(b) (i) Describe how the lock and key model can be used to explain how an enzyme breaks down a substrate molecule.

(*3 marks*)

(ii) Describe how the induced fit model differs from the lock and key model of enzyme action.

(*2 marks*)

Catalase is an enzyme found in many cells. It catalyses the breakdown of hydrogen peroxide:

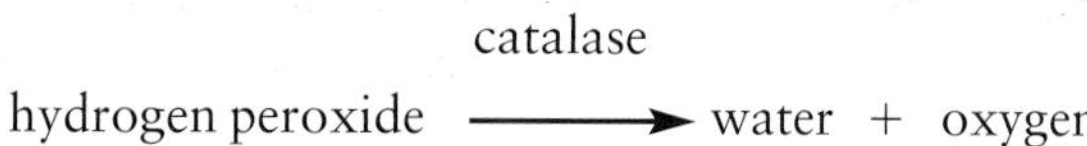

Cylinders of potato were cut with a cork borer. The cylinders were then sliced into discs of the same thickness and put into a small beaker containing 50 cm^3 of hydrogen peroxide. The mass of the beaker and its contents was recorded over a period of 15 minutes. The results are shown in Figure 4.23.

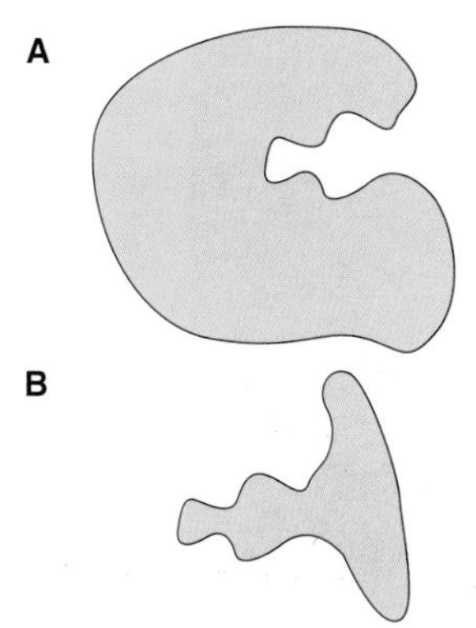

Figure 4.24

(c) Explain why the mass of the contents of the beaker fell as the reaction progressed.

(*1 mark*)

(d) Explain, in terms of collisions between enzyme and substrate molecules, why the rate of the reaction changed over the period of time shown on the graph.

(*2 marks*)

2 (a) Figure 4.24 shows an enzyme, and B is the substrate of this enzyme. By drawing on this diagram, show how a competitive inhibitor would affect the activity of the enzyme.

(*2 marks*)

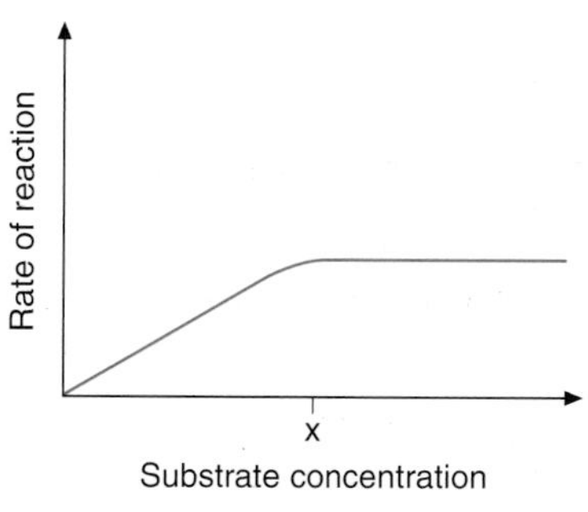

Figure 4.25

(b) Figure 4.25 shows the effect of changing substrate concentration on the rate of an enzyme-controlled reaction.

Explain why increasing the substrate concentration above the value shown at X fails to increase the rate further.

(*2 marks*)

(c) Explain how adding excess substrate could overcome the effect of a competitive inhibitor.

(*2 marks*)

Assignment

Most, if not all, biologists would describe their subject as a science and they would probably define biology as being the science of living organisms. But what is science? It is much more than the study of facts. Science is about solving problems by suggesting possible explanations or hypotheses and testing these with experiments. It is this that really defines science and makes it unique.

Biologists must obviously be able to design experiments, and you will be assessed on your ability to do this in your course work. Although every investigation is different, the principles adopted in designing an experiment are the same. In this assignment we will look at these basic principles and then apply them to one particular investigation.

Figure 4.26

We will start by taking an overall view. Designing and planning an experiment requires you to do three things.

- Make sure that you have a clear idea of what it is you are trying to test. Look at the simple sentence below. Filling in the gaps in this sentence should help you with this.
- As .... is changed, ... will change as a result.
- Identify the factor you need to change. This is the independent or manipulated variable. Decide how you will change it.
- Identify the factor that will change as a result. This is the dependent variable. You will need to decide how you will measure changes in the dependent variable so that you get quantitative results.

To summarise, make sure you know what it is that you are trying to do, decide how to change the independent variable and decide how to measure the dependent variable. Now, let us look at a particular problem.

When fruit like the tomatoes in Figure 4.26 ripen, several changes take place. The skin goes from green to red; they taste sweeter and they get softer. What causes them to get softer? We can use our knowledge of biology to suggest an explanation. Pectin is a substance found in and between plant cell walls which helps to bind cells together. As a fruit such as a tomato ripens, more of the enzyme pectinase is produced. Pectinase breaks down pectin into smaller, soluble molecules. As a result the cells separate from each other and the fruit gets softer.

Figure 4.27

The first step is to go back to the incomplete sentence in the first bullet point and make sure that we know exactly what we are trying to do. Filling in the gaps gives us the idea that:

As *the softness of the fruit* is changed, *the amount of pectinase in the tomato* will change as a result.

What we need to do then is to collect some tomatoes, ranging from hard and firm to soft and squashy, and find a way of measuring their squashiness. You can't buy a squashometer so you will need to design one for yourself.

1 (a) Think about tomatoes. If you apply a constant force to tomatoes of different ripeness, a very ripe tomato would respond differently to an unripe tomato. How?

(b) Use your answer to (a) to design a simple device that can be used to measure the squashiness of a tomato. You should describe it in sufficient detail for another member of your class to construct it without further help.

The next step is to isolate the pectinase enzyme from the tomatoes and measure its activity. You will need the information below to do this.

- Enzymes are soluble. The juice from a tomato will contain pectinase.
- Pectin is sold in many chemists and supermarkets. It is called Certo and is used by people who make jam because it helps the jam to set. Certo is a thick liquid. As it is broken down by pectinase it becomes thinner and flows more easily.

2 (a) Describe how you will get your samples of tomato juice from tomatoes of different softness.

(b) Suggest how you could measure how easily a Certo solution flows.

(c) Now draw a flowchart summarising the main steps in your procedure.

Another of the experimental skills you will need to develop is the ability to evaluate your ideas. You should be able to identify sources of error that are likely to affect the reliability of your results.

3 What are the steps in your procedure that are likely to make your results unreliable? Is there anything you can do about them at this stage?

Once we have the outline in place, we must ask ourselves if there are any other factors that could influence the results of this experiment and should therefore be kept constant. Remember, an enzyme is involved and there are a number of factors that influence the rate of enzyme-controlled reactions. If any of these are allowed to vary they will affect the reliability of the results.

4 List the factors that could influence the rate at which pectinase breaks down pectin. Suggest how you will keep each of these constant in your experiment.

There is a final experimentation step and it concerns setting up controls. The best approach to this is to look at the experiment and predict the results you would expect if it supports your hypothesis. In this case, we would expect the pectin to be broken down more quickly in the riper, squashier tomatoes because there is more pectinase in these tomatoes. But we must be careful.

5 Can you think of an explanation for your predicted results other than an increase in the amount of pectinase? If the answer is yes, you need to design a control to eliminate this possibility.

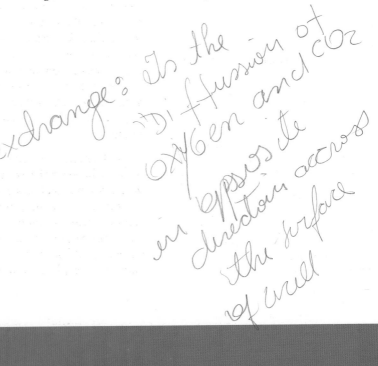

Gas exchange: Is the diffusion of oxygen and CO_2 in opposite direction across the surface of cell

5 Gas Exchange

Throughout much of human history, people have been obsessed with climbing tall mountains. One small but important branch of biology deals with the effect of high altitude on humans, as well as other animals. All tissues and organs, but especially the brain, are affected by a lack of oxygen (Figure 5.1). Thomas Hornbein, an anaesthetist, who climbed Mount Everest, concluded that at extreme altitude or low oxygen levels 'the brain...rather than the exercising muscle is the organ ultimately limiting function.'

Figure 5.1
To survive at high altitude humans normally require a supply of oxygen.

Extreme altitude is defined as a height greater than 5800 metres above sea level. At such heights, an organism is besieged by bitter cold, high winds, low humidity and high levels of solar and ultraviolet radiation. However, by far the most important physical challenge is hypoxia, or low levels of oxygen reaching the body tissues. The percentage of oxygen in the air actually remains the same, 21%. However, the atmospheric pressure drops steadily with increasing height, so that on top of a 3500 m mountain, for example Ben Nevis, the amount of oxygen present is only 65% that at sea level and at the summit of Mount Everest (8850 m) it is only 32% (see Figure 5.2).

At extreme altitude, most people experience a sharp decrease in appetite, breathlessness at rest, muscular fatigue, exhaustion on attempting any sort of exertion and a falling off of mental capacity. They have a significant reduction in the speed of their reactions and in their ability to make decisions. Also memory is also noticeably impaired. As a climber presses on to greater heights, brain tissue is further deprived of oxygen and bizarre mental states often occur.

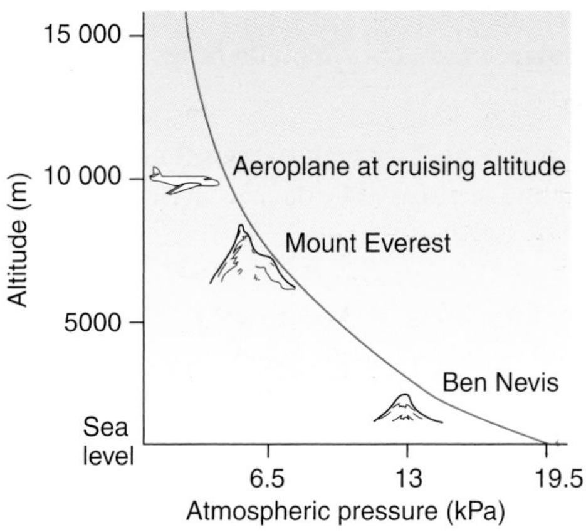

Figure 5.2
At higher altitudes, the atmospheric pressure is lower and as such the partial pressure of oxygen is less.

All living cells require a source of energy in order to survive. The energy is used for growth and for the processes which maintain each cell. All cells obtain their energy by the process of respiration. Respiration occurs in every cell within the body, generating adenosine triphosphate (ATP), using oxygen and producing carbon dioxide as a waste product. As a result, oxygen is depleted inside the cells whilst carbon dioxide accumulates. The diffusion of oxygen and carbon dioxide in opposite directions across the surface of a cell is called **gas exchange**.

In small animals, oxygen and carbon dioxide travel the entire distance between the external environment and mitochondria, where respiration takes place, by diffusion alone. In larger and more active animals, gas exchange over the general body surface cannot satisfy the animals' needs. One reason for this is that larger animals do not have a large enough surface area to exchange sufficient amounts of gases for the increasing demand made by a very large number of cells.

If the size of any organism increases, the relationship between its surface area and its volume changes. As Figure 5.3 shows, if you increase the size of an organism by doubling the length of each side, keeping the shape constant, the surface area (and hence the oxygen it can absorb) is increased by four times. However, its volume (and therefore the oxygen it needs) is increased by eight times. Thus by increasing body length the ability to supply oxygen by diffusion through the surface is reduced. Conversely, with a small organism, its volume decreases faster than its surface area. Thus the surface area:volume ratio is greater in small organisms.

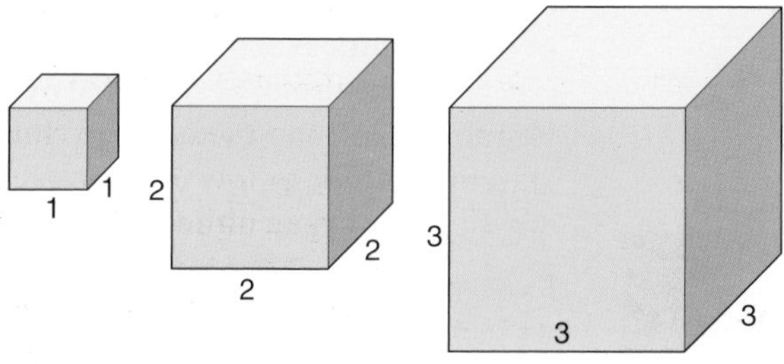

Figure 5.3
The relationship between the size of an organism and its surface area to volume ratio

Volume	1	8	64
Surface area	6	24	96
Surface area: volume ratio	6:1	3:1	1.5:1

There are three ways in which larger animals can obtain sufficient oxygen to allow respiration to occur in all cells:

- they can have a body shape that provides a large surface area relative to its volume. Some organisms develop flat bodies, which increase their surface area.
- they can develop a specialised gas exchange system, respiratory organs, which have a large surface area. These organs vary and are suited to the environment in which organisms live. They include lungs in humans and gills in fish.
- they can have a blood system to transport oxygen to cells that are far from the body surface. Oxygen and carbon dioxide are carried in the blood to and from a gas exchange organ.

Structure of the lungs

As air enters the body it is filtered, moistened and warmed by different parts of the upper respiratory tract (Figure 5.4). The nostrils are lined with small hairs that act as filters to prevent large airborne objects from penetrating the lungs. The nasal cavities warm and moisten the air and collect airborne particles on a mucous layer. Tiny hair-like structures called **cilia** (see Figure 5.5) protrude from the epithelial cells that line the nasal cavities and sweep the mucous and trapped particles back towards the throat.

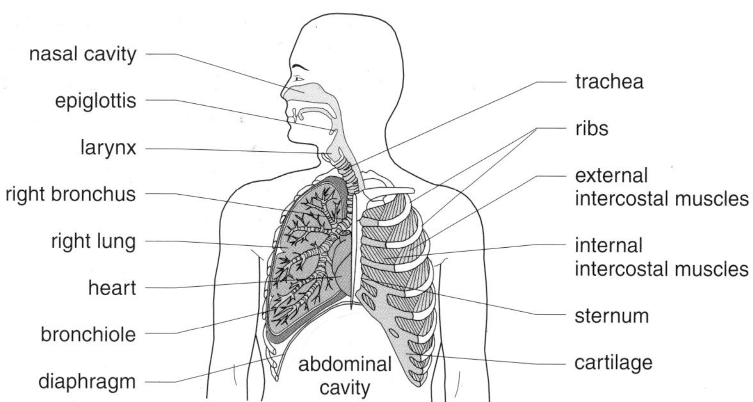

Figure 5.4
The human respiratory system.

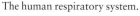

The nasal and buccal cavities are separated from each other; chewed food passes towards the oesophagus, while air entering the nasal cavities passes towards the trachea. Air, liquids and food pass through a chamber behind the tongue, the **pharynx**. To prevent food or liquid from entering the **trachea** during swallowing, a flap of tissue, the **epiglottis**, temporarily seals off the larynx, the box-like entrance to the trachea and lung system, so that food and liquids cannot enter the air passages.

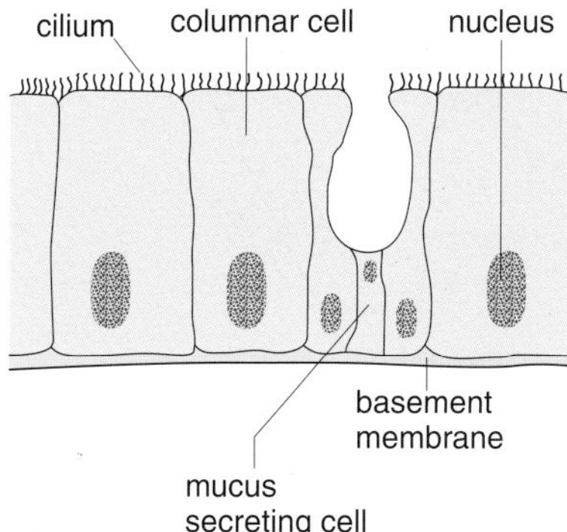

Figure 5.5 (above)
The ciliated epithelium.

The trachea is a muscular tube leading to the lungs. The inner surface of the trachea is lined with ciliated epithelial cells that also produce mucus, while the outer wall contains incomplete or C-shaped rings of cartilage that keep the trachea open at all times. The trachea divides into two **bronchi**, similar to trachea but with a smaller diameter. Each bronchus is supported by irregular plates of cartilage and, like the trachea, is lined with ciliated epithelia, which provide protection against microorganisms. The smallest bronchi divide still further into **bronchioles**. These contain no cartilage and are held open by the elasticity of the surrounding tissue. They have smooth muscles in their walls which enable their diameter to be controlled. These bronchial tubes form a system of hollow air ducts that end in clusters of tiny, thin-walled, blind-ending air sacs called **alveoli** (Figure 5.6). These average about 100 μm in diameter.

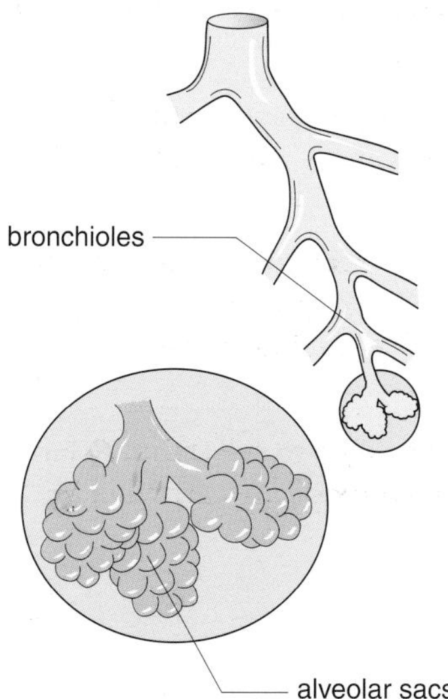

Figure 5.6 (above)
The branching network of airways leading to the alveoli.

Adult human lungs contain about 750 million alveoli, with a total surface area for gas exchange of about 80 m^2. Within the walls of each alveolus are numerous capillaries, which receive blood from the pulmonary arteries and drain it into the pulmonary veins (Figure 5.7). So dense are the capillaries that they form an almost continuous 'pool' of blood with an area estimated to be about 87% of the alveolar surface.

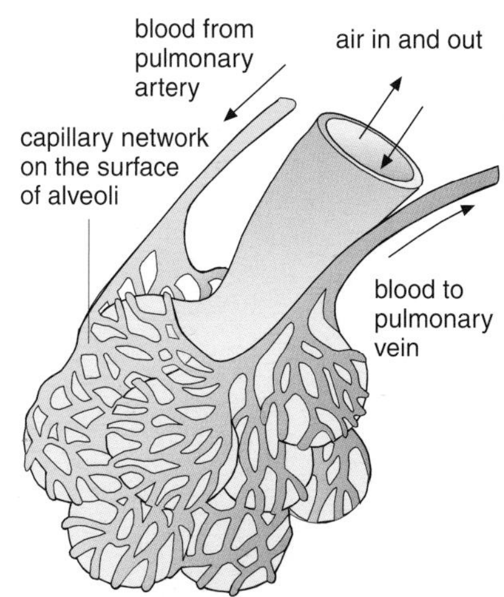

Figure 5.7 (right)
The numerous alveoli, with a dense capillary network, where gas exchange takes place.

Gas exchange in the lungs

The alveoli are the sites of gas exchange and may be regarded as the functional unit of the lungs. The walls of the alveoli are extremely thin, usually only one cell thick, and consist of squamous epithelia. Each alveolus is surrounded by a dense bed of blood capillaries (Figure 5.8).

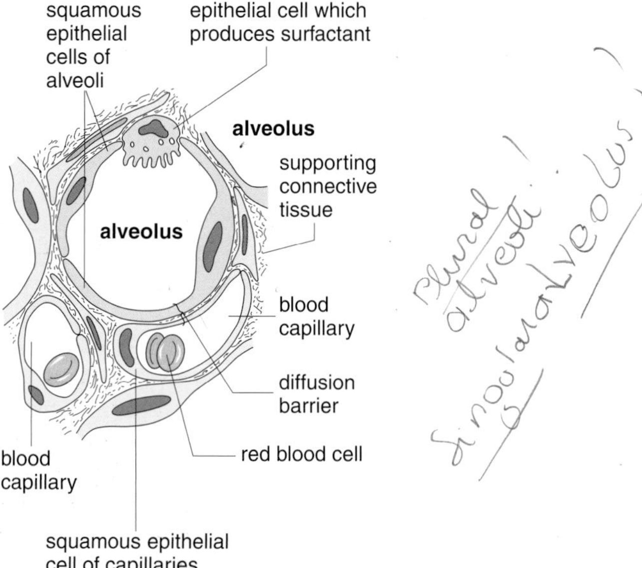

Figure 5.8
Photomicrograph of a section through several alveoli and capillaries.

Table 5.1 shows the amount of oxygen and carbon dioxide in the air moving in and out of the lungs. Other gases, such as nitrogen and water vapour, are also present. The percentage of nitrogen hardly varies, but the amount of water vapour in exhaled air is usually greater than that in inhaled air.

	Oxygen	Carbon dioxide
Inspired air	20	0.04
Alveolar air	13	5
Expired air	15	4

Table 5.1 The percentage of oxygen and carbon dioxide in inspired air, alveolar air and expired air.

Oxygen entering the alveolus dissolves in the film of water on its wall and then moves by diffusion across cells to the blood. As you can see from Figure 5.9, the blood flowing to the alveolus has a higher concentration of carbon dioxide and a lower concentration of oxygen than alveolar air. Under these circumstances oxygen and carbon dioxide diffuse down their respective concentration gradients: oxygen from the alveoli diffuses into the blood and carbon dioxide from the blood diffuses into the alveoli.

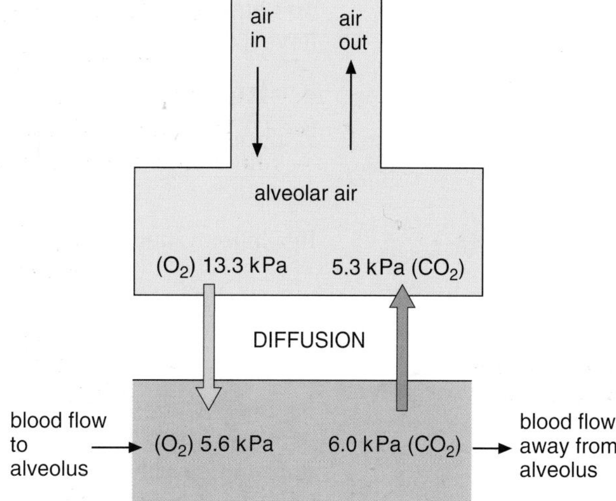

Figure 5.9
The exchange of oxygen and carbon dioxide that takes place as blood flows past an alveolus.

The movements of these gases are cases of simple diffusion and require no energy. There are actually few instances where the process of simple diffusion takes place within the body, but this is the case in the lungs. As a result, when the partial pressure of oxygen in the atmosphere falls below normal, oxygen deprivation rapidly develops. This is because the reduced concentration gradient slows the rate of diffusion and passive transport is insufficient to meet oxygen demands.

Once in the blood plasma, most oxygen passes into red blood cells, where it binds to haemoglobin and is transported through the body.

An important feature of multi-cellular animals is the presence of a gas exchange organ with a large enough surface area for sufficient intake of oxygen and release of carbon dioxide to occur to meet the demands of all body cells. However, as the body's requirement for energy and therefore oxygen increases, the rate of gas exchange must also increase. The large gas exchange organ with resulting large moist surface area has its limitations because the larger the surface area, the more water the body loses by evaporation. A compromise must be reached so that the gas exchange surface remains moist and permeable, and so these surfaces tend to be internal. Larger organisms have developed an impermeable outer layer and do not use their body surface for gas exchange.

Q

1. **Suggest why the gas exchange structures in air breathing animals tend to be internal?**
2. **Suggest why gas exchange surfaces are moist?**

Fick's law

Fick's law of diffusion states that the rate of diffusion of a gas is directly proportional to the area of the diffusion surface and to the difference in its concentration, but is inversely proportional to the thickness of the exchange surface or the distance over which it occurs:

$$\text{diffusion rate} \propto \frac{\text{surface area} \times \text{difference in concentration}}{\text{thickness of surface}}$$

This law provides an effective framework for considering how the maximum rate of diffusion of oxygen and carbon dioxide in the lungs is achieved.

To obtain the maximum rate of diffusion, a gas exchange system must have the following features:

- a large surface area
- the smallest possible diffusion pathway
- a large difference in concentration.

In a human lung these are achieved in the following way.

A large surface area

It may seem puzzling that the large surface area of the lung is achieved by having several million small alveoli rather than simply increasing the size of a few. Since the alveolar cell surface is the site of gas exchange between the air and the blood, a larger relative surface area allows a more rapid rate of diffusion of oxygen and carbon dioxide.

As explained earlier, when the length of each side of a cube doubles, the volume increases more rapidly than does the surface area. Hence the surface area:volume ratio goes down.

However, if a large cube (4 × 4 × 4) is cut into 64 small cubes (1 × 1 × 1) the total surface area becomes four times as large (see Figure 5.10). If these were alveoli, the 64 small air sacs would be better able to diffuse oxygen and carbon dioxide between the blood and the air in the lung than one large air sac.

Figure 5.10
The effect on surface area and surface area to volume ratio of dividing a large cube into smaller ones

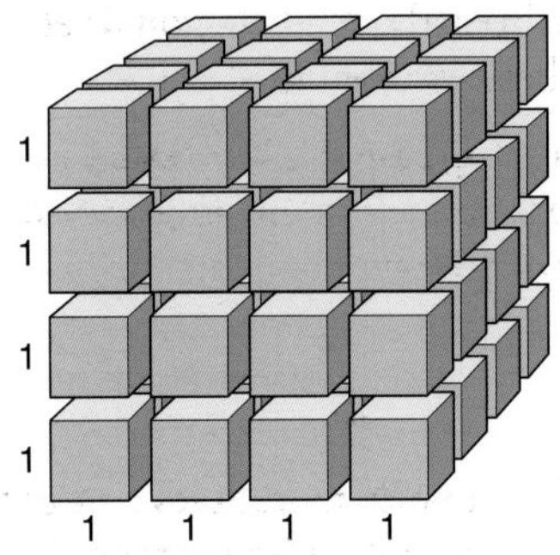

Volume	64	64
Surface area	96	384
Surface area: volume ratio	1.5:1	6:1

A small diffusion pathway

This is caused by the presence of a special type of cell, **squamous epithelium**, found in both the alveolar wall and the capillary. These cells are extremely thin. A small diffusion pathway is achieved because both the alveolar and capillary cells are so thin that the distance between the air in the lung and the blood is very short (Figure 5.11). In fact the average distance the gases travel is only about 0.5 μm

Q 3 What would happen to the oxygen concentration gradient across the gas exchange surface separating your lungs from your blood if the air in your lungs were not replaced frequently?

A large difference in concentration

This is maintained in two ways. Firstly, the partial pressure of oxygen in the alveoli is always greater than that of oxygen in the capillaries because of breathing, whereby a fresh supply of oxygen is brought to the alveoli. Secondly, the blood in the capillary is always moving, removing oxygenated blood and replacing it with deoxygenated blood. Thus it can be seen that the fine structure of the lungs is superbly adapted for gas exchange.

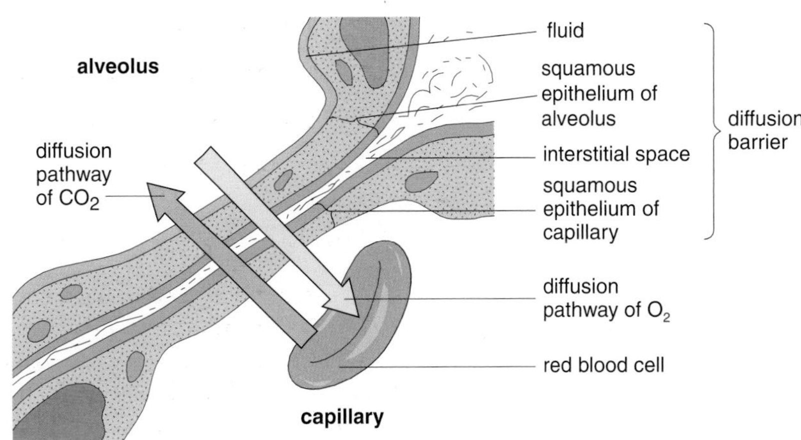

Figure 5.11
The diffusion pathway. Oxygen diffuses from the alveolus into the capillary. Carbon dioxide diffuses from the capillary to the alveolus.

The breathing mechanism

Humans fill and empty their lungs by means of a group of bones and muscles that work together. The human lungs lie within the **thoracic cavity**, a space enclosed by the ribs and separated from the abdominal cavity, which lies below by the **diaphragm**. In a human, the ribs curve forward from the backbone and meet at the breastbone. A diagonal set of muscles, the **intercostal muscles**, stretch between the ribs.

Breathing involves the alternate increase and decrease of air pressure in the lungs relative to that outside. A fall in air pressure in the lungs causes **inspiration** or breathing in; a rise in pressure in the lungs causes **expiration**, or breathing out.

Front view of rib cage and diaphragm

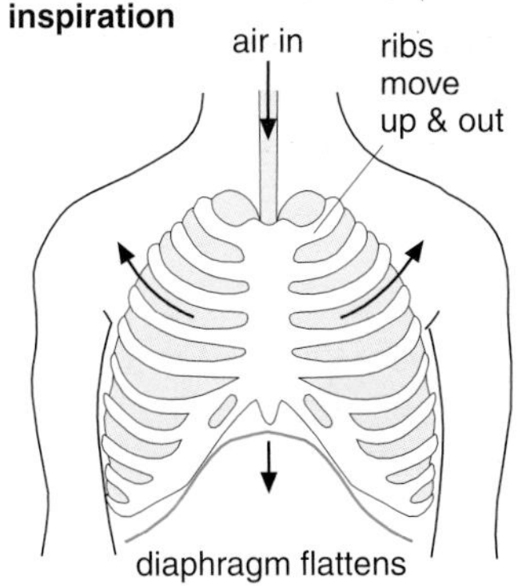

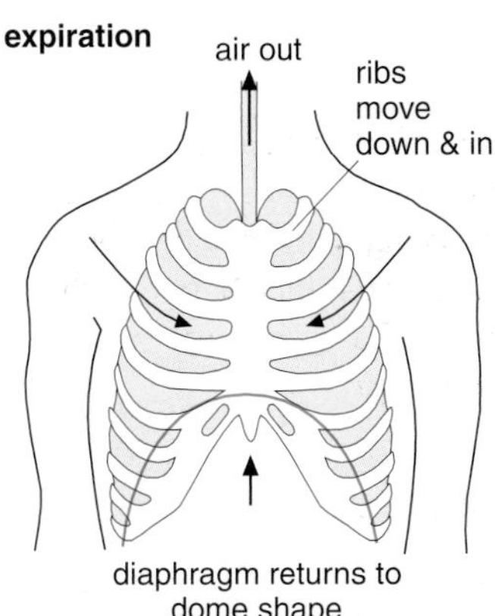

Figure 5.12
Diagram showing the movement of the ribs and diaphragm during inspiration and expiration.

Inspiration

Both the **external intercostal muscles** and the diaphragm muscles contract during inhalation of air (Figure 5.12). As the external intercostal muscles shorten, the rib cage moves up and out. At the same time, the diaphragm, which is dome-shaped at rest, flattens and pulls downward.

During inspiration, these muscular actions expand the volume of the thoracic cavity, lowering the air pressure inside the cavity below atmospheric pressure. As a result, air flows in through the nostrils, down the trachea, bronchi and bronchioles and into the alveoli.

Expiration

Expiration is largely a passive process brought about by relaxation of the external intercostal muscles and the diaphragm. The energy for expiration thus comes from energy stored as elastic tension generated during the previous inspiration. This relaxation is called elastic recoil. It results in the rib cage moving down and in, and the diaphragm returning to its original dome shape. These movements decrease the volume of the thoracic cavity, increasing the air pressure inside the cavity above atmospheric pressure and as a result air flows out.

During active exercise, however, a second set of muscles between the ribs, the **internal intercostal muscles**, contract and forcibly lower the rib cage. This expels more air from the lungs, making room for a larger volume of air with each inhalation.

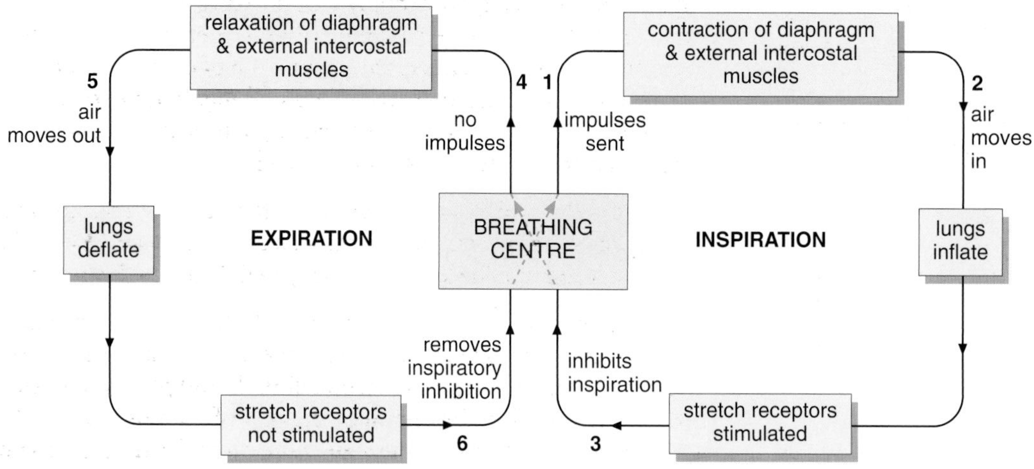

Figure 5.13
The coordination of breathing.

Coordination of breathing

Although we can breathe in or out voluntarily, breathing is largely automatic. The control centre for these automatic movements lies in the **breathing centre,** which is in the portion of the brain called the medulla. It is divided into two regions: an **inspiratory centre** and an **expiratory centre** (Figure 5.13).

When the inspiratory centre sends impulses to the external intercostal muscles via the intercostal nerves and to the diaphragm via the phrenic nerves, these muscles contract and cause inhalation.

As the lungs expand, stretch receptors in the walls of the bronchial tree are stimulated. These send impulses via the vagus nerve to the expiratory centre, which automatically cuts off inspiratory activity. This results in the diaphragm and external intercostal muscles relaxing, causing expiration. When the muscles have relaxed, the stretch receptors in the lungs are no longer stimulated therefore the expiratory centre becomes inactive and inspiration begins again. The inspiratory centre can once more send impulses to the external intercostal muscles and the diaphragm, and inhalation occurs again. This cycle is repeated again and again.

Q

4 **Explain why it is not possible to kill yourself by holding your breath?**

5 **Explain how hyperventilation is able to suppress the urge to breathe.**

Control of breathing

It is well known that both the rate and depth of breathing increase during exercise and decrease during sleep. The ventilation rate must therefore match the oxygen needs of the body. The breathing centre must have information concerning the activity of the body, but what is being monitored and where are the receptors?

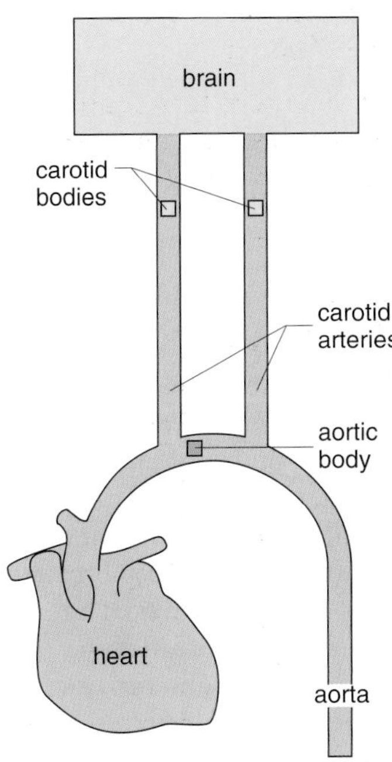

Figure 5.14
The aortic and carotid bodies.

There are two possible stimuli:

- carbon dioxide concentration: levels in the blood go up when the rate of respiration increases and more carbon dioxide is produced as a waste product
- oxygen concentration: levels in the blood go down as it is used in respiration to produce extra ATP as an energy source for exercise.

It has been shown experimentally that the oxygen level in the air has to fall from 20% to 12% before an increase in breathing rate is stimulated. In contrast, a tiny increase in the proportion of carbon dioxide in inhaled air strongly stimulates an increase in breathing. It appears therefore that the body is much more sensitive to changes in carbon dioxide concentration than to changes in oxygen levels.

In air, oxygen is normally abundant and therefore it is not likely to be a limiting factor, however, carbon dioxide removal is a problem. Thus, in humans carbon dioxide level in the body mainly drives the breathing centre.

The most important site for carbon dioxide detection is the breathing centre itself, situated in the medulla of the brain. A small group of cells, chemoreceptors, monitor the carbon dioxide and pH levels of the fluid within the brain. Two further small clusters of chemoreceptors, located in the arch of the aorta and in the wall of the carotid arteries, are known as the **aortic and carotid bodies** (Figure 5.14). These are stimulated by a rise in the level of carbon dioxide, a fall in pH and a fall in the level of oxygen in the blood. The breathing centre receives information as a nerve impulse from these chemoreceptors and uses this to regulate breathing rate.

In response to an increase in the concentration of carbon dioxide or a fall in pH (caused by carbon dioxide dissolving in the blood plasma, producing a weak acid) the inspiratory centre increases the rate of nerve impulses sent to the inspiratory muscles. This results in an increase in both the rate and depth of breathing.

The regulation of breathing is an example of **homeostasis**. When the concentration of carbon dioxide rises this stimulates an increase in breathing rate, which in turn reduces the concentration of carbon dioxide, leading to a decrease in breathing rate (Figure 5.15).

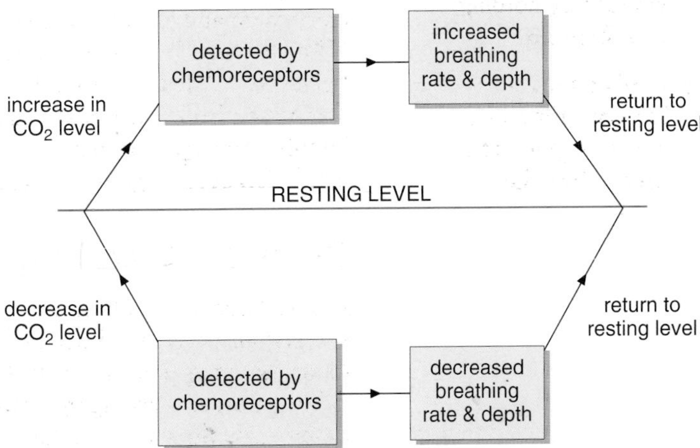

Figure 5.15
The homeostatic control of carbon dioxide.

Ventilation cycle

Q 6 **If the average breathing rate is 15 breaths per minute and the tidal volume is 0.5 litres. Calculate the ventilation rate.**

A person at rest who is breathing normally takes in and expels about half a litre of air during each ventilation cycle. This is known as the **tidal volume** and it can be measured and recorded using a spirometer.

The rate at which a person breathes is called the **ventilation rate.** This is usually expressed as the volume of air breathed per minute:

ventilation rate = tidal volume × number of breaths per minute

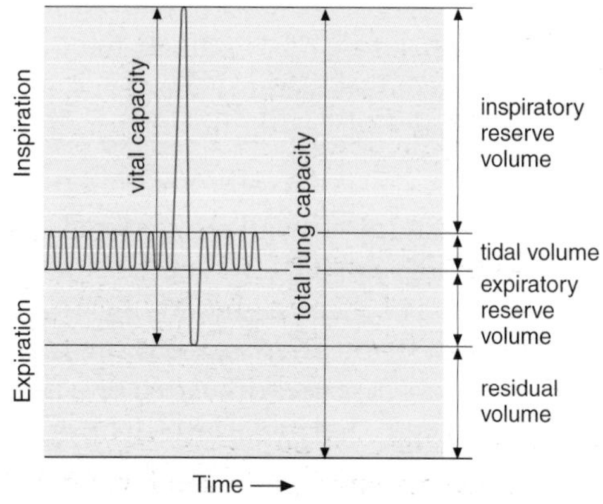

Figure 5.16
The typical lung volumes of humans.

The ventilation rate changes according to the circumstances: in muscular exercise, for example, both the frequency and depth of breathing increase, resulting in a higher ventilation rate. Lungs have a much greater potential volume than is ever used in resting conditions and this allows the ventilation rate to adapt to changing needs. If you take a deep breath, you can take into your lungs an extra 3 litres of air over and above the tidal volume. This is called the **inspiratory reserve volume** and is brought into use when required. If at the end of a normal expiration you expel as much air as you possibly can, the extra air expired is approximately 1 litre and is called the **expiratory reserve volume.** The total volume of air that can be expired after maximum inspiration is known as the **vital capacity.** The vital capacity of an average person is between 4 and 5 litres but a fit athlete may exceed 6 litres. Even after maximum expiration, approximately 1.5 litres of air still remain in the lungs. This is known as the **residual volume**. The various lung volumes are shown in Figure 5.16.

Of the half litre inspired in quiet breathing, only about 350 cm^3 gets into the parts of the lung where gas exchange takes place. The rest remains in the trachea and bronchial tubes, collectively known as the 'dead space', where no gas exchange takes place.

Extension box 1

Consider how cigarette smoke affects the lungs. Chemicals in the smoke from just one cigarette immobilise cilia in the bronchioles for several hours. The particles in smoke also stimulate mucus secretions that in time block the airways. Other chemicals in smoke can kill protective white blood cells present in the respiratory tract. What starts as 'smoker's cough' can end in bronchitis and emphysema. One in five of all smokers develop the crippling disease emphysema, which cannot be cured as the damage caused to the lungs cannot be reversed.

Obstructive lung disease, which includes bronchitis and emphysema, is the name given to a number of lung conditions in which the passage of air is obstructed. These two conditions frequently exist together, bronchitis being a disease of the bronchi and emphysema a disease of the bronchioles and alveoli. In emphysema the alveoli become permanently and abnormally enlarged due to destruction of alveolar walls.

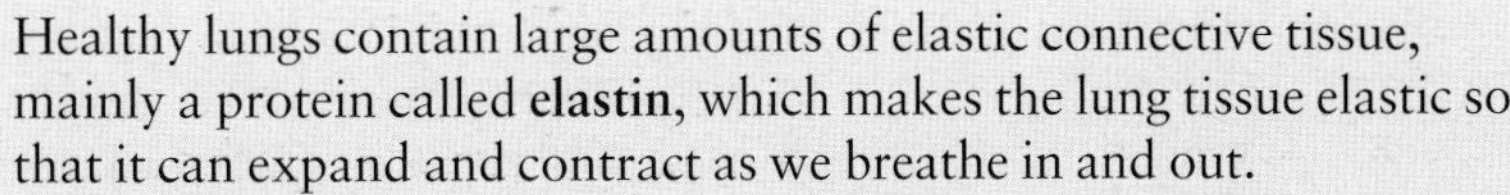

Healthy lungs contain large amounts of elastic connective tissue, mainly a protein called **elastin**, which makes the lung tissue elastic so that it can expand and contract as we breathe in and out.

The development of emphysema appears to be related to the amount of the enzyme **elastase** that acts on lung tissue. The greater the amount of elastase, the more elastin is broken down, and large holes appear in the lungs.

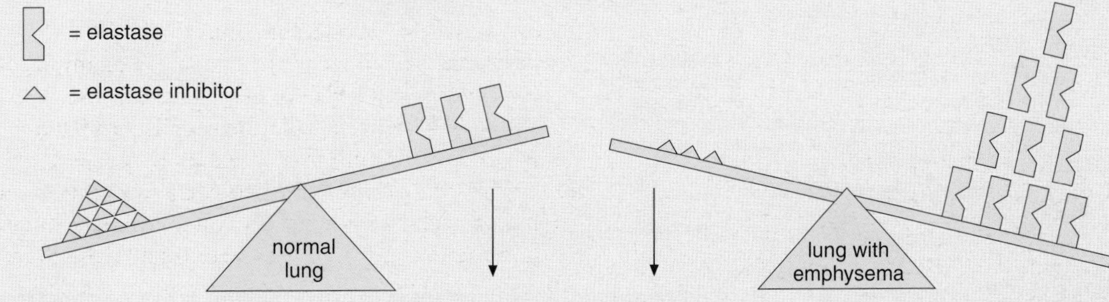

Figure 5.17
The balance between elastase and the elastase inhibitor in a normal lung and a lung with emphysema.

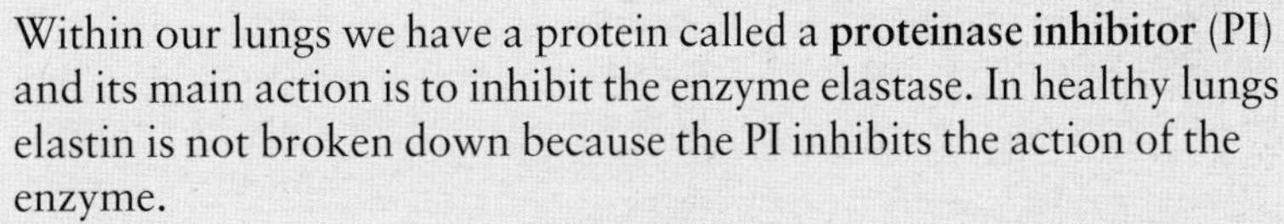

Within our lungs we have a protein called a **proteinase inhibitor** (PI) and its main action is to inhibit the enzyme elastase. In healthy lungs elastin is not broken down because the PI inhibits the action of the enzyme.

In emphysema, lung tissue is destroyed because the normal balance between elastase activity and anti-elastase activity is shifted in favour of elastase (Figure 5.17).

In lungs with emphysema, the elastin, which can be compared to elastic bands, has perished and become permanently stretched. The lungs are no longer able to force out all the air from the alveoli. The residual air left in the alveoli prevents fresh air, taken in with each breath, from reaching the alveolar walls and passing through to the blood.

The stretched and damaged alveoli are often referred to as 'dead space' because little if any exchange of gas can take place across them. As the enzyme elastase can degrade many other proteins besides elastin, there is breakdown and loss of other lung tissue too until, in extreme cases of the disease, the lung is composed of rather large holes that are non-functional (Figure 5.18).

The loss of alveoli has two important consequences. There is a decrease in surface area, resulting in reduced gas exchange, and with the loss of the elastic fibres in the alveolar walls, the lungs have a decreased ability to recoil and expel air.

The only way to minimise the chance you have of developing emphysema is not to smoke at all or give up; lung function cannot be restored to smoke-damaged lungs but giving up smoking does significantly reduce the rate of further deterioration of lung function.

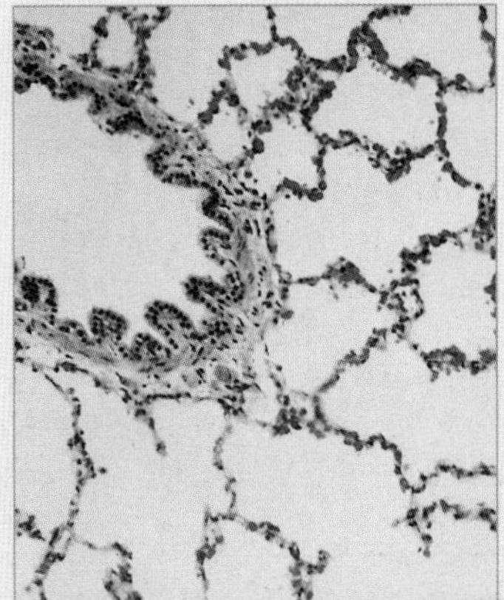

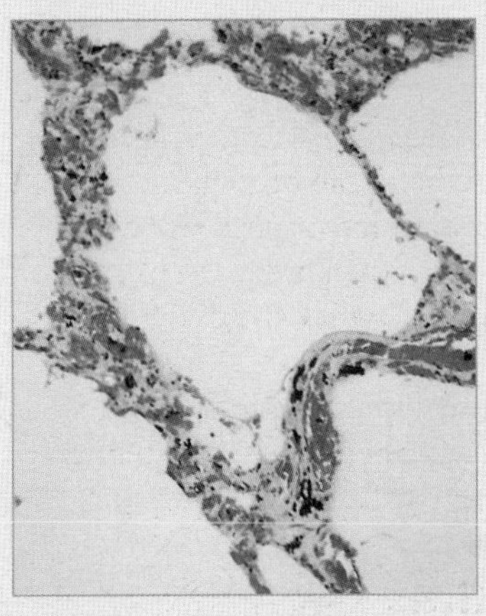

Figure 5.18
Photograph of a normal lung (top) and a lung with emphysema (bottom).

Extension box 2

We breathe air, which is a mixture of gases. Atmospheric air at sea level contains approximately 20% oxygen and 80% nitrogen by volume and exerts a total pressure of about 100 kPa. Of this 100 kPa only 20 kPa is due to oxygen molecules; the remaining 80 kPa is due to nitrogen. These are the partial pressures of the individual gases and are symbolised as pO_2 and pN_2 (Figure 5.19).

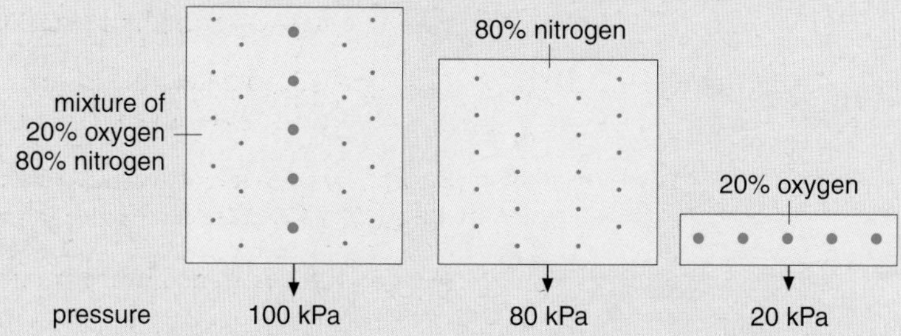

Figure 5.19 The partial pressures of oxygen and nitrogen in atmospheric air.

Partial pressure is the driving force causing the transfer of oxygen from air to blood in the alveoli of the lungs and from blood to mitochondria in the tissues of the body.

The partial pressure falls at each stage of transfer: from atmosphere to lung air, to blood, to tissue fluid, to cytoplasm to mitochondria (Figure 5.20).

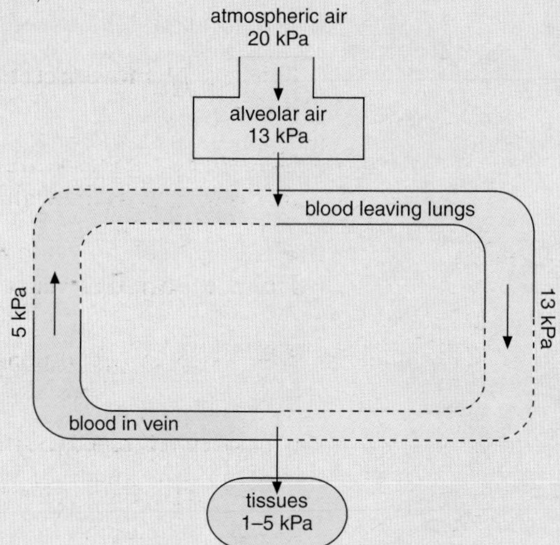

Figure 5.20 The fall in partial pressures during the transfer of oxygen from air to blood.

Why partial pressure and not just percentages?

This is because oxygen uptake is affected by atmospheric pressure. Even at sea level, atmospheric pressure is not always exactly 100 kPa and at the top of Mount Everest the pressure can vary greatly from this. Here the air still contains 20% oxygen by volume but due to a very low atmospheric pressure, the partial pressure of oxygen entering the lungs is only 5.8 kPa compared with 20 kPa at sea level. Thus the concentration gradient of oxygen between the lungs and the blood is greatly reduced.

Summary

- The human gas-exchange system consists of the nasal cavity, trachea and lungs. Movement of the lungs is effected by the intercostal muscles.
- Gas exchange takes place across the alveolar epithelium.
- The alveoli provides a large surface area for gas exchange.
- In the alveoli, oxygen diffuses from the alveolar air to the blood, and CO_2 diffuses from the blood into the alveolar air.
- Fick's law provides a framework to show how the maximum rate of diffusion of respiratory gases is achieved.
- The medulla and the phrenic nerves generate a basic breathing rhythm
- Pulmonary ventilation is the product of tidal volume and breathing rate.
- Exercise increases pulmonary ventilation.

Examination questions

1 Figure 5.21 shows a section through some alveoli in the lung of a human.

Give **three** features visible in the figure that help to increase the rate of diffusion across the wall of the alveoli into the blood.

(3 marks)

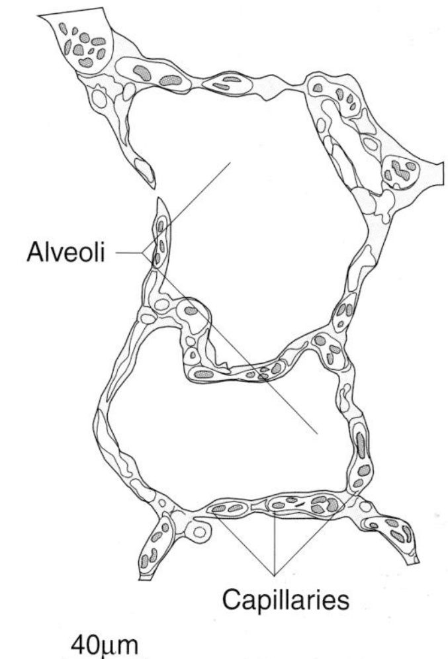

Figure 5.21

2 Fick's law can be given as:

$$\text{diffusion rate is proportional to } \frac{\text{surface area} \times \text{difference in concentration}}{\text{thickness of surface}}$$

(a) With reference to gas exchange in the human lung:

(i) describe two features that ensure a large surface area for gas exchange.

(2 marks)

(ii) give two processes that ensure that a difference in concentration is maintained.

(2 marks)

Figure 5.22 (below)
Figure 5.23 (right)

(b) Figure 5.22 shows the use of the mouth-to-mouth method of resuscitation (ventilation). Suggest how it is possible for exhaled air to be effective when this method is used.

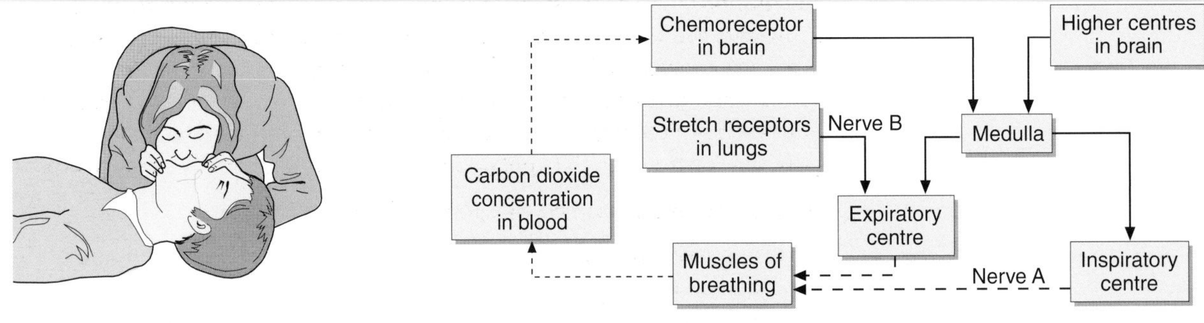

(2 marks)

3 Figure 5.23 shows some parts of the mechanism that controls breathing.

(a) Which muscles are stimulated by Nerve A?

(2 marks)

(b) Explain the role of Nerve B in the control of breathing when a subject is at rest.

(1 mark)

(c) Give one example of the role of higher centres of the brain in the control of breathing.

(1 mark)

(d) Table 5.2 shows the effect of variations in breathing rate and tidal volume on alveolar ventilation (fresh air reaching the alveoli per minute). The dead space of the respiratory system, i.e. the volume of the trachea and bronchi through which no respiratory exchange can take place, is 150 cm^3.

	Column X	**Column Y**
Breathing rate (breaths per minute)	30	10
Tidal volume (cm^3)	200	600
Pulmonary ventilation rate (cm^3 per minute)	6000	6000
Alveolar ventilation rate (cm^3 per minute)	$(200 - 150) \times 30 = 1500$	$(600 - 150) \times 10 = 4500$

Table 5.2

(i) Describe how a person would be breathing to produce the figures in column X.

(1 mark)

(ii) During exercise, both the breathing rate and tidal volume increase. Use the information above to explain why an increase in the breathing rate alone would not be able to satisfy the demands of the body during exercise.

(3 marks)

Assignment

Biologists are like other scientists, they need to be able to carry out simple calculations in order to summarise and analyse data, and interpret the results. In this assignment we shall look at the relationship between size and surface area, and the way in which it affects the diffusion of respiratory gases and the transfer of heat. In carrying out these exercises you will have the opportunity to gain evidence for your Key Skills portfolio concerning the application of number. You will be required to:

- carry out multistage calculations to do with:
  - amounts and sizes
  - scales and proportions
  - handling statistics
  - rearranging and using formulae
- interpret the results of calculations, present findings and justify methods.

It is not easy to find the surface area of an animal such as a rabbit or a cow. Various attempts have been made by measuring skins that have been removed from the animals concerned or by comparing them with a series of cylinders and cones. Perhaps the most imaginative approach involved using a roller to paint a cow and calculating the surface area from the number of revolutions of the roller!

1 Explain why there are likely to be errors in finding an animal's surface area by:

(a) removing its skin and measuring this

(b) comparing the animal with a series of cylinders and cones.

The simplest way to investigate the relationship between size and surface area is not to consider an animal but to look instead at a simple geometrical shape such as a cube.

2 (a) Copy and complete the table below. This will allow you to compare a number of cubes which have sides of different lengths. For each cube, you will need to calculate the total surface area (remember, a cube has six faces), volume and the ratio of surface area to volume. For simplicity, units have been omitted.

Length of one side	Total surface area	Volume	Ratio of surface area to volume
1			
2			
3			
4			
5			
6			

Table 5.3

(b) Which of the following statements are true and which are not true about the information in the table:

A The larger the cube, the greater its surface area.
B Small cubes have small surface areas in relation to their volumes.
C As cubes get larger, their volumes increase faster than their surface areas.
D An increase in the size of a cube is associated with a decrease in the ratio of surface area to volume.

The graph in Figure 5.24 shows measurements of the body mass and surface area of a large number of different species of vertebrate animals.

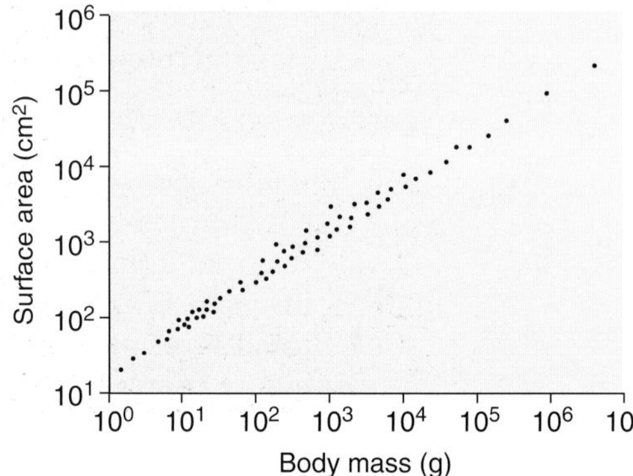

Figure 5.24
Graph showing the relationship between surface area and body mass in different species of vertebrate.

3 (a) A log scale was used to plot the data in Figure 5.24. This means that each point marked on the axes is 10 times larger than the previous point. What is the advantage of using a log scale when plotting these data?

(b) Use the graph to estimate:
(i) the surface area of an animal that weighs 1 kg
(ii) the body mass of an animal that has a surface area of 10 m^2.

The relationship between surface area and the size of an animal, measured in terms of either its volume or its mass, is important when considering the transfer of heat. The two animals shown in the photographs in Figures 5.25 and 5.26 both live in tropical regions. Use the information in your table and that in Figure 5.24 to answer the questions under each photograph.

Figure 5.25

4 What is the advantage of large ears to an African elephant?

5 Large reptiles such as this crocodile are found in the tropics. Use your knowledge of heat transfer and body size to suggest why large reptiles could not live in colder regions.

Figure 5.26

The relationship between size and surface area is also important when considering the diffusion of substances into and out of cells. Figure 5.27 shows a photograph of a cell from the lining of the small intestine.

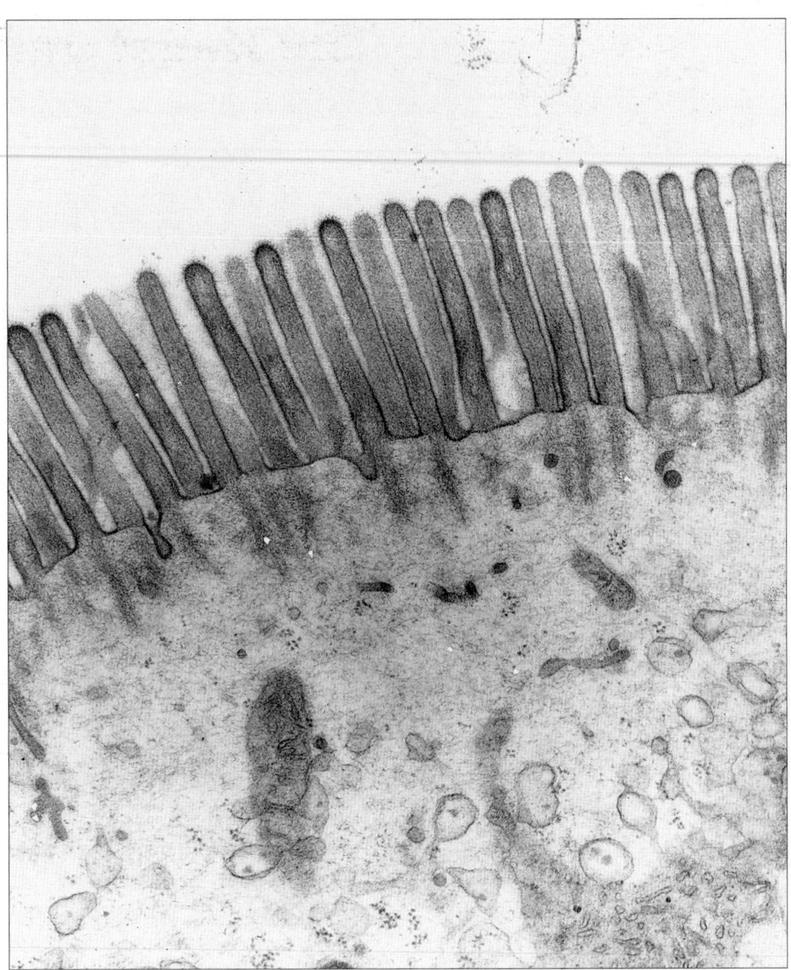

Figure 5.27

6 The relationship between real size in a photograph and the magnification is given by the formula:

$$\text{real size} = \frac{\text{size in photograph}}{\text{magnification}}$$

Use this formula to calculate the magnification of the photograph. Show your working taking the length of one microvillus to be 3μm.

7 This cell absorbs molecules from digested food. By how many times do the microvilli increase the surface area of the cell that is in contact with digested food?

Note:

- There are no instructions to tell you how to do this. However you go about this you should remember that all you can really do is to make a good estimate. Because of this you are not justified in calculating your answer to several decimal places.
- Set out your calculations as clearly as you can, giving your reasons for carrying out each step. Describe any assumptions you need to make.

6 The Heart and Circulation

The human heart is a remarkable organ. It beats over 100 000 times a day, that's around two and a half thousand million times in a lifetime and it never stops for a rest. To do this, the heart muscle must maintain a high rate of respiration and receive sufficient oxygen and glucose.

A heart attack occurs when one of the coronary arteries becomes blocked. Part of the heart muscle is starved of oxygen and dies. Sometimes a heart attack affects the pacemaker and other parts of the system by which the heartbeat is controlled. This results in heart block: a condition which, if not treated, is fatal. An artificial pacemaker may be fitted. This consists of a thin wire that is threaded along one of the veins in the neck or arm into the right side of the heart where its tip is embedded in the wall of the ventricle. The other end of this wire is attached to an artificial pacemaker, which is adjusted so that the ventricle beats at between 60 and 80 beats per minute.

There are some much more advanced pacemakers. They measure changes in breathing rate, body temperature and vibration, and alter the rate at which the artificial pacemaker is causing the heart to beat. When the person takes some exercise, the pacemaker makes sure that the heart beats at an appropriate rate.

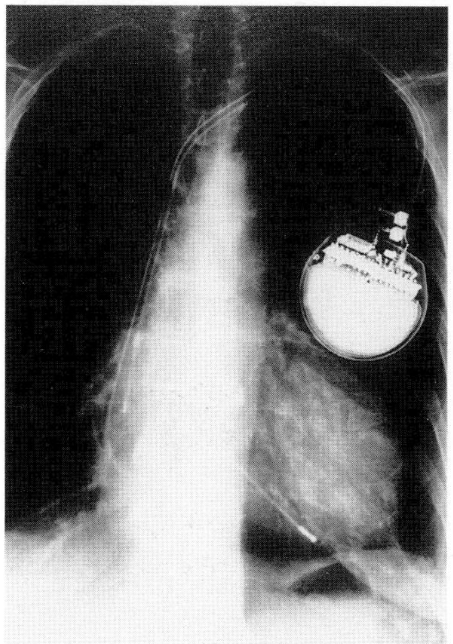

Figure 6.1
This person has been fitted with an artificial pacemaker. You can see the pacemaker and the wires leading to the heart. Apart from regular visits to the hospital to check that the pacemaker is still working, the patient can expect to live a normal life.

key term

The **coronary arteries** supply the heart with oxygen and glucose.

In Chapter 2, we looked at ways in which different substances move in and out of cells. One of these was diffusion. Diffusion is very efficient at transporting substances over short distances but it is nowhere near as effective when greater distances are involved. In a mammal, there are many surfaces where substances are exchanged between organs and their surroundings: respiratory gases are exchanged in the alveoli of the lungs, nutrients obtained from the digestion of food are absorbed by the cells lining the gut, and muscle cells exchange respiratory gases, obtain their glucose and get rid of various waste products. These processes take place over very short distances and frequently involve diffusion. However, we need a system that will link the different exchange surfaces and allow the rapid bulk movement of substances around the body. This is the role of the blood. Because of what it does, it is sometimes described as a **mass transport** system, transporting large quantities of different substances from a **source** to a **sink**.

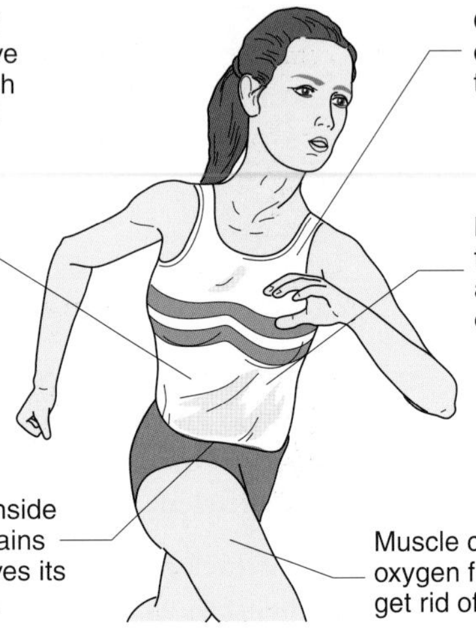

One of the functions of the kidneys is to remove excretory products such as urea from the blood

Oxygen and carbon dioxide are exchanged in the alveoli of the lungs

Nutrients obtained from the digestion of food are absorbed through the wall of the small intestine

The fetus developing inside its mother's uterus obtains its nutrients and removes its waste via the placenta

Muscle cells obtained glucose and oxygen from the blood. They also get rid of their waste products

Figure 6.2
The many exchange surfaces in the body of a human are linked by the blood system, which acts as a mass transport system.

Q 1 **Oxygen and glucose are transported around the body in the blood system. For each of these two substances, name one organ that acts as a source and one that acts as a sink.**

In this chapter, we shall look at the role of blood cells and different sorts of blood vessels in transporting substances between exchange surfaces. We shall also consider the ways in which the structure of the heart is related to its function, knowledge of which has helped us to develop new ways of treating heart disease, such as the use of artificial pacemakers. Finally, we shall see that the blood system is much more than a series of pipes running between different organs. Flow through the different vessels is constantly being adjusted to meet the needs of the body.

The circulation of blood

A mammal's heart serves as a pump. It pumps blood out through arteries. These branch into smaller arterioles and then into capillaries. Capillaries are tiny vessels with very thin walls that allow the exchange of substances between the blood and the cells of the body. Blood is collected from the capillaries by a network of venules. It flows from these venules into the larger veins and back to the heart.

The larger blood vessels in the body share the same basic structure. Their walls have three layers:

- a thin inner lining layer, or endothelium, that is very smooth and enables blood to flow through vessels without causing friction
- a middle layer containing muscle and elastic fibres
- a tough outer layer that helps to protect the vessel from the pressure exerted by other organs rubbing against it.

Table 6.1 shows how the relative thickness of these layers varies and depends on the type of blood vessel involved. The blood contained in the larger arteries has just left the heart. It is at a high pressure and there is a surge every time the ventricles of the heart contract. As the blood is forced along the artery, the elastic fibres in the middle layer are stretched, allowing the wall to bulge outwards. Between the surges in pressure, these elastic fibres recoil. This helps to force the blood along the artery and gradually evens out the flow of blood.

Vessel	Appearance	Inner layer	Middle layer	Outer layer
Artery		Present	Present as a thick layer containing muscle and many elastic fibres	Present
Capillary		Present	Absent	Absent
Vein		Present	Present but thinner than in an artery Contains muscle and a few elastic fibres	Present

Table 6.1

2 Explain why you can feel a pulse if you place your fingers over a large artery in the wrist or neck but not if you place them over a vein.

Figure 6.3
A number of different features allow blood to be returned from the feet of a giraffe to its heart.

The amount of blood required by different parts of the body varies. If you take some form of fairly vigorous exercise, the blood supply to your muscles and skin will increase, while that to your digestive system will decrease. The presence of muscle in the middle layer of the walls of the smaller arteries and arterioles allows the diameter of these blood vessels to change and the blood supply to various organs to be constantly adjusted to meet the needs of the body.

In the capillaries, substances are exchanged between the blood and the surrounding cells. Capillary walls consist only of the thin endothelium and, as a result, they are permeable, that is water and many substances with small molecules are able to pass through them. Some of the white cells in the blood can squeeze between the endothelial cells and leave the blood, particularly at sites of infection.

Once the blood has drained back into the veins it is at a relatively low pressure. Look at the giraffe in Figure 6.3. There are a number of aspects of the blood system which enable the blood to be returned to the heart. These include:

- valves in the veins (see Figure 6.4) that prevent backflow
- muscles that surround the veins. As an animal such as this giraffe walks, these muscles contract and squash the veins. This squeezes the blood along them. The presence of the valves ensures that blood can only be squeezed in one direction – towards the heart.
- breathing in creates a negative pressure in the chest. This helps to draw blood into the heart from the veins.

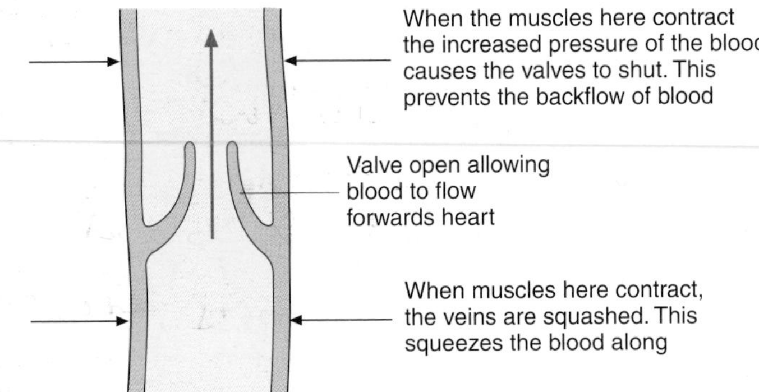
When the muscles here contract the increased pressure of the blood causes the valves to shut. This prevents the backflow of blood
Valve open allowing blood to flow forwards heart
When muscles here contract, the veins are squashed. This squeezes the blood along

Figure 6.4
The valves in veins are rather like the hip pocket in a pair of jeans. If you run your hand upwards, it squashes the pocket flat and your hand moves smoothly upwards. If you run your hand down, it pushes the pocket open and stops your hand moving further.

Q 3 **A person has an accident and cuts through a major blood vessel in the wrist. Explain how you could tell if the escaping blood was coming from an artery or from a vein.**

Extension box 1

Figure 6.5
Oryx are antelope found the semi-desert conditions of Africa and Arabia. When oryx were exposed to daytime temperatures of 40°C and night-time temperatures of 22°C their body temperatures showed a 7°C fluctuation.

Some mammals live in extremely inhospitable conditions. In places such as the deserts of Arabia and Africa, daytime temperatures can rise as high as 45°C. Yet these regions are not devoid of life. Far from it. There are many mammals found in such regions. Some are small and can either rest in the shade or remain underground during the heat of the day. Larger animals such as antelopes find this solution impossible. They have complex physiological mechanisms that allow them to tolerate high temperatures.

Suppose we look at what happens when an antelope such as an oryx (see Figure 6.5) is exposed to very hot and dry conditions. It doesn't sweat as we do. Instead of staying constant, its body temperature rises. This would be fatal in a human. The advantage to the antelope of this behaviour is that valuable water is not lost as sweat in trying to keep a constant body temperature. Not all organs in the body, however, can tolerate such high temperatures. The animal cannot survive if the temperature of the brain is allowed to rise in this way. So how is the blood supply to the brain cooled? This is where the special pattern of circulation in the nose comes in.

An antelope exposed to these very hot conditions does not sweat but it pants. This results in moisture evaporating from the membranes in the nose. Heat energy is transferred from the body in turning the water on these surfaces into vapour. The inside of the nose will therefore be cooler than the rest of the body. This means that the blood flowing back through the veins may be several degrees cooler than that in the rest of the body.

At the base of the brain there is a branching network of blood vessels. In this network, the warm arterial blood flows past the cooler venous blood. Heat is lost from the blood in the arteries to that in the veins. This serves to cool the blood supply to the brain. The blood flow in the two vessels is in opposite directions and this makes the transfer of heat more efficient. It has been shown that the blood supplying the brain may be cooled by as much as 3°C in this way.

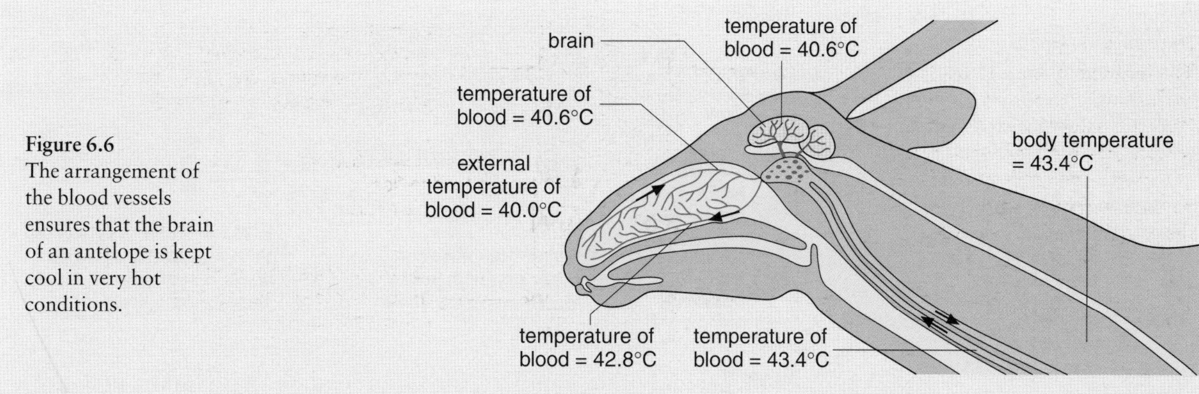
brain
temperature of blood = 40.6°C
temperature of blood = 40.6°C
external temperature of blood = 40.0°C
body temperature = 43.4°C
temperature of blood = 42.8°C
temperature of blood = 43.4°C

Figure 6.6
The arrangement of the blood vessels ensures that the brain of an antelope is kept cool in very hot conditions.

So, do mammals which live in deserts have big noses? This certainly ought to be the case, as the larger the nose, the greater will be the evaporative surface that cools the venous blood. It seems to be true. The surface area of the nasal sinuses of breeds of cattle and sheep that live in hot dry conditions has been found to be much greater than those of animals living in less dry conditions.

Capillaries and exchange surfaces

There are over 80 000 km of blood vessels in the body of an average human adult, and most of them are capillaries. In order to function efficiently, the individual cells in the organs that make up the body must be supplied with the oxygen, glucose, amino acids and other substances they need, and waste products such as carbon dioxide must be removed. Within almost every organ is a network of capillaries that supply or remove these substances (Figure 6.7).

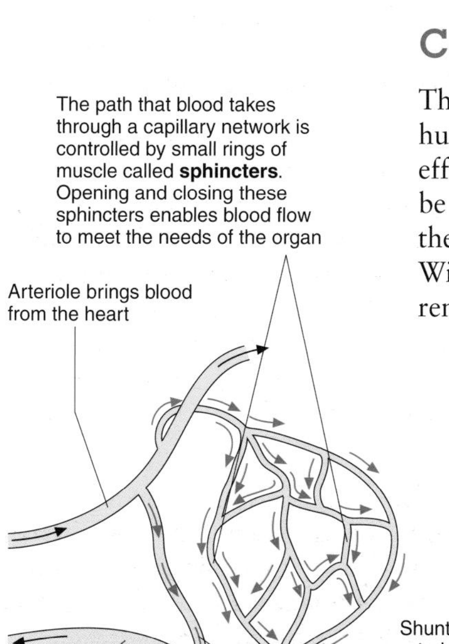
The path that blood takes through a capillary network is controlled by small rings of muscle called **sphincters**. Opening and closing these sphincters enables blood flow to meet the needs of the organ
Arteriole brings blood from the heart
Shunt vessel joins arteriole directly to venule. Allows blood to bypass capillary network
Venule returns blood to the heart

Figure 6.7
The capillary network is like a road system in a city. If you want to travel from one side of a city to the other, you can bypass the city on a motorway (the green arrows through the shunt vessel), you can use the ring road (the blue arrows through the capillaries) or you can go through the side streets (the red arrows through the smaller capillaries).

The cells in the body are surrounded by **tissue fluid.** This tissue fluid is continually formed from and reabsorbed into the blood in the capillary network. Figure 6.8 shows part of a single capillary. Blood flowing into this capillary from an arteriole has a high **hydrostatic pressure.** This is the pressure brought about by the pumping action of the blood and it tends to force water and substances with small molecules out through the permeable walls of the capillary. At the same time the blood contains soluble globular proteins, called **plasma proteins**, and other dissolved substances. These lower the water potential of the blood and tend to cause water to be drawn back into the capillary by osmosis. At the arteriole end of the capillary, the effect of the hydrostatic pressure is greater than that of the water potential so fluid tends to be forced out. This fluid is very similar in composition to blood plasma but it lacks the plasma proteins, which are too large to escape.

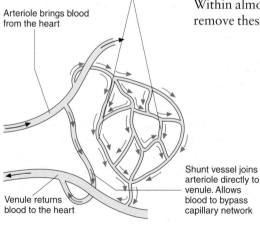

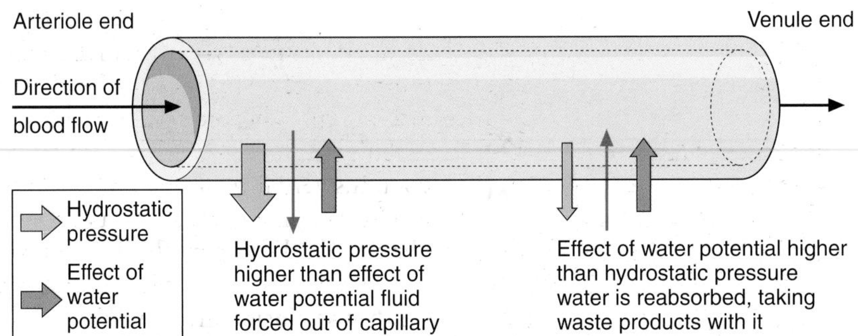

Figure 6.8
The formation and reabsorption of tissue fluid. At the arteriole end, the high hydrostatic pressure results in fluid being forced out of the capillary. At the venule end water is reabsorbed as a result of the water potential of the blood plasma.

At the far end of the capillary, the venule end, the hydrostatic pressure is lower. This is mainly the result of some of the fluid being forced out of the blood. As the volume falls, so does the hydrostatic pressure. However, the water potential has changed very little because the plasma proteins are too large to get through the capillary walls. The effect of the water potential is now greater than that of the hydrostatic pressure and water tends to be drawn back into the blood, taking with it waste products produced by the cells.

Q 4 **Explain how the following may affect the formation and reabsorption of tissue fluid:**
(a) **high blood pressure**
(b) **a fall in the amount of plasma protein as a result of prolonged illness.**

Over the course of a day, more fluid is forced out of the capillaries than is reabsorbed back into them. This accumulated tissue fluid is returned to the blood by the lymphatic system. Excess tissue fluid drains into small, blind-ending tubes called **lymphatic capillaries.** These lead into larger **lymph vessels,** which finally empty the lymph that they contain into the blood in the veins in the neck. Lymph vessels possess valves that ensure that their contents only flow in one direction.

Extension box 2

Sometimes more tissue fluid is formed than can be reabsorbed or removed by the lymphatic system. It accumulates and the tissues swell. This is known as **oedema.** There are many factors that may cause oedema. Some of them, like the slight swelling that takes place in the fingers in very hot weather, are of no concern in an otherwise healthy person. However, oedema may suggest that there is something much more seriously wrong with the body.

Figure 6.9
A child suffering from kwashiorkor.
Note the typical swelling. This is oedema.

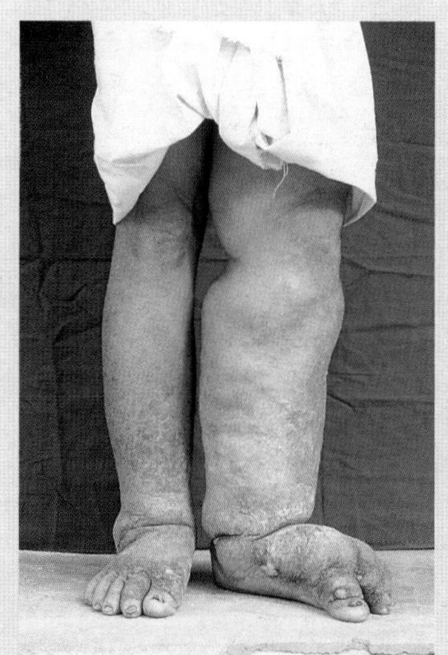

Figure 6.10
This person is suffering from filariasis. Parasitic worms have blocked the lymph vessels in his leg. Tissue fluid has accumulated and has produced permanent oedema. Over a period of time other changes occur, including the thickening of the skin which gives the condition its common name of elephantiasis – the affected leg looks like that of an elephant.

In many parts of the world malnutrition is all too common. The symptoms of **kwashiorkor** often become apparent in the second year of a child's life when he or she is weaned on to a diet consisting mainly of carbohydrate (cassava, plantain or yam) mashed up with water. This diet contains very little protein. One of the effects of the low level of protein is that there is a low concentration of amino acids in the blood. They are almost all taken up by the muscles, where they form some new protein. The liver receives very few of these amino acids and so can no longer synthesise plasma proteins. This affects the water potential of the blood and less of the tissue fluid is reabsorbed. Accumulation of this tissue fluid gives rise to the oedema that is a characteristic of kwashiorkor.

Plasma proteins can also be lost from the body in patients with some forms of kidney disease. In these people there is also a fall in the plasma protein concentration as the plasma proteins escape from the body in the urine and an accumulation of tissue fluid.

A heart attack or disease of the heart valves can lead to **heart failure**. This is a general term used to describe conditions where the heart is unable to pump out enough blood to meet the demands of the body. It is often associated with complex changes involving the nervous and hormonal systems. Some of these changes result in less sodium being excreted by the kidneys and an increase in the sodium concentration in the blood. Sodium ions pass out of the blood through the capillary walls during the formation of tissue fluid. As a result of the higher sodium concentration in the tissue fluid, there is a smaller water potential gradient between the tissue fluid and the blood plasma. Less tissue fluid is absorbed and again the characteristic swelling caused by oedema will be evident.

Blood and blood cells

We saw in Chapter 1 that living organisms are made from cells. In complex organisms such as animals, similar cells are grouped together to form **tissues**, each of which has a particular function. Epithelial tissue, for example, forms the lining of blood vessels, nervous tissue conducts impulses within the nervous system and muscle tissue is able to contract and bring about movement. An organ consists of a number of different tissues. An **organ** may be defined as a structure that has a particular function in the organism. Arteries, for example are organs. They are made up of different tissues and their function is the transport of blood from the heart. Together organs make up a **system**.

Q 5 **An artery is an organ. Name two tissues found in the wall of a large artery.**

Blood is a liquid but it is still an example of a type of tissue known as a connective tissue. Other connective tissues, such as those that make up the outer layer of the walls of arteries and veins, have cells which are separated from each other by a **matrix** that does not contain cells. In

key term

Plasma is blood with all the cells removed.

blood this matrix is a liquid, the **plasma**. Within the plasma are the red blood cells, whose main function is the transport of respiratory gases, and the white cells, which are involved in protecting the body from disease. We will look at each of these.

Plasma

If we take a sample of blood and spin it in a centrifuge, the cells will sink to the bottom and leave a clear, straw-coloured liquid at the top. This liquid is the plasma. It contains many of the substances that are transported round the body: glucose and amino acids, hormones, mineral ions and urea. One important feature of plasma is that it is not constant in composition. Figure 6.11 shows the blood system supplying blood to and taking it away from the liver.

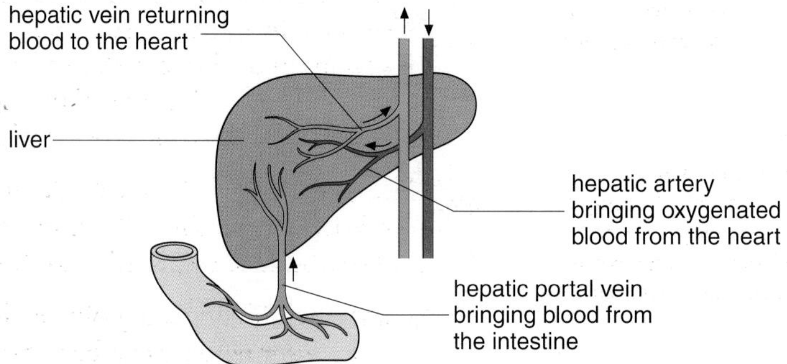

Figure 6.11
The liver receives its blood supply from two sources: the hepatic artery brings oxygenated blood from the heart, while the hepatic portal vein brings blood from the intestine and some other abdominal organs. The blood is returned from the liver to the heart by the hepatic vein.

Immediately after a meal, the blood in the hepatic portal vein has a very high concentration of glucose as it is transporting glucose from the intestine to the liver, where it is stored. The concentration of glucose in this blood vessel will gradually fall as time goes by and the glucose is absorbed. If we eat too much protein, the body cannot store the excess amino acids formed from the digestion of the protein. The liver breaks down the excess amino acids and produces urea. Blood in the hepatic vein has a higher concentration of urea than blood in the other two vessels.

Q 6 **In which of the three blood vessels shown in Figure 6.11 would you expect:**
(a) **the concentration of amino acids to vary most**
(b) **the concentration of carbon dioxide to be highest?**

Red blood cells

key term

Erythrocytes (red blood cells) transport respiratory gases.

The cells that collect at the bottom of the tube when the blood is centrifuged make up what is called the **packed cell volume** (Figure 6.12). This is the percentage of the blood volume taken up by blood cells. In humans, it is approximately 40% and most of the cells involved are red blood cells or **erythrocytes**. Their main function is the transport of respiratory gases.

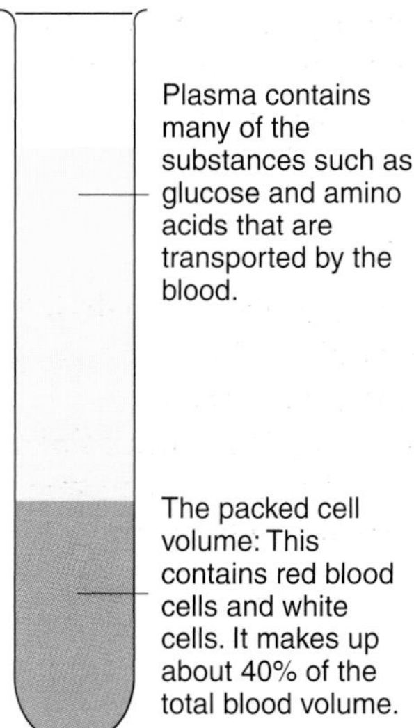

Plasma contains many of the substances such as glucose and amino acids that are transported by the blood.
The packed cell volume: This contains red blood cells and white cells. It makes up about 40% of the total blood volume.

Figure 6.12
The cells that collect at the bottom of a tube when a sample of blood is centrifuged form the packed cell volume. It is slightly higher in men than in women.

If you have studied the assignment at the end of Chapter 1, you will already know a lot about red blood cells. They have a number of adaptations that make them very efficient when it comes to absorbing and transporting respiratory gases. These include:

- **small size:** human red blood cells are only about 7.5 microns (μm) in diameter. This is very much smaller than most other cells in the body. A small size means that all of the haemoglobin molecules that these cells contain are close to the surface and this allows oxygen to be picked up and released rapidly.
- **shape:** a sphere is the shape with the maximum volume. A spherical blood cell can therefore hold a large amount of haemoglobin and transport a lot of oxygen. On the other hand, a blood cell with a flat, disc-like shape would have the maximum surface area, important for the efficient diffusion of oxygen into and out of the cell. Red blood cells are biconcave discs. This is a compromise between having a large volume and a large surface area. It allows the cell to contain a lot of haemoglobin while still allowing efficient diffusion through the large surface area provided by the plasma membrane.
- **organelles:** red blood cells do not contain either nuclei or mitochondria. This allows more space inside the cell for haemoglobin, the substance responsible for transporting oxygen.
- **haemoglobin:** haemoglobin is the oxygen-transporting pigment. Much more haemoglobin can be packed into red blood cells than could be dissolved in the plasma. It is also kept in chemical conditions that allow it to function as efficiently as possible in loading and unloading respiratory gases. There would also be a number of disadvantages in having haemoglobin carried in solution in the plasma. These include making the blood so thick that it would have difficulty flowing through the blood system, affecting the water potential and losing the haemoglobin from the body as the blood passed through the kidneys.

Q 7 **Explain why red blood cells:**
(a) **cannot respire aerobically**
(b) **cannot synthesise proteins.**

For more about red blood cells, look at the assignment in Chapter 1.

White cells

key term

Leucocytes (white cells) are concerned with protecting the body from disease.

White cells are usually larger than red blood cells but they are found in much smaller numbers. Some of these cells are able to escape from the blood system and are found elsewhere in the body, so white cell is a more accurate name for them than white blood cell. They are colourless so they have to be stained in order to be seen clearly with a light microscope. Staining allows us to see details such as the shape of the nucleus and the presence of granules in the cytoplasm. Table 6.2 shows the important differences between the main types of white cell.

Type of leucocyte	Appearance	Function
Lymphocyte	Has a large, round nucleus and a relatively small amount of cytoplasm	Some types of lymphocyte secrete antibodies Others have different functions, such as killing infected cells and controlling aspects of the immunological process
Monocyte	Has a large, kidney-shaped nucleus	These cells engulf bacteria
Granulocyte	Possess a lobed nucleus and granular cytoplasm	Granulocytes have a number of different functions: some engulf bacteria and others are involved with allergies and in inflammation

Table 6.2

Extension box 3

How long do red blood cells live?

In an adult man there are approximately 5.5 million red blood cells in a single cubic millimetre (mm^3) of blood; in an adult woman the figure is a little less, around 4.7 million per mm^3. With the blood volume of an average adult around 5 litres, this is an enormous number of red blood cells. One interesting fact, however, is that in about 120 days' time, none of the red blood cells that are in your blood now will still be there, they will all have been replaced. So how did we find out for how long red blood cells live?

If we want to investigate what happens to a molecule in a biochemical pathway or an individual organism in its natural environment, we need to label it in some way. This principle is also used to find the length of life of a red blood cell. We remove and label some of a person's red blood cells, put them back and then take blood samples at regular intervals until all the labelled cells have disappeared from the blood. A number of modifications of this approach have been tried out.

You will probably know that human blood can be divided into four main groups: A, B, AB and O. Besides this system, however, there are many other ways of grouping blood and these can be used to estimate a red blood cell's life span. The subject of the experiment is given a transfusion of compatible but slightly different blood. Although the red cells introduced will not harm the subject in any way, the slight differences can be detected with suitable antibodies. The number of foreign cells present in the subject's blood can be counted at regular

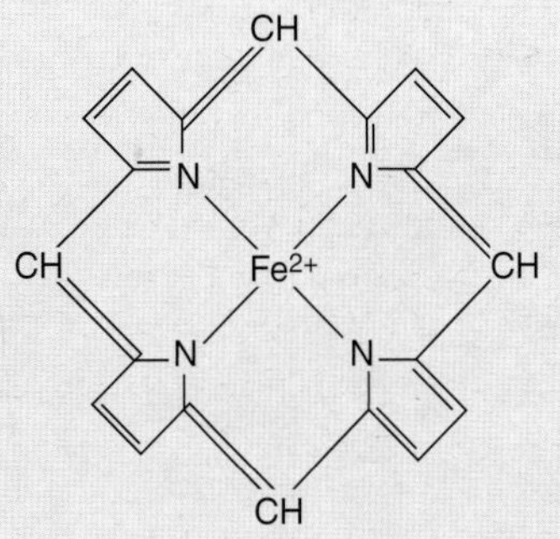

Figure 6.13
The chemical structure of a haem group. Note that it consists of four nitrogen-containing rings grouped round an iron atom.

intervals, allowing us to determine the maximum time for which they can live. This is quite a useful method but it has a problem in that it really only measures for how long one person's blood cells can live in the blood of another person.

If we can use the subject's own blood, we might be able to overcome this problem. A good idea would be to label haemoglobin in some way since this substance is found in large amounts in red blood cells. A molecule of haemoglobin consists of four polypeptide units. Each of these so-called globin chains is associated with an iron-containing haem group. The chemical structure of this haem group is shown in Figure 6.13.

Perhaps the first thing that you might notice is that haem contains iron. Could we label the haemoglobin with a radioactive isotope of iron? Unfortunately this will not work. The red cells disappear from the blood because they are broken down in the liver. The iron is removed from the haemoglobin in these cells and used to make more haemoglobin in more red blood cells so the radioactivity will not disappear from the blood as the labelled cells reach the end of their lives.

We can, however, label the nitrogen in the haem group. We give the subject a small amount of the amino acid glycine containing the radioactive isotope of nitrogen, ^{15}N. We know that the nitrogen in the haem group is derived from glycine so this technique allows us to attach a radioactive label to the red blood cells being formed immediately after the subject was given the glycine. The graph in Figure 6.14 shows the changes in the level of ^{15}N in the blood of one person treated in this way.

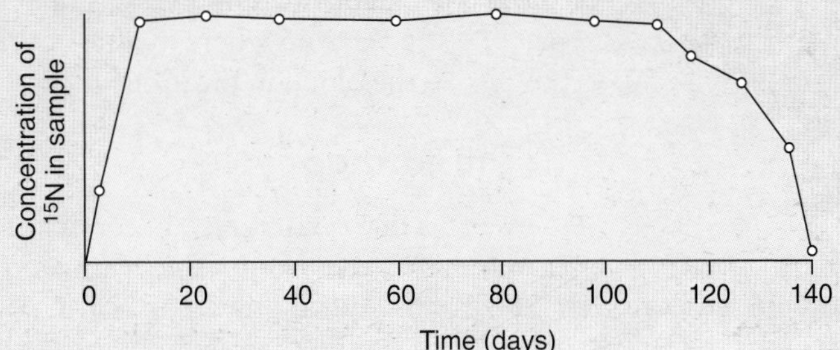

Figure 6.14
The concentration of ^{15}N in samples of blood from a person who has been fed with ^{15}N-labelled glycine. The upward slope represents the time taken for labelled haemoglobin to be formed and incorporated in red blood cells. The steep fall in the amount of labelled haemoglobin allows us to calculate the length of life of the red blood cells. This ranges from 110 to 130 days.

The heart

Humans, like all mammals, have a **double circulation** (see Figure 6.15). The heart has four chambers. Blood flows into the right atrium and out of the right ventricle into the **pulmonary circulation**. Here it is oxygenated as it passes through the lungs. It returns to the left atrium and then goes from the left ventricle into the **systemic circulation**. This is the blood supply to the rest of the body. Most mammals are very active and it is important that oxygen reaches their tissues rapidly. When the blood passes through the many tiny blood vessels in the lungs it loses pressure. By passing through the left side of the heart before being sent round the body, the pressure can be increased, allowing oxygenated blood to reach the organs of the body rapidly.

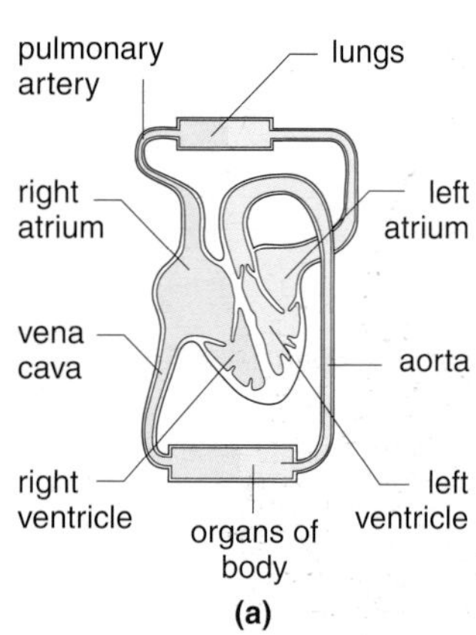

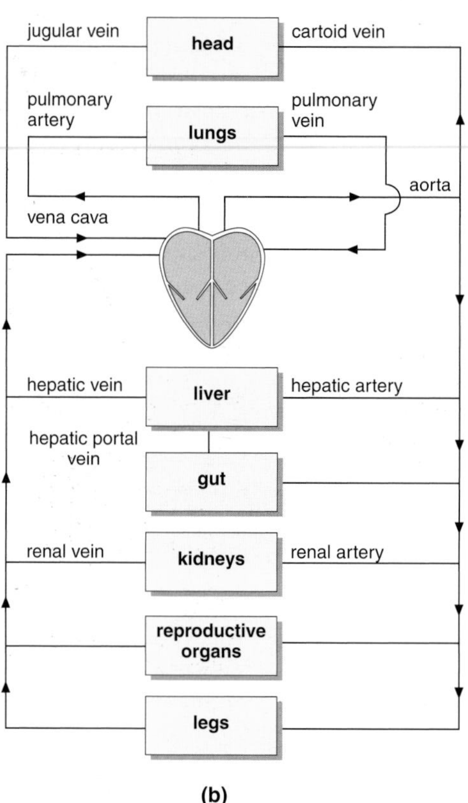

Figure 6.15
(a) The double circulation in a human. Blood is passed from the right side of the heart to the lungs, where it is oxygenated. This oxygenated blood is returned to the heart. It is pumped to the rest of the organs of the body by the left side of the heart.
(b) The main blood vessels that make up the human blood system.

Q 8 Name the blood vessels through which a molecule of urea will pass from where it enters the blood in the liver to where it leaves the blood in the kidney.

Q 9 If you could measure a pulse in the pulmonary artery would it beat at the same rate, slower or faster than a pulse in the wrist?

The cardiac cycle

When a person is at rest, his or her heart beats approximately 70 times a minute. Each beat of the heart represents a cardiac cycle in which the heart fills and empties. The beating of the two sides of the heart is synchronised. When the left side is filling, so is the right; when the left ventricle contracts, so does the right. The left side of the heart is responsible for pumping the blood all the way round the body; the right side only pumps it to the lungs, which lie relatively close to the heart. The wall of the left ventricle is much thicker than the right and produces much greater pressure when it contracts.

Figure 6.16 shows that, although the cardiac cycle is continuous, we can recognise three main stages: atrial systole, ventricular systole and diastole.

Atrial systole

Some of the blood that enters the atria from the veins flows straight through into the ventricles. The rest is pumped through as the muscle in the walls of the atria contracts. Each atrium is separated from its corresponding ventricle by a valve, the **atrioventricular** (AV) **valve**. The

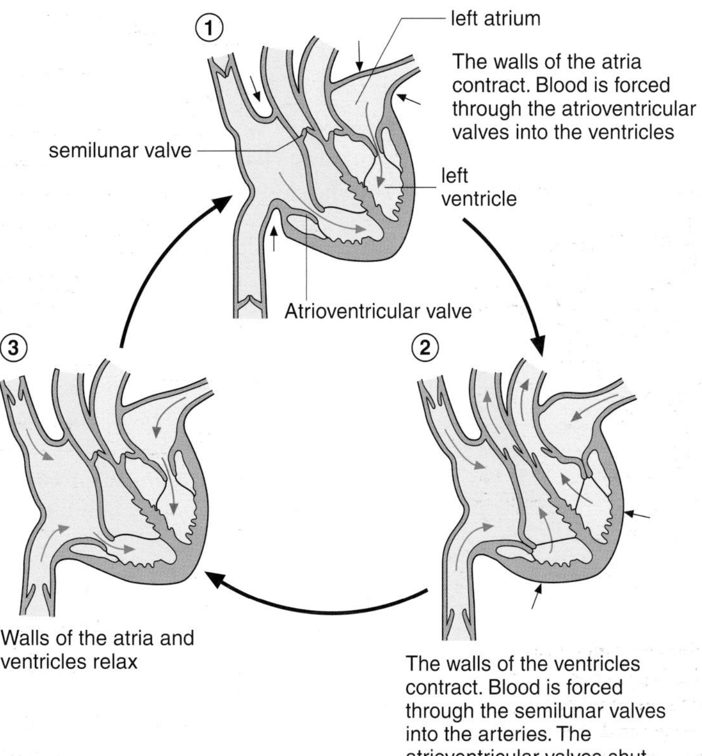

Figure 6.16
There are three main stages in the heart cycle. The atria contract (atrial systole), the ventricles contract (ventricular systole) and finally both atria and ventricles relax (diastole).

atrial walls are thin, so they do not produce much pressure when they contract but it is enough to open the AV valves and force blood through into the ventricle. Atrial systole takes approximately 0.1 seconds.

Ventricular systole

During ventricular systole, the thick muscular walls of the ventricles contract. The blood in the ventricles is squeezed and its pressure rises. As soon as the pressure of the blood in the ventricles is higher than that in the atria, the flaps of the AV valves are pushed shut. This prevents blood from flowing back into the atria. When a doctor listens to a patient's heart with a stethoscope, the sound made by these valves closing can be heard: it is often described as sounding rather like 'lub'. The pressure of the blood in the ventricles is now higher than that in the arteries leaving the heart. This results in the semilunar valves opening and blood flowing out of the right ventricle into the pulmonary artery, and out of the left ventricle into the aorta. Ventricular systole lasts about 0.3 seconds.

Diastole

The final stage of the cardiac cycle involves relaxation of the atrial and ventricular walls. As the walls of the ventricles relax, the pressure of the blood they contain drops rapidly. The arterial blood pressure is now higher and forces the semilunar valves to shut. This can also be heard with a stethoscope and produces a sound described as 'dub'. With the semilunar valves shut, blood is unable to flow back into the ventricles.

As the muscle in the atrial walls also relaxes, blood is able to flow from the veins into the atria. The walls of the atria now start to contract again, starting another cardiac cycle.

Figure 6.17 summarises the changes in pressure that occur in the left atrium, the left ventricle and the aorta during one complete cardiac cycle. You should be able to explain the main features of these curves. Look first at the curve representing the changes in pressure that take place in the left ventricle. As the muscles in its wall contract, the pressure increases very steeply to reach a maximum value of approximately 16 kPa. When the muscles relax, the pressure falls. There is also a rise in the pressure of the blood in the atria as the atrial walls contract at the beginning of the cycle. The second rise, between about 0.2 and 0.45 seconds, is due to blood flowing into the atria from the veins. Flow of blood from the left ventricle into the aorta results in an increase in the pressure of blood in the aorta. The graph also shows how differences in pressure result in the AV valves and the semilunar valves opening and shutting.

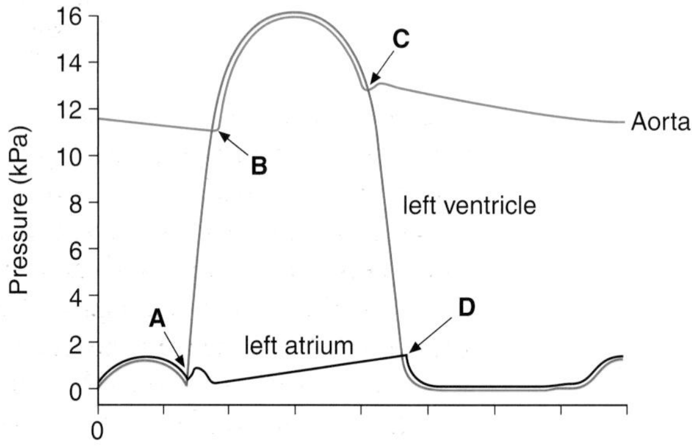

Figure 6.17
Change in pressure during one cardiac cycle. This graph shows figures relating to the left side of the heart. The pressures on the right side are generally a little lower.

Q 10 At which of the points A, B, C or D in Figure 6.17 would you expect to hear:
(a) the first heart sound (lub)
(b) the second heart sound (dup)?

Coordinating the heartbeat

Suppose you wanted to squeeze all the toothpaste out through the nozzle of a toothpaste tube as quickly as possible. There would be no point in simply squashing the middle of the tube. Some of the toothpaste might come out through the nozzle, but you would be likely to split the tube so that toothpaste leaked out from various places. It is the same with the heart. For the blood to flow from atria to ventricles and then out through the arteries requires the heartbeat to be coordinated. If parts of the heart beat independently, it will not function properly as a pump.

Cardiac or heart muscle differs from other muscles in the body because it is **myogenic**. This means that it beats on its own. It does not need a nerve impulse to make it contract. A heartbeat starts with an electrical signal from an area of muscle in the wall of the right atrium called the **sinoatrial node** (SAN) or pacemaker. The position of the SAN is shown in Figure 6.18.

This electrical signal sets the rate at which the heart beats. Every time the muscle cells in the SAN beat they send out a wave of electrical activity which spreads over the surface of the atria, causing the muscles in the atrial wall to contract. This excitation wave starts at the top of the atria and spreads towards the ventricles, ensuring that the atrial wall contracts in a way that will force blood into the ventricles.

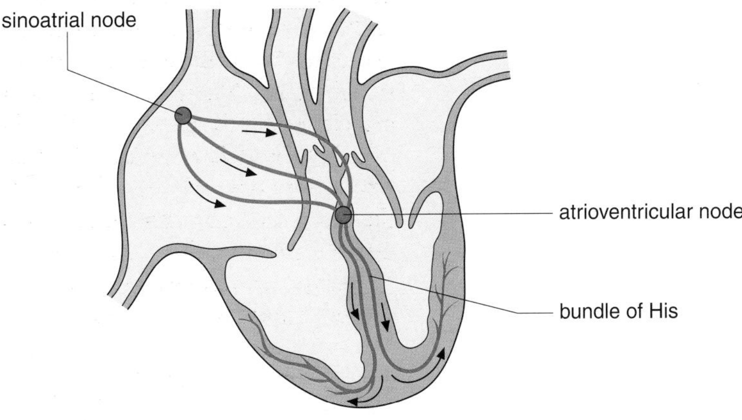

Figure 6.18
Every time the heart beats a wave of electrical activity spreads over the heart from the sinoatrial node. This acts as a signal and causes the heart to beat in a regular, organised way.

The ventricle muscle must not start contracting until the atrial muscles have finished contracting and squeezed all the blood into the ventricles. The delay in ventricle contraction is brought about by delaying the passage of the excitation wave through to the ventricle. A ring of fibrous tissue between the atria and the ventricles prevents its spread and it can only pass through in one region, the **atrioventricular node** (AVN). After a short delay here, the wave passes down specialised conducting fibres in the wall or septum between the right and left ventricles. These fibres, which form the **bundle of His**, conduct the excitation wave very rapidly to the base of the ventricles. It then spreads upwards through the muscle in the wall of the ventricle. This ensures that the ventricles contract from the base upwards, squeezing blood into the arteries.

Q 11 Explain the importance in the cardiac cycle of:
(a) delaying the spread of the excitation wave in the atrioventricular node
(b) conducting the excitation wave very rapidly to the base of the ventricles.

key term

The **cardiac output** is the amount of blood pumped out of the left side of the heart in one minute.

The effect of exercise on the heart

The cardiac output is the product of the **stroke volume**, the amount of blood that the left ventricle pumps out each time it beats, and the **heart rate**, the number of times the heart beats in one minute. This can be represented as a simple equation:

$$\text{cardiac output} = \text{stroke volume (heart rate)}$$

Q 12 A man's resting cardiac output is 5500 cm^3. His heart is beating at 70 beats per minute. Calculate the volume of blood that his left ventricle pumps out each time it contracts.

If you take some exercise, one of the things you are likely to notice is an immediate increase in heart rate. This is only part of the story, however, as both the rate and force with which the heart beats are continuously adjusted so that cardiac output is matched to the needs of the body. In this way, enough oxygen can be supplied and carbon dioxide can be taken away.

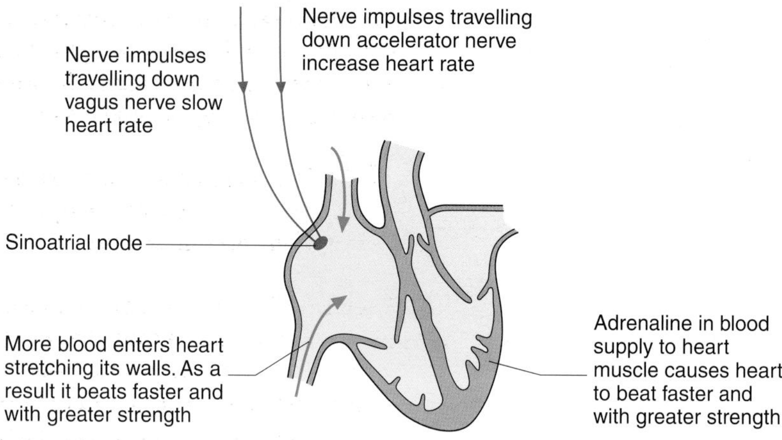

Figure 6.19
Cardiac output can be adjusted to meet the needs of the body. The heart responds to hormones, to electrical impulses passing down nerves from the brain and to changes in the amount of blood returning to the heart through the veins.

Figure 6.19 summarises the three ways in which cardiac output can be adjusted. These involve the heart responding to hormones, nerve impulses from the brain or changes in the volume of blood returning to the heart.

Hormones

When you are under stress or there is a need for action, the hormone adrenaline is secreted by the adrenal glands, which are situated at the top of your kidneys. Adrenaline travels in the blood to all areas of the body. It has many effects and one of them is to increase the heart rate by increasing the rate at which the SAN sends out its waves of electrical activity.

key term

The **cardiovascular centre** in the brain is involved in controlling the cardiac output.

Nerve impulses from the brain

There are two nerves from the cardiovascular centre which go the heart. These are:

- the **accelerator nerve**: impulses travelling down this nerve to the SAN increase the heart rate and impulses going to the heart muscle increase the strength of the contractions
- the **vagus nerve**: impulses travelling down this nerve to the SAN slow down the heart rate.

When you start exercising, the rate of respiration in your muscles increases. More carbon dioxide is produced as a result. Receptors monitor the concentration of carbon dioxide by detecting changes in pH. Nerves from these receptors send impulses to the cardiovascular centre, which responds in an appropriate way. The cardiovascular centre also responds to other signals. It slows the heart rate in response to high blood pressure and increases it in response to signals from other parts of the brain, such as, for example, in anticipation of activity.

Changes in the volume of blood returning to the heart

During a period of exercise, muscles respire faster and use up more oxygen. A fall in the concentration of oxygen causes the veins bringing blood back to the heart from the body to get wider. More blood therefore enters the heart and stretches its walls more than normal. The heart responds to this by beating faster and with greater strength.

Q 13 If a person has a heart transplant, the new heart has no nerves going to it. Explain why a transplanted heart can still beat faster if the person exercises.

Exercise affects the rest of the circulation as well as bringing about an increase in cardiac output. The energy required for increased muscle contraction comes from respiration. More exercise means a faster rate of respiration and the need for more oxygen to be supplied to the muscles. This can be achieved by increasing the amount of blood flowing through the capillaries. When a person is resting, many of the capillaries in muscles are closed. When the muscle is active, most of these capillaries have blood flowing through them. The graph in Figure 6.20 shows the effect of different amounts of exercise on the rate of blood flow through muscle capillaries.

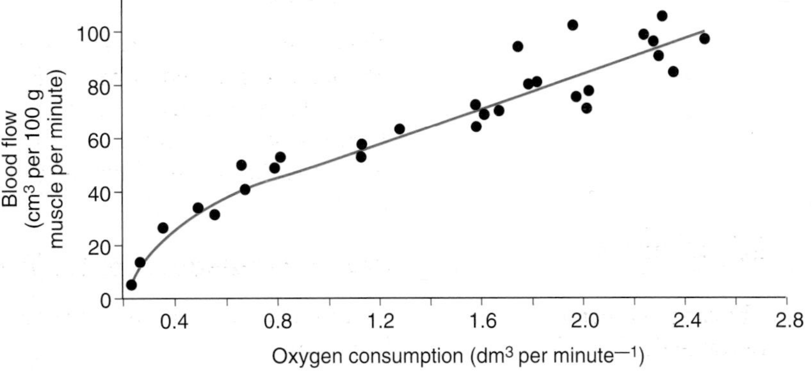

Figure 6.20
This graph shows the effect of exercise on the flow of blood through muscle capillaries. The x-axis shows the oxygen taken up in dm^3 per minute. The more vigorous the exercise, the greater the amount of oxygen consumed.

A large increase in blood flowing to one part of the body must be met by a reduction in the amount of blood supplying other parts of the body. During a period of exercise, there is an increase in the amount of blood flowing to the muscles and to the skin but a decrease in the supply to other organs such as the kidneys and those that make up the digestive system. The brain needs a constant supply of oxygen so its blood supply is not affected however severe the exercise.

Summary

- Mass transport in mammals occurs through the arteries, arterioles, and veins.
- Capillaries are important in metabolic exchange and in the formation of tissue fluid.
- Red cells (erythrocytes) in the blood transport oxygen.
- White cells in the blood are involved in protecting the body from disease.
- The heart cycle consists of the atrial systole, the ventricular systole, and the diastole.
- The heart is myogenic. The sinoatrial node (SAN) begins each heartbeat by generating an electrical signal.
- Exercise increases cardiac output and affects the distribution of blood to the different organs in the body.

Examination questions

1 The table shows the pressures measured at two different places in a blood capillary.

	Arterial end of capillary	Venous end of capillary
Blood pressure/kPa	4.6	1.3
Osmotic pressure of blood plasma/kPa	3.3	3.3

(a) Using the data given in the table, explain why fluid leaves a capillary at the arterial but not at the venous end.

(2 marks)

(b) Under normal conditions the total volume of fluid leaving the capillaries is greater than that returned. Explain why this extra fluid does not accumulate in the tissues.

(2 marks)

2 The graph shows the changes in pressure which take place in the left side of the heart.

(a) Use the graph to calculate the heart rate in beats per minute. Show your working.

Heart rate = ____________ beats per minute

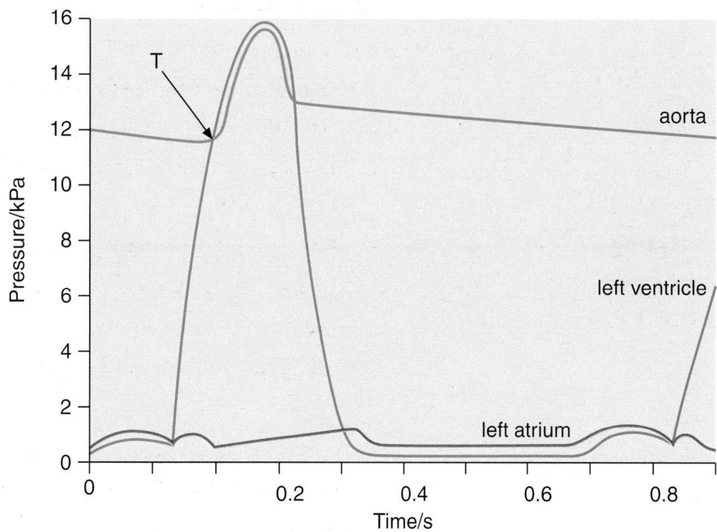

(b) (i) Explain, in terms of pressure, why the valve between the left ventricle and the aorta opens at time **T**.

(1 mark)

(ii) For how long is the valve between the left atrium and the left ventricle closed? Explain how you arrived at your answer.

(2 marks)

(c) (i) How would you expect the pressure in the right ventricle to differ from that in the left ventricle?

(1 mark)

(ii) Explain what causes this difference in pressure.

(1 mark)

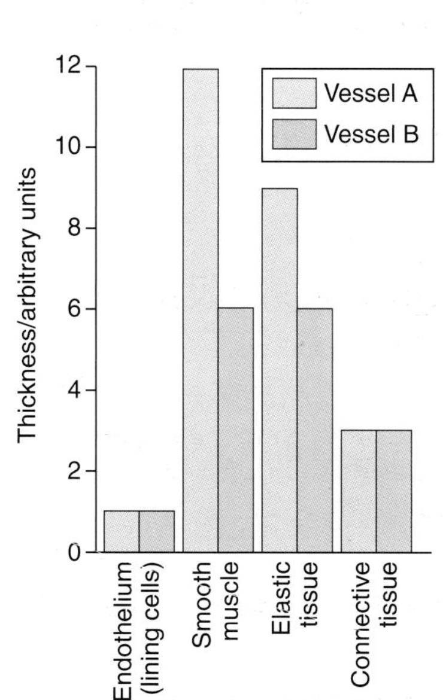

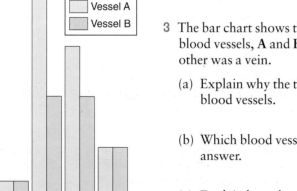

3 The bar chart shows the relative thickness of parts of the walls of two blood vessels, **A** and **B**. One of these blood vessels was an artery, the other was a vein.

(a) Explain why the thickness of the endothelium is the same for both blood vessels.

(1 mark)

(b) Which blood vessel is the artery? Explain the reasons for your answer.

(2 marks)

(c) Explain how the structure of veins ensures the flow of blood in one direction only.

(2 marks)

Assignment

The assignments in the first five chapters in this book have introduced you to some of the skills that a biologist needs. The assignment in this chapter involves using all of these skills. You will need to answer some questions based on a short written passage, apply your knowledge to new situations and handle and interpret data.

Many people living in more developed countries survive to middle and old age. In view of this, it is not surprising that diseases involving the degeneration of organs and systems account for so many deaths. At the top of the list is heart disease, which kills around a third of all people in the UK and leads to the chronic ill-health of many more. One of the most valuable tools we have for studying the heart and detecting heart disease is the electrocardiogram (ECG). Read the following passage, which explains the principles underlying electrocardiography.

The cardiac cycle is controlled by a series of electrical events. A wave of electrical activity spreads over the surface of the heart. The body tissues also conduct electricity, so voltage changes that affect the heart can also be detected on the body surface. Taking an ECG involves attaching a series of electrodes to the surface of the body. They are self-adhesive and coated with a jelly that ensures a good electrical contact. The signals from these electrodes are processed and displayed to give a picture of the electrical events taking place as the wave of electrical activity spreads over the surface of the heart.

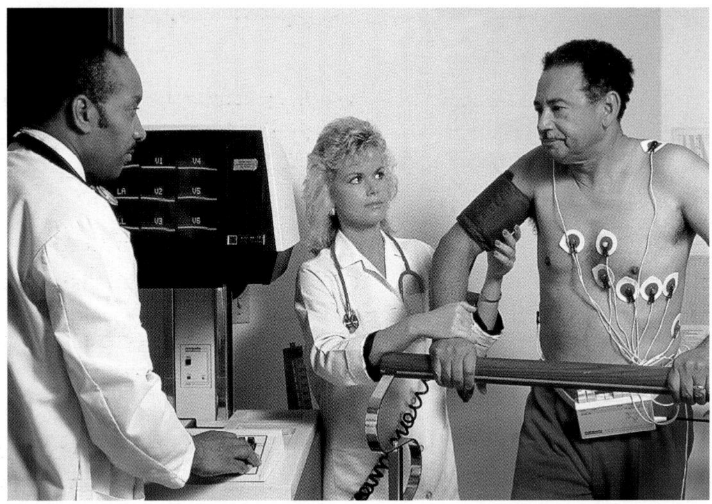

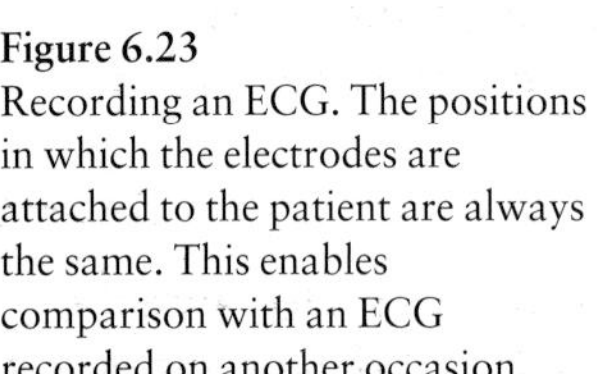

Figure 6.23
Recording an ECG. The positions in which the electrodes are attached to the patient are always the same. This enables comparison with an ECG recorded on another occasion.

An ECG shows the passage of an electrical signal in a particular direction. If we compare the picture we get from a pair of electrodes, we can see the changes that occur as the electrical activity spreads from one electrode to the other. By attaching the electrodes to different places, we can see what is happening to different parts of the heart. In this assignment we will only be looking at the pattern we see from a single pair of electrodes, showing the spread of electrical activity from the atria to the ventricles.

Figure 6.24 shows an ECG of the events associated with a single heartbeat in a healthy person. Letters are used to identify particular features:

- the P wave shows the spread of electrical activity over the surface of the atria
- the QRS complex shows the spread of electrical activity over the surface of the ventricles
- the T wave shows the electrical recovery of the ventricles; this is the time when the muscle is relaxing and the ventricles are starting to fill with blood.

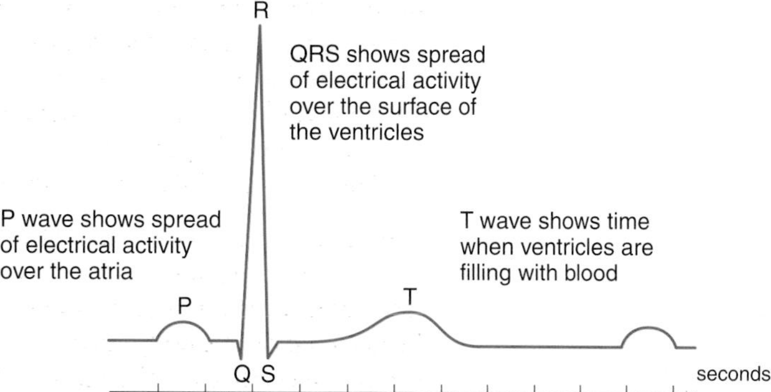

Figure 6.24
Interpreting an ECG.

Now use the information in the passage and Figure 6.24 to answer these questions.

1 Describe the path taken by the wave of electrical activity as it spreads over the surface of the heart.

(2 marks)

2 An ECG shows a number of traces which, although they show recordings of the same heartbeats, differ in appearance from each other. Explain why.

(2 marks)

3 Pregnancy has many effects on the body. One of these is to cause the heart to twist slightly so that it is lying in a slightly different position in the chest.

(a) Suggest why a woman's heart may twist and lie in a slightly different position during pregnancy.

(2 marks)

(b) Why might an ECG be misinterpreted if it was not known that the woman from whom it was recorded was pregnant.

(1 mark)

4 Look at the ECG shown in Figure 6.24.

(a) How long does one heartbeat take?

(b) What is this person's heart rate in beats per minute?

(2 marks)

5 Explain why no electrical activity can be detected between the end of the P wave and the beginning of the QRS complex.

(2 marks)

Patient **A**

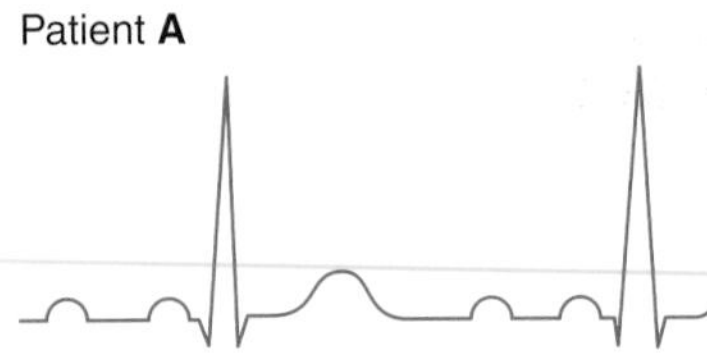

Patient **B**

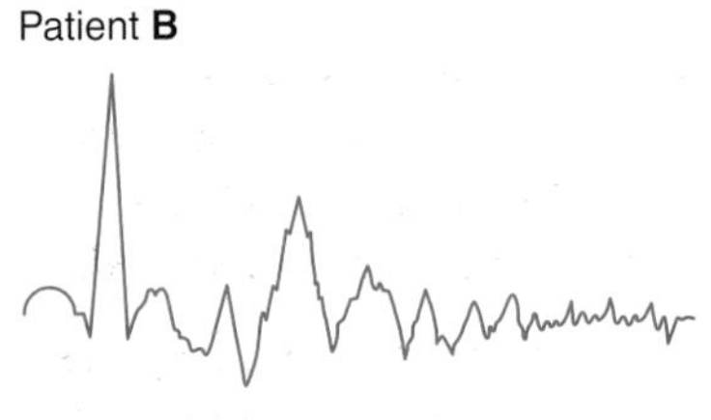

Figure 6.25

Figure 6.25 shows ECGs obtained from patients whose heartbeats are abnormal. Compare them with Figure 6.24 and answer the question below.

6 (a) Describe how the heartbeats of Patient A differ from those of a healthy person.

(b) Patient B has a condition called ventricular fibrillation. When a patient is being kept in intensive care and ventricular fibrillation is detected, a warning is given and emergency treatment has to be given immediately. Suggest why ventricular fibrillation may result in the death of the patient.

(2 marks)

Look at Figure 6.24 again. The interval between the beginning of the QRS complex and the peak of the T wave is the time when the heart is emptying. The interval between the peak of the T wave and the beginning of the next QRS complex is the time when the heart is filling.

In an investigation, an adult male had his ECG recorded while his heart was beating at different rates. The results are shown in Table 6.3.

Heart rate (beats minute^{-1})	Emptying time (s)	Filling time (s)
55	0.43	0.66
60	0.43	0.57
70	0.42	0.44
80	0.42	0.33
90	0.43	0.24

Table 6.3

7 (a) Plot these data as a suitable graph using a single pair of axes.

(4 marks)

(b) What do these data show about the way in which heart rate increases?

(2 marks)

Enzyme Technology

Enzymes have enormous potential in the commercial world. Since they can catalyse specific reactions at relatively low temperatures, they are more versatile and much cheaper than inorganic catalysts. Once a suitable enzyme has been found, it has to be made on a large scale and then purified for the task involved.

There is a vast world-wide demand for sweeteners, mainly for confectionery and soft drinks. Traditionally, sucrose, from sugar beet or sugar cane, has been used, but in recent years a sucrose substitute known as high fructose syrup has become cheaper to produce. Fructose is sweeter than sucrose and can be made from starch, a relatively abundant and cheap foodstuff, often a waste product of the food industry.

Four enzymes are involved in the production of high fructose syrups: bacterial α-amylase, fungal amyloglucosidase, pullulanase and bacterial glucose isomerase. The first three catalyse the conversion of starch to glucose. Glucose isomerase then converts the glucose to a 50:50 mixture of glucose and fructose. The end product is high fructose syrup, a mixture of about 42% fructose and 55% glucose. In these proportions it has the same sweetening power as sucrose.

An alternative sweetener is aspartame, sold as *Canderel*™ and *Nutrasweet*™, a dipeptide, which is 180 times sweeter than sucrose. The commercial production of aspartame again involves enzymes, particularly aspartase.

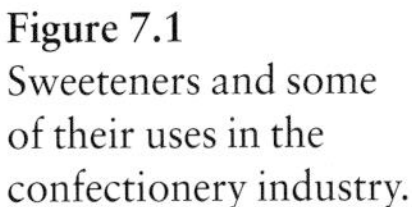

Figure 7.1
Sweeteners and some of their uses in the confectionery industry.

Chocolate lovers everywhere may be surprised to learn that the soft centres inside chocolates are possible because of the action of an enzyme. To start with the centre is solid, containing polysaccharides and an enzyme. Once the chocolate coating has set, the enzyme breaks down the polysaccharide filling, turning the hard centre into the familiar runny filling.

Enzymes are described as biological catalysts, that is they speed up the rate of chemical reactions. The essential role of the enzyme is to provide an active site, which has a complementary shape to the molecules that bind there. Enzymes are proteins and they are extremely specific, generally reacting with only one substrate. Substrates are converted to products through the formation of an intermediate enzyme–substrate complex. To be effective in a production process the enzyme molecules must be brought into maximum contact with the substrate molecules.

Enzymes have three main advantages when used in industrial processes and these are directly related to their properties:

- they are specific in their action, reacting with only one substrate to produce a specific product, and they are therefore less likely to produce unwanted by-products
- they are biodegradable and therefore cause less environmental pollution
- they are able to work at moderate temperatures, neutral pH and normal atmospheric pressure and are therefore energy saving.

Q 1 What is the advantage of using enzymes compared with inorganic catalysts in manufacturing industries.

The main disadvantages of enzymes are that they are:

- highly sensitive to changes in the physical and chemical environment surrounding them. Thus they may be denatured by even a small increase in temperature and changes in pH. This means that the conditions in which they work must be stringently controlled. In particular the enzyme–substrate mixture must not be contaminated with other substances that might affect the reaction so the equipment used must be scrupulously clean
- expensive to purify
- often unstable in the purified state.

Enzymes have been used, often unknowingly, in the food industry since ancient times. Today enzymes play an important role in numerous industrial applications and are important agents in biotechnology in general (Table 7.1).

The traditional use of enzymes in the production of beer and in cheese manufacture involves the whole microorganism. However, recent developments use isolated enzymes. This has several advantages:

- when whole organisms are used, some of the substrate is converted to microbial biomass therefore using isolated enzymes is far less wasteful
- only one enzyme is present, so there will not be wasteful side reactions
- only one chemical process needs to be considered, so it is much easier to set up the optimum environmental conditions for that process
- a single product is formed, so it is easier to isolate and purify the desired product.

Industry	Enzyme	Use
Dairy industry	Lipases	Blue cheeses ripened by extracellular lipases from mould, e.g. *Penicillium roquefortii*
Brewing industry	Amyloglucosidase	Breaks down sugars in production of low-calorie beer
Baking industry	Proteases	Lowers protein content of flour for biscuit production
Agricultural forestry	Lignases	Most wood waste is not very useful as it contains lignocellulose and few organisms can utilise this: lignases from *Sporotrichum pulverulentum* make cellulose available for animal feed
Leather industry	Trypsin	Removes hair and excess tissue from hides and skin to make the leather more pliable (trypsin used to be obtained from dung)
Medical	Trypsin	Dissolves blood clots and cleans wounds
Biological washing powders	Mostly bacterial extracellular proteases	Remove organic stains from washing (factory workers may become allergic; immobilisation of enzymes overcomes this)

Table 7.1 Industrial uses of enzyme technology.

The development of enzyme technology is comparatively recent. There is now large-scale production of enzymes to catalyse a range of reactions at lower temperatures and pressures than would otherwise be required.

Since enzymes are proteins, they may be denatured by extremes of temperature and pH. However, industrial enzymes must be very robust and able to withstand a wide range of working conditions. They must be able to tolerate a wide temperature range, 10–55°C. They must also have a wide pH tolerance as they may need to work in the presence of chemicals, such as sulphur dioxide, which usually inhibit enzyme action.

The organisms grown must therefore be selected carefully to produce enzymes with these qualities. Indeed some enzymes have exceptional properties, such as α-amylase, which can degrade starch at temperatures in excess of 100°C.

Industrial production of enzymes

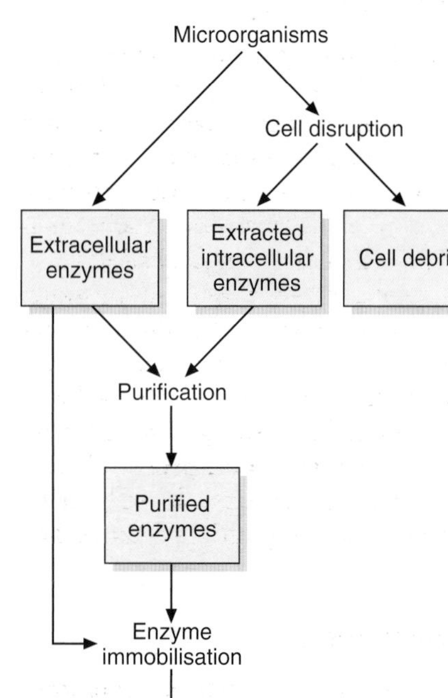

Figure 7.2
Flow diagram showing the industrial production of enzymes.

The majority of enzymes used commercially are obtained from microbial sources, usually fungi and bacteria. Relatively few come from plants or animals.

Most industrial enzymes are **extracellular enzymes**, that is they are secreted by the microorganism into their surroundings. These enzymes can easily be extracted from the contents of the fermenter by filtration and can then be concentrated. If **intracellular enzymes** are required, that is those that function within the cell, recovery is more complex. The cells must be broken up and the desired enzyme extracted from a mixture of many enzymes and cell debris.

Microorganisms have been found to be a valuable source of enzymes for several reasons:

- they produce more enzyme molecules in relation to their mass than most other organisms
- product yield can be increased by means of strain selection, mutation and optimisation of growth conditions
- they are easy to manipulate genetically and can be subjected to gene transfer techniques
- they can be grown in suitable laboratories anywhere and thus are not influenced by climate: this ensures independence of supply from world markets, which may be subject to political influence
- they can occupy a great variety of habitats and extremes of conditions, so their enzymes function in an enormous range of pH and temperature.

Industrial microbiologists have many factors to consider when selecting suitable microorganisms for enzyme production, including:

- the nutritional requirements must be simple
- they must not produce toxins or offensive odours
- they must be non-pathogenic
- they must have a high growth rate
- they must have the optimum growth temperatures for the process required.

When the organism has been selected, genetic engineering may be used to produce a strain that gives a high enzyme yield combined with other useful properties such as thermal stability. Any less favourable characteristics, such as odour, can be selected out.

key term

Large-scale fermentation, both aerobic and anaerobic, is carried out in large cylindrical steel containers called **bioreactors** or **fermenters**.

Methods of producing enzymes

Fermenters

Fermenters are required for the large-scale growth of microorganisms, which produce enzymes, and the substrates on which the microorganisms feed should be cheap, plentiful and non-toxic. Commonly used substrates include whey, molasses and waste from flour milling.

Stainless steel is used to make fermenters because apart from being able to withstand the high internal pressures resulting from gas production, it provides a smooth, hard and corrosive-resistant inner surface, from which microorganisms can be removed after washing. After the inner surface has been disinfected and washed with distilled water, the central cavity is filled to a specified level with a sterile nutrient solution. This nutrient solution is then inoculated with a pure culture of a bacterium or fungus. Paddles rotate the mixture so that the suspension of microorganism and substrate is always well mixed. As nutrients are used up, more can be added. Probes monitor the formation of foam at the surface of the mixture and changes in pH, temperature and oxygen concentration within it. A water jacket surrounding the fermenter normally contains fast-flowing cold water to cool the fermenter since fermentation is a heat-generating process. Most of the air, together with carbon dioxide and other gases produced by cell metabolism, leave the fermenter by an exhaust pipe positioned above the liquid level.

Q 2 Why are bacteriophages a serious problem if they infect the contents of an industrial fermenter?

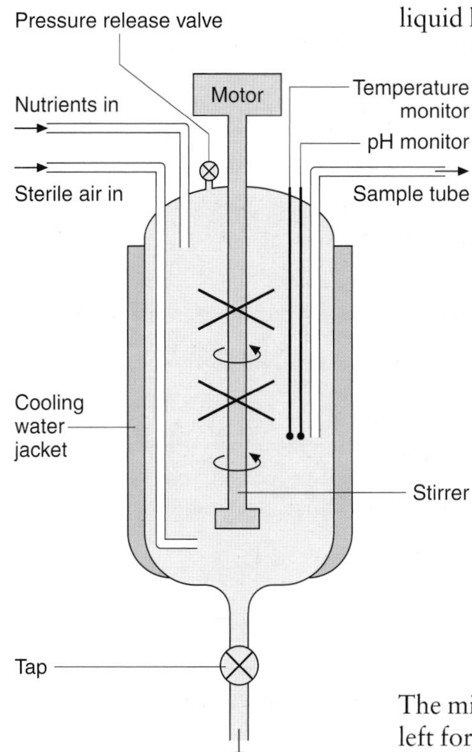

Figure 7.3
Simplified diagram of a batch fermenter.

Method of fermentation

Enzymes are usually produced by **batch** fermentation, although it is sometimes found that better yields are obtained if different substances are added at various times during the fermentation in a **fed-batch process**.

There are several stages involved in the extraction and preparation of extracellular enzymes:

- filtering off the microorganism
- concentrating the enzyme by reducing the water content of the liquor; often done by reverse osmosis
- addition of antibacterial agents to prevent contamination
- quality control, to ensure uniformity of the product
- packaging.

This is carried out in a **closed** fermenter (Figure 7.3).

The microorganism is put into the fermenter with a nutrient medium and left for the fermentation to take place. The product is separated from the rest of the mixture at the end. While the process is going on, nothing is added to the vessel and nothing is removed, except for the venting of waste gases. All nutrients are used up by the end of the fermentation.

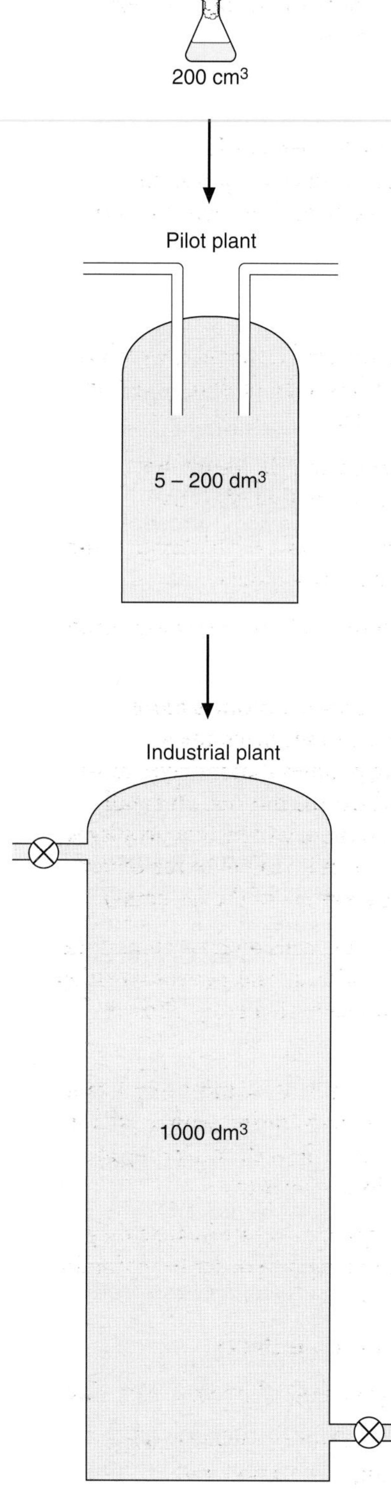

Figure 7.4
Diagram showing the relative scale of fermenters at each stage in scaling up.

During batch fermentation environmental factors are constantly changing, although the temperature is usually controlled. The phase of exponential growth of the microorganisms lasts only a short time.

A batch culture has a number of advantages compared with other industrial procedures:

- it is easy to set up and easy to control the environmental factors
- vessels are versatile: they may be used for different processes at different times, enabling a manufacturer to meet market demands more easily
- should a culture become contaminated, only one batch is lost, so the cost to the manufacturer is minimised.

Large-scale production

Industrial and research laboratories have a strict set of guidelines to follow for microbial work:

- laboratory surfaces must be smooth and easy to clean
- some areas are designated 'clean zones', where full protective clothing must be worn to avoid the escape of pathogens or the introduction of contaminating organisms into pure cultures
- laminar flow cabinets, where contaminated air is continuously removed, are often used to create a sterile working area
- the fermentation vessel should be designed to keep contaminating organisms out
- the sampling outlets must be carefully designed to contain any spillage
- all effluent from the fermenter must be passed through a continuous steriliser.

Scaling up

Modifying a laboratory procedure so that it can be used on an industrial scale is called 'scaling up' (Figure 7.4).

It is not always straightforward to scale up a laboratory process to produce a profitable and efficient industrial process. Laboratory procedures are normally scaled up via intermediate models of increasing size. To do this three stages are involved:

- basic screening
- a pilot plant
- industrial scale.

Basic screening

The researcher uses a small laboratory flask with a volume of approximately 200 cm^3 to grow a microorganism that has been found to make a useful product. The optimum conditions for growth of the organism are determined and if the process works smoothly, the next stage is to scale up the process into a bigger vessel.

Pilot plant

The microorganism is cultured in a small-scale fermenter of approximately 5–200 litres in volume. This is to find the optimum operating conditions, which may not be the same as they were for very small-scale growth.

Industrial scale

The microorganism is grown in a massive industrial fermenter, which is thousands of litres in volume. The fermenter has to be built to very high specifications. However, there are many problems:

- contamination of the desired microorganism and therefore the exclusion of any contaminating organism, is the first objective
- the need to install highly sensitive controls that allow pH, temperature and fluid volume to be maintained within narrow limits
- foam formation at the surface of the culture broth must be monitored and controlled
- as cells multiply in the culture broth, the mixture becomes more viscous. This increasing viscosity can pose serious mechanical problems. Both the motor and the mixing paddles that it drives must be large enough to continue to rotate and stir the mixture. A small vessel is stirred by a small electric mixer so there will be few problems in cooling the vessel sufficiently, but when this is scaled up the mixer can generate enough heat to vaporise the contents of the fermenter
- at this stage, when the mixture becomes thick, cells tend to stick to the sides of the vessel and the surfaces of the paddles. These surfaces either have to be non-adhesive themselves or coated with a non-adhesive substance
- in cultures of aerobic organisms, oxygen supply is the most important limiting factor. For large-scale cultures, forced aeration with small air bubbles with a large surface area to volume ratio is the only way in which the high oxygen demand of the cells can be met
- microorganisms contained in a confined space generate heat as they grow. This must be removed via a heat exchanger in order to maintain a constant temperature.

Other problems involved in scaling up the process include:

- the heating up and cooling down of a large volume of liquid is very slow
- special equipment needs to be designed to allow all the processing after fermentation, such as filtering, drying, distillation and extraction, to be carried out on a large scale

Nutritional requirements

Organisms need nutrients to generate ATP and synthesise cell materials. These nutrients are normally obtained from their environment. An artificial environment such as a culture medium must thus contain all the necessary nutrients in sufficient amounts. However, different microorganisms have different requirements.

Element	Function
Carbon	Component of organic molecules, e.g. proteins, carbohydrates, lipids and nucleic acids
Nitrogen	Component of proteins and nucleic acids
Sulphur	Component of some amino acids
Phosphorus	Component of nucleic acids, phospholipids and ATP

Table 7.2 Function of some elements within microorganism.

Carbon

Some microorganisms can use carbon dioxide as their carbon source, others need organic carbon. Some microorganisms may be able to use a range of sources to fulfil their carbon requirements, while others may be highly specialised.

Nitrogen

Some microorganisms can fix atmospheric nitrogen. Many take up nitrogen in the form of nitrates or as ammonium ions. However the nitrogen requirement is often met by organic nutrients, such as amino acids or peptides which can also provide the carbon and energy source.

Sulphur

Some microorganisms take up sulphur as sulphates. The sulphur requirement is often met by organic nutrients, such as amino acids or peptides. These can also form the carbon, nitrogen and energy source.

Extra growth factors

These are substances that some microorganisms cannot make, which are required in small amounts for use in the synthesis of cell materials. For example:

- amino acids (needed for protein synthesis)
- vitamins.

Oxygen

Many organisms need molecular oxygen, O_2, for aerobic respiration.

Culture media

The culture media must contain a balanced mixture of the required nutrients at concentrations to allow a good growth rate. If a nutrient is in excess, it can inhibit growth or may even be toxic. Usually a mineral base is made which contains all the possible nutrients in inorganic form. To this can be added a carbon source, an energy source, a nitrogen source and, if required, growth factors to suit the organism being cultured.

A particular medium may be satisfactory for initial growth, but microorganisms change the pH of the medium as they grow because their metabolic waste products accumulate. To counteract this, buffers are often added.

As well as the correct culture medium, other factors, such as light, are often necessary to meet the needs of individual microorganisms.

Aseptic conditions

For successful fermentation, it is vital that there should be no contamination of the culture. For this reason the entire fermenter, all ancillary equipment and the growth medium must be sterile before inoculation. The air supplied during the fermentation must also be sterile and there must be no mechanical breaks in the fermenter that might allow microorganisms to enter.

All equipment is cleaned with hot water before use and then sterilised in situ with steam. The steam must be able to reach all parts of the fermenter assembly, so the design of the fermenter is vital to avoid pockets of air, which may occur if valves are incorrectly sited. All the interior surfaces should be polished since rough surfaces can act as a reservoir for contaminating microorganisms.

The culture medium, which can be a volume of several thousand litres, may be heat sterilised in the fermenter itself by passing steam through the cooling coils and jacket. Alternatively, the medium may be prepared in a separate vessel and passed through a continuous steriliser.

All additives, for example antifoaming agents, must also be sterilised, although strong acids or alkalis added to adjust the pH do not need sterilisation. The air supply to the culture medium is sterilised by filtration. Many types of filter materials have been developed to trap microorganisms. However, viruses such as bacteriophages in the air are too small to be removed in this way and their presence could result in the loss of an entire culture. In some cases therefore the air is heated to inactivate any bacteriophages and then cooled again before it enters the vessel.

Air leaving the fermenter must also be sterilised, particularly if recombinant organisms are being grown. Not only is this a safety requirement, stopping their release into the environment, but it also prevents key microbial strains from becoming available to competitors.

Downstream processing

This is the name given to the extraction and purification of the desired end product or products of the fermentation processes. The procedure adopted depends on whether it is the cells, or the solution surrounding them, that contains the desired end product. Downstreaming to extract cells from liquid cultures may involve the following stages:

- flocculation to precipitate the cells
- filtration to separate the cells from the solution
- centrifugation to concentrate cell masses
- drying to prepare the cells for packaging.

If the solution surrounding the cells is required, the following downstream process may be used:

- flocculation to precipitate the cells
- ultrafiltration or centrifugation to obtain a cell-free solution
- distillation to concentrate the solvent
- drying to obtain a crude solute
- purification to obtain a pure solute.

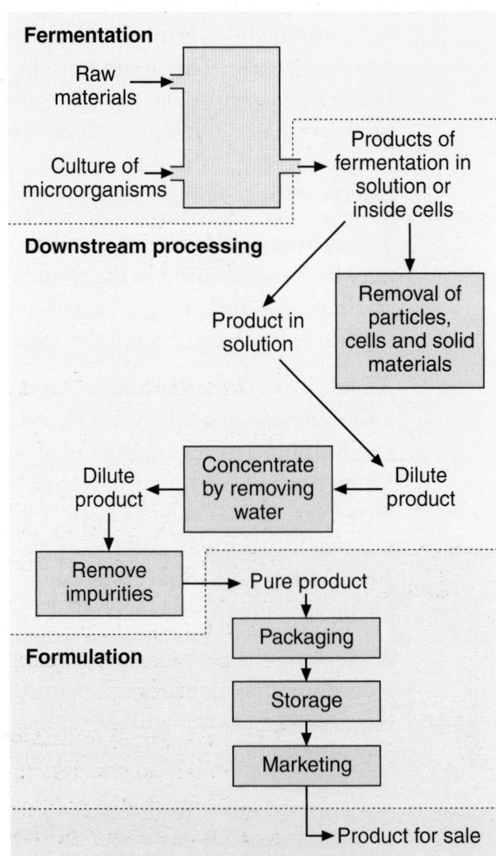

Figure 7.5 Summary of stages involved in enzyme production

key term

Sensory systems using biological processes can be extremely sensitive and are often referred to as **biosensors**.

key term

The **transducer** is an electrical device that can register chemical changes and convert them into an electrical signal. This signal can be amplified and the strength of the signal is directly proportional to the amount of the particular substance being measured.

Q 3 Explain how a biosensor might be used to control the amount of insulin given to a person with diabetes

Enzymes used as analytical reagents

Biosensors for analysing biologically active compounds in solution could simplify the workload of doctors, the police and analytical chemists. Biosensors would be of value to the police as they could obtain rapid estimates of blood alcohol levels in drink-driving suspects.

There are three main types of enzyme-based biosensors:

- enzyme transducer sensors
- enzyme thermistors
- enzyme biochips.

In a biosensor there is a **biological recognition layer**, which often contains an enzyme. This layer is used to recognise a particular substance by producing a biochemical signal when it is present. This biological component is the part that ensures sensitivity and is responsible for the high degree of specificity shown by biosensors. The basis of the sensor is that when two biological molecules interact, there are measurable chemical changes. These chemical changes can be converted into an electrical signal by a transducer. The strength of the signal is directly proportional to the concentration of the particular substance being measured. (Figure 7.6).

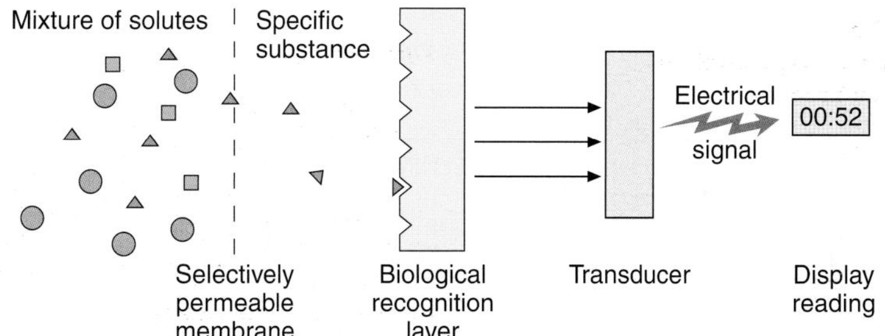

Figure 7.6 (right)
Diagram showing the general features of a biosensor.

Figure 7.7 (below)
Diagram showing an enzyme electrode for the detection of urea.

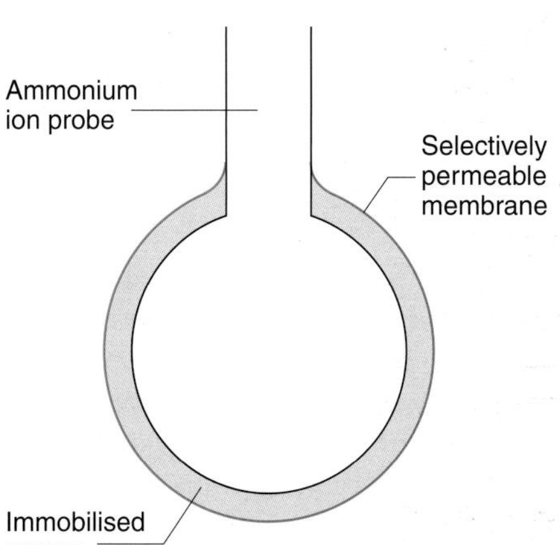

Enzyme transducer sensors

An early example of a biosensor was developed to allow surgeons to continuously monitor blood glucose levels during surgery. In this sensor the enzyme used is **glucose oxidase**. This catalyses the reaction between glucose and oxygen in solution to form **gluconic acid** and **hydrogen peroxide**. The enzyme is associated with a platinum oxygen electrode, which measures the changing oxygen concentration as blood glucose levels change. Increasing glucose levels are proportional to decreasing oxygen levels.

The same principle can be used for measuring the amount of urea in blood or urine. Immobilised **urease**, surrounded by a membrane that is permeable to urea, is packed around the tip of an ammonium ion probe (Figure 7.7).

When the probe is dipped into a solution containing urea, molecules of the solute diffuse through the membrane and are broken down to carbon dioxide and ammonium ions by the immobilised enzymes.

$$\text{urea} + H_2O \rightarrow CO_2 + NH_4^+$$

The volume of ammonium ions released in a given time is proportional to the concentration of urea in the solution. An appropriate transducer converts levels of the ammonium ion into an electrical signal (Figure 7.8).

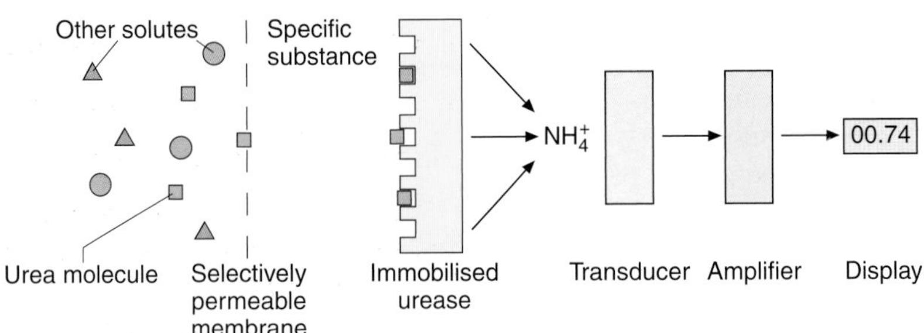

Figure 7.8
Specific example of a biosensor using immobilised urease for the detection of urea.

Enzyme thermistor

This is a biosensor that measures minute changes in temperature during exothermic reactions. Such devices are sensitive to temperature changes between 0.004 and 1.0°C, the amount of heat generated during a reaction being directly proportional to the concentration of the reactants.

Enzyme biochip

Q 4 When alcohol and cholesterol are broken down they produce organic acids. Design, in outline only, a biosensor which might be used to monitor the levels of cholesterol or alcohol in human blood.

The miniaturisation of biosensors has led to the development of probes where the enzyme is immobilised on the surface of a silicon chip, making a **biochip**. Such biosensors are tiny, some only 2 mm across, and one has been developed to detect blood glucose levels, again using the immobilised enzyme glucose oxidase but this time measuring the amount of gluconic acid produced. This reaction generates an acid, or more specifically positively charged protons (H^+), which collect. Negatively charged electrons (e^-) from the silicon chip are attracted to the positive charges. The flow of electrons through the silicon terminals generates a current, which is proportional to the amount of glucose in solution. These biochips are small and self-contained and it may soon be possible to implant them under the skin.

Immobilised enzymes

Until quite recently many enzyme-catalysed reactions were carried out on a commercial scale using enzyme/substrate/water mixtures. This system led to enzymes being wasted. It was often quite difficult to remove enzymes from the end-products, and even more difficult to purify the recovered enzyme so that it could be used again. An increasing demand for pure enzymes and their products led to **enzyme immobilisation.** This is a technique whereby an enzyme is bound immovably to a surface and not allowed to dissolve in a solution containing its substrate. The enzymes can thus be held in place during the reaction, removed afterwards and used again.

There are various methods of enzyme immobilisation, which may be used to bind either the enzymes themselves or cells capable of producing them. Some current examples are enzymes that are:

- adsorbed onto an insoluble matrix, such as collagen (Figure 7.9a)
- held inside a gel, such as silica gel (Figure 7.9b)
- held within a semi-permeable membrane (Figure 7.9c)
- trapped in a microcapsule, such as alginate beads (Figure 7.9d)
- attached to cellulose fibres (Figure 7.9e).

All these processes involve physical bonding of the enzyme, which is not easy to carry out and generally results in lower enzyme activity. Covalent bonding of an enzyme to a support medium results in higher enzyme activity, although preparing the enzyme in this way is difficult (Figure 7.9f).

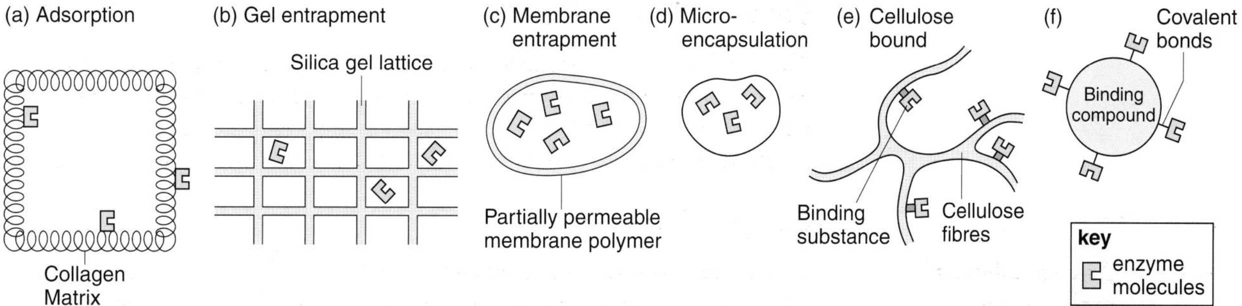

Figure 7.9 (a-f)
Diagrams showing different methods of immobilising enzymes.

Q 5 What are immobilised enzymes?

Immobilisation may affect the shape and properties of the enzyme molecules involved. The enzyme substrate, for example, may not bind so effectively, which causes the rate of reaction to fall. This disadvantage, however, may be offset by the fact that immobilisation may increase the temperature and pH range over which an enzyme remains active.

The advantages of using immobilised enzymes are that:

- enzymes can be recovered and used over and over again, which is particularly useful when the enzyme is expensive or difficult to produce
- the product will not be contaminated by the enzyme because the enzyme is held in a matrix
- the matrix protects the enzyme with a physical barrier, so it is more stable at extremes of temperature and pH.

Immobilised enzyme technology is still developing rapidly and there are likely to be many new applications for immobilised enzymes in industry, medicine and waste disposal.

Extension box 1

Biological washing powders

Although approximately 2000 enzymes have been identified and of these 150 are used in industrial processes, their best-known use is in biological washing powders. In 1913 Dr Otto Rohn manufactured the world's first enzyme detergent. The powder consisted mostly of a mixture of washing soda (sodium carbonate) and the enzyme trypsin, extracted from the pancreas. This formula was, however, far from successful. The soda made the solution very alkaline, thus reducing the activity of the enzymes.

After the Second World War, in 1947, fermenters were developed for the large-scale manufacture of penicillin; however it was not until 1963, in Denmark, that Novo Industries used this ferment technology to produce a protease enzyme from *Bacillus licheniforms* that was active and stable in alkaline washing conditions. Today enzymatic detergents account for 85% of the market in Western Europe. But how are they made? The main steps are shown in Table 7.3.

Process	Description of each step
Fermentation	The protease enzyme is produced by bacteria in a fermenter released into the culture medium
Centrifugation	The bacterial cells are separated from the culture medium by centrifugation
Evaporation	The culture medium is concentrated by removing water
Ultrafiltration	Any bacterial cells still present are filtered out
Precipitation	The enzyme is separated from the remaining liquid by precipitation with salt
Salting	The product is granulated with salt to act as a preservative
Paste formation	The product is mixed with water to form a paste
Shaper	The paste is shaped into spheres
Drying	The spheres are dried
Waxing	The spheres are coated with a layer of wax and cooled

Table 7.3 The main steps in the production of enzymes for biological washing powders.

Q 6 **Biological washing powder manufactuers claim their products 'remove biological stains'. What do they mean by a biological stain?**

The finished particles are added to the other ingredients of the washing powder and packaged. The wax subsequently melts in the wash, releasing the enzyme.

One of the initial problems of this formulation was an allergic response to the free enzyme; making the enzyme dust free by encapsulating it in wax, from which it is released only when in the wash, has reduced this.

The advantage of modern biological washing powders is that they are effective at relatively low temperatures, and are therefore energy saving as well as gentler on clothes.

Since proteases currently in use work best at around 55°C, the search is on for new enzymes with an optimum temperature of around 20–30°C.

It may be possible to transfer genes, which code for the production of suitable low temperature enzymes, into existing harmless microorganisms.

Extension box 2

Figure 7.10
An industrial plant producing ammonia by the Haber process.

Nitrogenase, a remarkable enzyme

Fertiliser is big business. Every year, over one and a half million tonnes of nitrogen-containing fertiliser are produced in the UK alone. It is made industrially from ammonia, produced by combining nitrogen and hydrogen in the Haber process:

$$N_2 + 3H_2 \rightarrow 2NH_3$$

In the Haber process, the nitrogen and hydrogen mixture is first compressed to a pressure of about 200 kPa and then passed over a catalyst heated to a temperature of 400 °C. Energy is needed for raising the pressure and temperature of the gas mixture. Without this input of energy, the atoms of the nitrogen molecule cannot be separated and the reaction will not take place. This is not the whole story, however, because the hydrogen used has to be produced from natural gas. A number of stages are involved, most of which again need energy in the form of heat.

Ammonia can be produced from nitrogen and hydrogen but the process needs large amounts of energy to provide the necessary high temperatures and pressures.

Living organisms can also fix nitrogen, that is, produce ammonia from nitrogen and hydrogen. The reaction involves nitrogenase and the remarkable thing about this enzyme is that, unlike the Haber process, it controls a reaction that works at everyday temperatures and pressures. However, the nitrogen molecule again has to be split, and this still requires a considerable amount of energy. It comes in this case from ATP and a lot is needed – for every mol of nitrogen fixed, 25 mols of ATP are

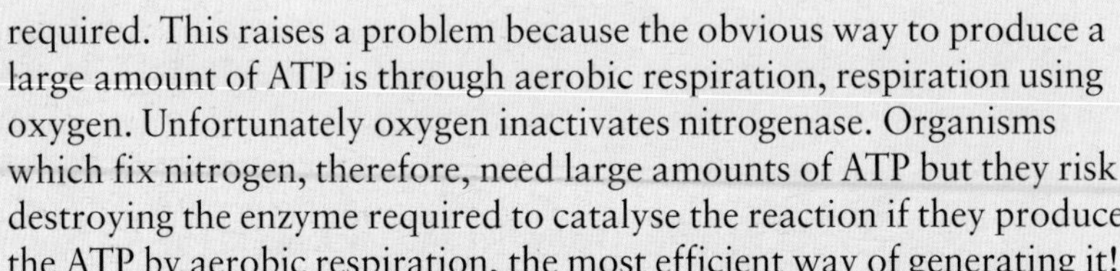

required. This raises a problem because the obvious way to produce a large amount of ATP is through aerobic respiration, respiration using oxygen. Unfortunately oxygen inactivates nitrogenase. Organisms which fix nitrogen, therefore, need large amounts of ATP but they risk destroying the enzyme required to catalyse the reaction if they produce the ATP by aerobic respiration, the most efficient way of generating it!

Despite this, many microorganisms can fix nitrogen. They often form close relationships with larger organisms which are beneficial to both species involved. The eukaryote respires aerobically providing the microorganism with its energy requirements. The microorganism is able to live in an oxygen-free environment within its host where it fixes nitrogen with the aid of its nitrogenase. Some of the ammonia and other nitrogen-containing compounds produced as a result may be used by the eukaryote.

(a)

(b)

(c)

Figure 7.11
Leguminous plants such as peas, beans and this Judas tree (a) form associations with nitrogen-fixing bacteria. Other less familiar organisms such as the fungi in this lichen (b) and a group of insects, the termites (c) form similar relationships.

Extension box 3

Enzymes are green

Leather is produced from animal skins. The fresh skin must be treated before the tanning process can be started. The first step is to remove the hair. At one time, this was done chemically by using a mixture of calcium hydroxide and sodium sulphide. The calcium hydroxide caused the tissues in the skin to swell and become softer while the sodium sulphide weakened the hair at its base so that it could be washed off. This treatment was effective but unfortunately produced considerable amounts of the poisonous gas, hydrogen sulphide.

The skin must also be treated to remove the fat and to make it more pliable, a process called baiting. The pliability of leather results from partly breaking down protein fibres in the skin. The amount of protein that is actually broken down depends on the use to which the leather is to be put. Soft leathers such as those used to make gloves have most of this protein removed while the tougher, firmer leathers used for the soles of shoes do not need so much baiting. The calcium hydroxide and sodium sulphide mixture also helped with this softening process while other solvents were used to remove the fat. Not surprisingly, waste from tanneries produced severe pollution problems.

Nowadays, increasing use is made of enzymes. For removing the hair from cattle and pig skins, a protease enzyme which works in alkaline conditions is used. This enzyme is produced by a species of bacterium. It partly digests keratin, the main protein present in hair, allowing the hair to be removed easily. Sheep skins have to be treated rather more gently because the wool that is removed is valuable. Protein-digesting enzymes extracted from the pancreas of slaughtered animals are applied to the underside of the skin. These enzymes diffuse into the skin and break down the proteins around the hair follicles, allowing the wool to be removed undamaged. Pancreatic enzymes are also used for baiting hides from other animals.

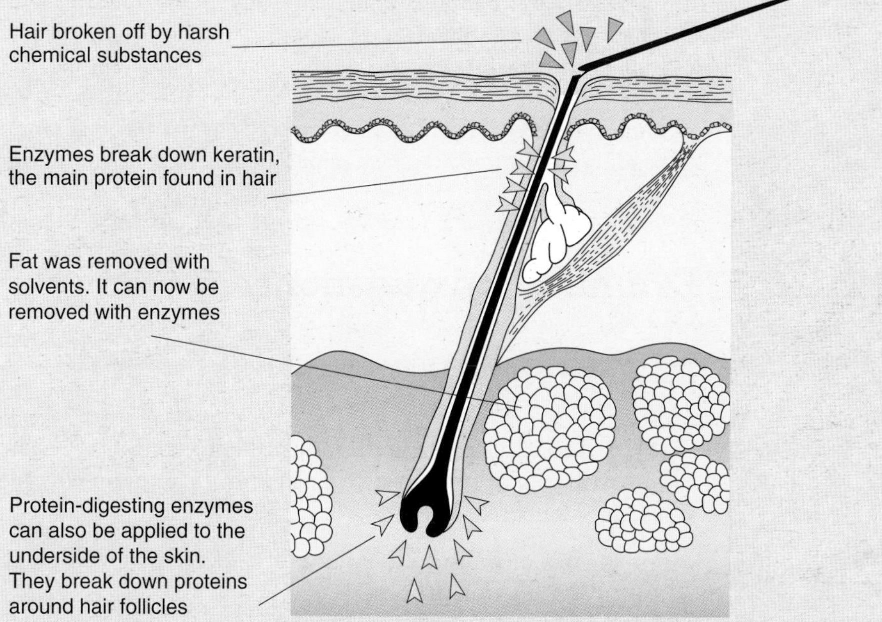
Hair broken off by harsh chemical substances
Enzymes break down keratin, the main protein found in hair
Fat was removed with solvents. It can now be removed with enzymes
Protein-digesting enzymes can also be applied to the underside of the skin. They break down proteins around hair follicles

Figure 7.12
Various treatments have been used to remove the hair from animal skins. The use of enzymes has allowed a reduction in the use of harsh chemical substances and has played a significant role in reducing the pollution from tanneries.

Tanning factories were for a long time a major source of pollution. Harsh inorganic substances were used in the process and large amounts of waste were produced. There are two ways in which increasing the use of enzymes may contribute to reduced pollution. First, it offers an alternative to the use of substances such as calcium hydroxide and sodium sulphide. A recent estimate suggested that widespread use of protein-digesting enzymes has reduced the consumption of sodium sulphide by some 40%. Obviously with less sodium sulphide being used, hydrogen sulphide production also becomes less of a problem. The other contribution of enzymes is that their use contributes to more effective treatment of tannery waste. It can be broken down naturally and recycled by the microorganisms in the nitrogen cycle.

Industries involving animal skins can lead to severe environmental pollution. For example, lanolin is a fatty substance produced from wool. At one time much of the lanolin used in producing cosmetics was extracted from Bradford sewage – the result of pollution from the large numbers of woollen mills in the area.

Summary

- Extracellular enzymes are secreted to function outside the organism's cell: intracellular enzymes function inside the cell.
- The commercial production of enzymes involves the growth of large numbers of microorganisms using specific media and aseptic conditions.
- The commercial production of enzymes involves the isolation and purification of the enzyme product by downstream processing.
- Enzymes may be used as analytical reagents because of their high sensitivity and specificity.
- Industrial processes require a high degree of thermostability.
- Immobilised enzymes can be separated easily from reactants and products and can thus give a higher degree of control; they are also more stable.

Examination questions

1 (a) What is an extracellular enzyme?

(1 mark)

(b) Bacteria are grown in the industrial production of certain extracellular protein-digesting enzymes. Explain why the bacteria are incubated:

(i) in a medium containing a lot of protein in the first part of the production process

(2 marks)

(ii) in a medium containing little protein in the second part of the production process.

(1 mark)

(c) This process is carried out in aseptic conditions, which prevent contamination with other species of bacteria. Give **one** reason why it is necessary to prevent contamination with other species of bacteria.

(2 marks)

2 Lactose is a sugar found in milk. Many adults cannot digest lactose. An industrial reactor, making use of immobilised enzymes, can be used to reduce the amount of lactose in milk.

(a) What is an immobilised enzyme? Explain the advantage of using an immobilised enzyme in this reactor.

(4 marks)

(b) The contents of this reactor are kept at 7°C. Use this information to suggest why it was necessary to pass the milk through the reactor several times in order to reduce the amount of lactose enough.

(2 marks)

Assignment

When you have read this chapter and studied the relevant part of the specification, you should appreciate that enzymes have many applications in industrial processes. In this assignment you will be working either on your own or as part of a group to produce a poster. Your poster should show how the features of enzymes, which you looked at in Chapter 4, make them particularly useful in industrial processes.

This assignment will also provide you with an opportunity to collect evidence for your Key Skills portfolio. Successful completion of this exercise will enable you to demonstrate communication skills, in particular your ability to:

- make a presentation of a complex subject using at least one image to illustrate complex points
- read and synthesise information from extended documents about a complex subject including documents containing images.

Start by reading the information in the box in this assignment. It tells you some things about designing a poster. You should keep this information in mind as you work.

Now look at Figure 7.13. You will see that there are three main ideas that you will need to research. Each of these corresponds to one of the Extension boxes in this book, although not necessarily in this chapter. The material in these Extension boxes should provide the basic information you require. In each case, however, you should try to obtain a little more detail about the particular topic by looking at other sources of information. The notes below should help you with this.

Enzymes are specific

Start with the Extension box **Restriction enzymes**, page 167. Look up: Specificity, Enzyme electrodes, Biosensors.

Enzymes are environmentally friendly

Start with the Extension box **Enzymes are green**, page 128. Look up: The nitrogen cycle, Tanning, Thermal pollution, Heavy metal pollution.

Enzymes do not require extreme conditions

Start with the Extension box **Nitrogenase**, a remarkable enzyme, page 127. Look up: Nitrogen fixation, The Haber process, Thermostable enzymes, Immobilisation.

Once you have collected this basic material, you can start to produce your poster. Your final task ought to be to write about the features of enzymes which form the links with the topics you have written about. You will find it easier to write this last so that you can be sure that what you plan to include is relevant.

Designing your poster

Your poster has one main purpose. It must communicate scientific information about a particular topic as clearly as possible. In designing it, you must not lose sight of this purpose. A good poster is one that conveys relevant information effectively to its target audience. As you work, try to keep the following points in mind.

- Think about this poster as being designed to provide information for other students following an AS Biology course. Make sure that it can be understood readily by explaining new ideas clearly and using scientific terms with which your readers will be familiar.
- The layout should be simple and uncluttered. In a good poster the main ideas are separated from each other but it is easy to see how they relate. Figure 7.13 shows one possible way of laying out these main ideas, but you should feel free to change this if you think you can improve on it.

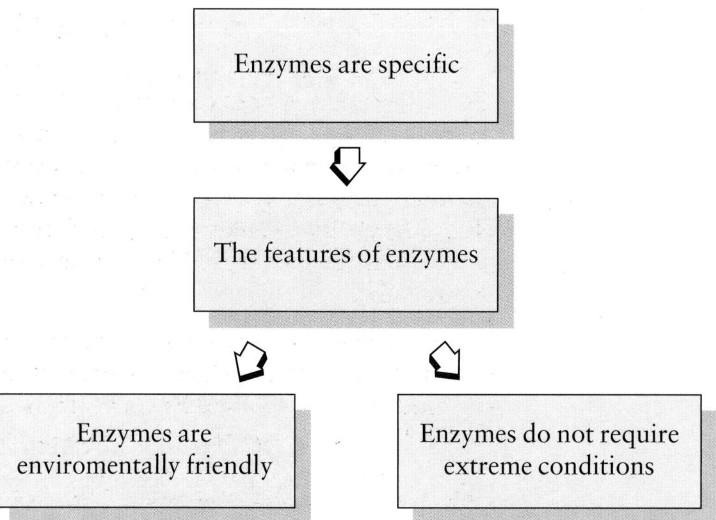

Figure 7.13
Laying out your poster.

- Large blocks of text can be difficult to understand. You can often make complex points easier to follow if you use images such as graphs, flow charts or diagrams. But, remember, images also have to be clear, relevant and easy to understand. They are not just pictures.
- There is very little room on the A2-sized sheet of paper that you will require as a backing sheet so you will have to make sure that what you write is relevant and concise. Writing the material for a poster is as much about what you can leave out as what you should include.

8 DNA and Protein Synthesis

Chromosomes contain the DNA that controls our characteristics. As you will learn in this chapter, each DNA molecule is a chain of nucleotides. On each nucleotide is one of four organic bases. It is the order of these bases that determines the genetic code.

In the 1990s, over 1000 laboratories around the world joined the Human Genome Project (HGP). One of the goals of this project is to determine the entire sequence of bases in human DNA. This means determining the sequence of about 3.2 billion bases. If you think this is a daunting task, imagine the problems in managing the resulting data: it is thought that a list of the base sequence of human DNA would fill 480 000 pages of single-spaced word-processed text.

Originally scheduled to be completed in 2005, completion of the first draft of the HGP was announced on 26th June 2000. The knowledge it will bring will undoubtedly be helpful. For example, knowing about the effects of DNA variations among individuals should be helpful in diagnosing, treating and eventually preventing many disorders. However, at the time of its completion the HGP had cost many millions of US dollars, mostly of taxpayers' money. Many people think the money could have been better spent on other research programmes.

Figure 8.1 represents chromosome 22, the first human chromosome to be decoded as part of the HGP. The short arm of the chromosome contains genes that encode the structural RNA of the ribosomes. The long arm of the chromosome has 33.4 million base pairs and contains the protein coding sequences. Chromosome 22 is implicated in the functioning of the immune system, congenital heart disease, schizophrenia, mental retardation, birth defects and several cancers, including leukaemia.

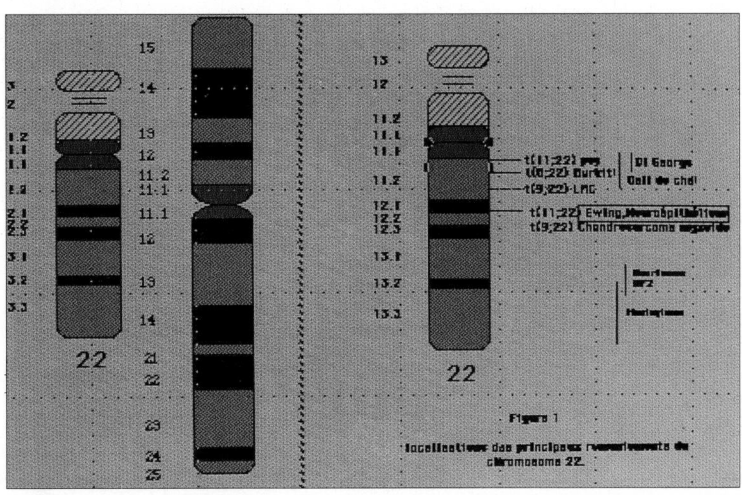

Figure 8.1 Chromosome 22 is the second smallest human chromosome. In December 1999 it became the first chromosome to be decoded in the Human Genome Project.

key term

Nucleic acid carries the **genetic code** that cells need to make proteins. Each **gene** codes for the amino acid sequence of one protein.

Nucleic acids

Nucleic acids carry the genetic code. This code determines the order of the amino acids in each protein that a cell makes. (Look back to Chapter 3 to remind yourself of how proteins are made from chains of amino acids linked by peptide bonds.) Cells can only make a particular protein if they have nucleic acid with the correct code. If the code for a particular protein changes (i.e. a **mutation** occurs), the cell might lose its ability to make that protein.

You learned in Chapter 3 that cell proteins:

- form part of the structure of membrane and cytoplasm
- can act as enzymes, speeding up chemical reactions in cells.

It is because they determine which proteins a cell can produce that nucleic acids are so important.

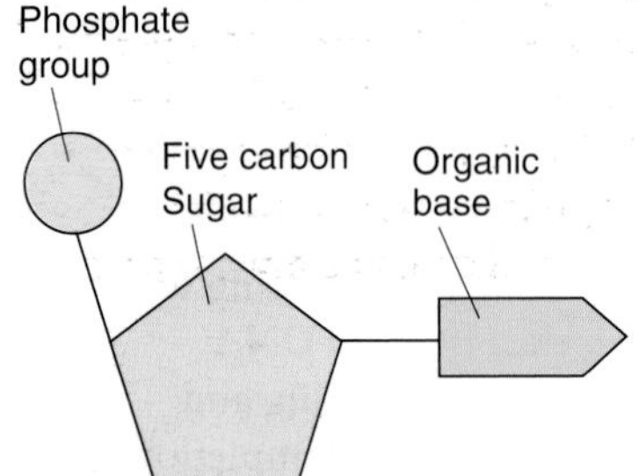

Figure 8.2
A single nucleotide. DNA and RNA are made of long chains of nucleotides. The compositions of nucleotides from DNA and RNA are compared in Table 8.1.

Nucleotides

There are two types of nucleic acid present in cells: **deoxyribonucleic acid** (DNA) and **ribonucleic acid** (RNA). Each is a long chain of smaller units, called **nucleotides**. Figure 8.2 represents a single nucleotide. Each nucleotide has:

- one five-carbon sugar: this sugar is deoxyribose in a molecule of DNA but is ribose in a molecule of RNA
- one phosphate group: these are the same in both DNA and RNA. Each has a negative charge, making nucleic acids highly charged molecules.
- one organic base: the organic bases fall into two groups, purines and pyrimidines. **Purines** have two rings of carbon and nitrogen atoms and **pyrimidines** have a single ring of carbon and nitrogen atoms.

The compositions of nucleotides from DNA and RNA are compared in Table 8.1.

Feature	Nucleotide from DNA	Nucleotide from RNA
Five-carbon sugar	Deoxyribose	Ribose
Phosphate	Phosphate group (PO_4^-)	Phosphate group (PO_4^-)
Purine (organic bases)	Adenine or guanine	Adenine or guanine
Pyrimidine (organic bases)	Cytosine or thymine	Cytosine or uracil

Table 8.1 A comparison of nucleotides in DNA and in RNA.

key term

There are two types of nucleic acid in cells: **DNA** and **RNA**.

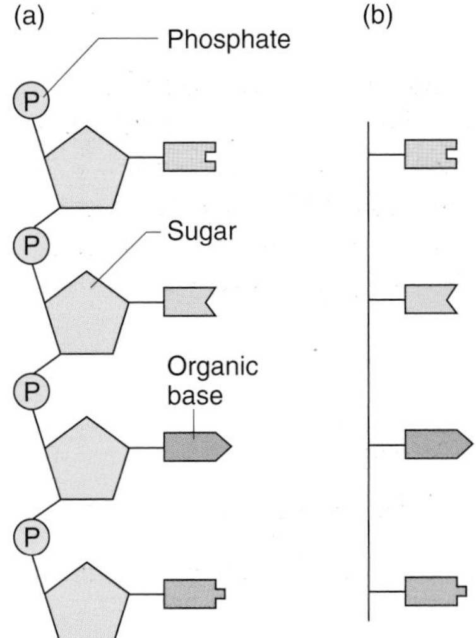

Figure 8.3
(a) Part of a chain of nucleotides.
(b) A simpler way of drawing a chain of nucleotides. This is an easier diagram to remember in a Unit Test.

Chains of nucleotides

A nucleic acid molecule is a very long chain of nucleotides. Figure 8.3a shows part of such a chain. The chain is held together because the phosphate group of each nucleotide is linked to the sugar of the next. Remember that the phosphate group and sugar are identical throughout the chain. The only variable is the organic base. This allows us to simplify our diagram. Figure 8.3b shows the simplified version. Only the bases are identified.

Q 1 **What is the difference between a nucleotide and a nucleic acid?**

Base pairing

When two nucleotide chains lie next to each other, purines and pyrimidines naturally link together by weak hydrogen bonds. We call this **base pairing**. Bases always pair together in a precise way. One member of each pair is a purine and the other is a pyrimidine. In DNA, adenine (A) always pairs with thymine (T) and cytosine (C) always pairs with guanine (G). In RNA, adenine always pairs with uracil (U) and cytosine always pairs with guanine (G).

key term

Bases link together by hydrogen bonds to form base pairs. In DNA, adenine and thymine always pair together (A=T base pair). Cytosine and guanine always pair together (C=G base pair).

Nucleic acid molecules

Figure 8.4a represents part of a DNA molecule. It has two separate chains of nucleotides. These chains are held together by base pairs (A=T and C=G). In a prokaryotic (bacterial) cell, a molecule of DNA can be up to 4.6 million nucleotides long. In a eukaryotic cell, it is much longer.

Figure 8.4
(a) Part of a DNA molecule showing its sugar-phosphate chains and base pairs.
(b) Part of a DNA molecule showing the double helix. This is too complex a diagram to attempt in a Unit Test.

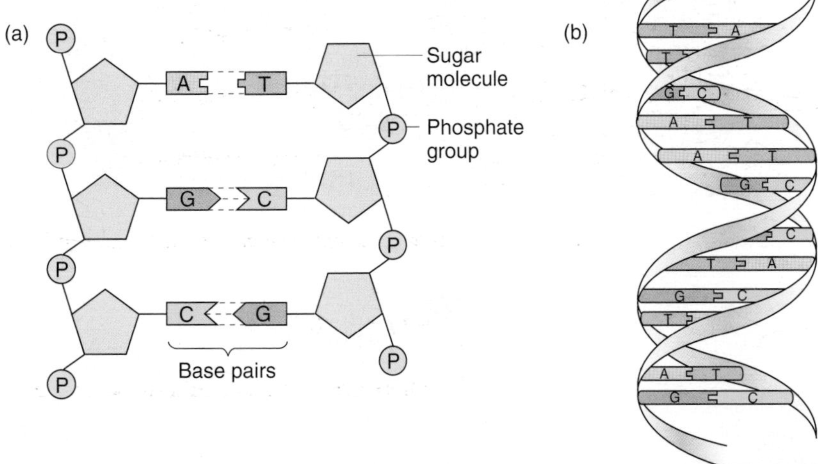

DNA molecules normally twist into a helix (coil). Figure 8.4b shows this helix in part of a DNA molecule. Because the molecule has two chains of nucleotides, it is often referred to as a **double helix**. Figure 8.4b is too complex to draw in a Unit Test. You should learn to draw Figure 8.4a.

A molecule of RNA has only one chain of nucleotides. As a result, it does not form a long regular helix like the two chains in DNA. A helix is formed at certain parts of a molecule of RNA.

Q 2 **Name *three* differences between the structure of a DNA molecule and that of an RNA molecule.**

Protein synthesis

We can use our knowledge of DNA and RNA to learn how proteins are made. Table 8.2 summarises the main processes involved in protein synthesis.

Structure	Function	Name of process
DNA	Carries code for all the proteins that a cell can make	None
Messenger RNA (mRNA)	Copies the code for a single protein from DNA Carries the code to ribosomes in the cell's cytoplasm	Transcription None
Ribosome	Reads mRNA code and assembles amino acids in their correct sequence to make a functional protein	Translation
Transfer RNA (tRNA)	Brings individual amino acids from the cytoplasm to the ribosome	None

Table 8.2 The main processes of protein synthesis. A DNA base sequence is copied on to a mRNA base sequence during transcription. This sequence is converted into an amino acid sequence during translation.

key term

Messenger RNA (**mRNA**) copies part of the DNA code. Transfer RNA (**tRNA**) carries specific amino acids to ribosomes as they make proteins.

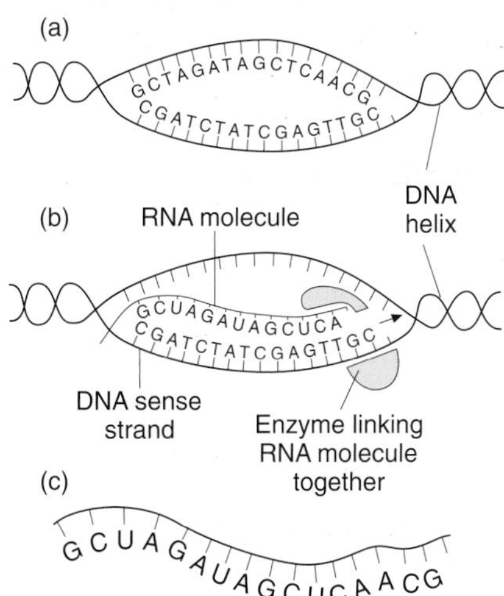

Figure 8.5 (above)
DNA transcription produces a mRNA copy of part of the DNA code. (a) Where the DNA is copied, its chains of nucleotides separate, leaving their bases unpaired.
(b) Free mRNA nucleotides pair with the exposed DNA bases.
(c) An enzyme joins the mRNA nucleotides and a mRNA molecule leaves the DNA it has copied.

Figure 8.6 (right)
DNA contains non-coding sequences of bases, called introns. These are copied onto mRNA but then removed before the mRNA is used.

Transcription (DNA to mRNA)

Figure 8.5 shows how a molecule of messenger RNA (mRNA) is made from part of the DNA code. The DNA helix unwinds where it is to be copied. This happens because the hydrogen bonds that hold together this part of the DNA break down. Figure 8.5a shows the separated chains of DNA nucleotides with their unpaired bases. Free mRNA nucleotides pair with the exposed DNA bases. This is shown in Figure 8.5b. As we saw earlier, the base pairs are precise. The bases A, C, G and T in DNA pair with U, G, C and A respectively in mRNA. Finally, an enzyme links the individual mRNA nucleotides together to form mRNA. Figure 8.5c shows the final mRNA molecule.

In Figure 8.5, only one of the DNA chains is being copied. This is what happens in cells. Only one of the exposed chains of DNA nucleotides is ever copied during transcription. The chain that is copied is called the **'sense' strand.** Figure 8.5 also shows that every DNA base is copied onto mRNA during transcription. Biologists now know that long sequences of the DNA are non-coding. These sequences are called **introns.** Although they have been copied onto mRNA, the non-coding sequences are of no use. Before mRNA is used, these non-coding sequences are removed from their molecules. Figure 8.6 shows how mRNA is modified before use.

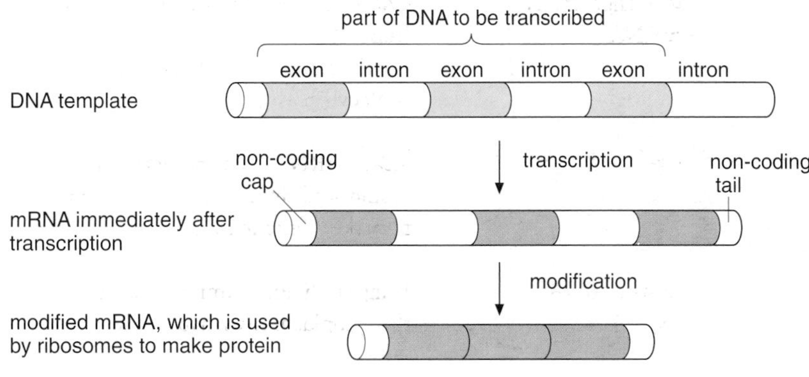

Q 3 The code on part of the sense strand of DNA is ACC GAC. What will be:

(a) the code on the complementary strand
(b) the code on the mRNA made from the sense strand?

key term

Introns are regions of non-coding DNA. They are copied onto mRNA during transcription. They are removed from the mRNA before it is used.

Genetic code

We can now see that the base sequence of part of the DNA has been copied to a base sequence in mRNA. This mRNA sequence is the genetic code for a protein. Each amino acid is coded by a combination of three mRNA bases. This mRNA code for a single amino acid is called a **codon**. Table 8.3 shows the genetic code. To read it, you need to select the first base from the left of the table. Then you select the second base from the middle of the table. Finally, you must select the third base from the right of the table. For example, the codon GAU codes for aspartic acid. The genetic code has two important features:

- the code is **non-overlapping:** a specific 'start' codon ensures that ribosomes 'read' the mRNA code in a unique way. For example the sequence AUGAAUTCGCCU will always be read as AUG AAU TCG CCU, not as A UGA AUT CGC CU
- the code is **degenerate:** this means that some amino acids are coded for by more than one codon.

First base	**Second base**				**Third base**
	Guanine (G)	**Adenine (A)**	**Cytosine (C)**	**Uracil (U)**	
G	GGG alanine	GAG glutamic acid	GCG alanine	GUG valine	G
	GGA glycine	GAA glutamic acid	GCA alanine	GUA valine	A
	GGC glycine	GAC aspartic acid	GCC alanine	GUC valine	C
	GGU glycine	GAU aspartic acid	GCU alanine	GUU valine	U
A	AGG arginine	AAG lysine	ACG threonine	AUG start	G
	AGA arginine	AAA lysine	ACA threonine	AUA isoleucine	A
	AGC serine	AAC asparagine	ACC threonine	AUC isoleucine	C
	AGU serine	AAU asparagine	ACU threonine	AUU isoleucine	U
C	CGG arginine	CAG glutamine	CCG proline	CUG leucine	G
	CGA arginine	CAA glutamine	CCA proline	CUA leucine	A
	CGC arginine	CAC histidine	CCC proline	CUC leucine	C
	CGU arginine	CAU histidine	CCU proline	CUU leucine	U
U	UGG tryptophan	UAG **stop**	UCG serine	UUG leucine	G
	UGA **stop**	UAA **stop**	UCA serine	UUA leucine	A
	UGC cysteine	UAC tyrosine	UCC serine	UUC phenylalanine	C
	UGU cysteine	UAU tyrosine	UCU serine	UUU phenylalanine	U

Table 8.3 The genetic code. You will not be expected to remember details of this code for a Unit Test. Notice that three mRNA bases code for one amino acid. Each unit of three mRNA bases is called a codon. Some amino acids are coded by more than one codon. Like AUG, UAG and UAA, some codons are 'reading instructions' for ribosomes.

key term

The genetic code is non-overlapping and degenerate.

Q 4 What sequence of amino acids would be coded by the mRNA bases AUGGGGCACUGCGUA?

Translation (mRNA to proteins)

A molecule of mRNA moves into the cytoplasm. Here, a ribosome becomes attached to its 'start' codon. Only a special part within the ribosomes attaches to codons. From the 'start' codon, the ribosome moves along the mRNA three bases at a time. Figure 8.7a shows a ribosome attached to a molecule of mRNA. In this position, it 'reads' the codon to which it is attached.

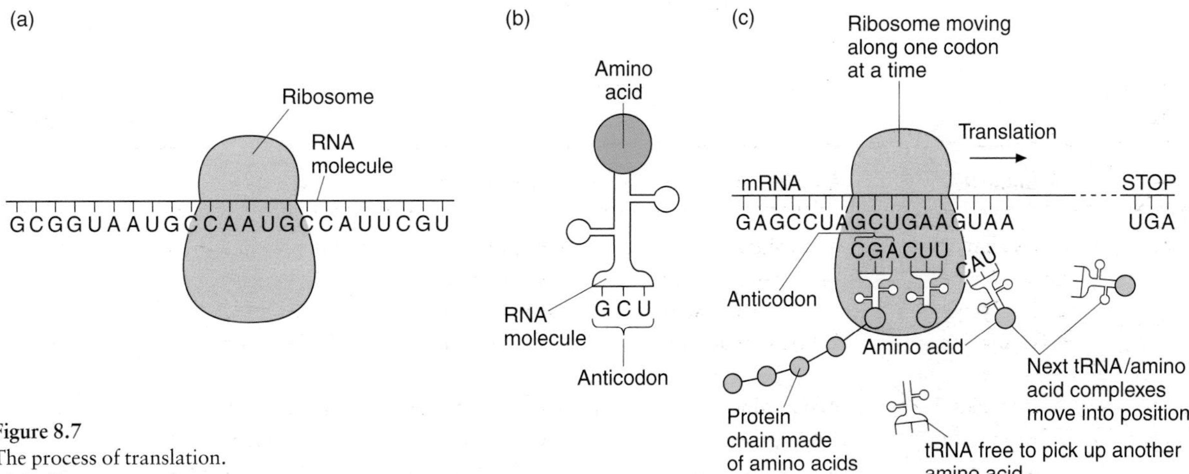

Figure 8.7
The process of translation. (a) A ribosome attaches to a molecule of mRNA in the cytoplasm. (b) A molecule of tRNA has three bases, called an anticodon, that complement the codon in mRNA. (c) A ribosome 'reads' the mRNA code three bases at a time. Appropriate amino acids are brought to the ribosome by different, specific molecules of tRNA, and the ribosome joins these together to make a protein.

A ribosome does not rely on the diffusion of amino acids. Instead, small molecules of transfer RNA (tRNA) bring individual amino acids to the ribosome. The combination of tRNA and amino acid is specific. Each tRNA molecule has three unpaired bases, called an **anticodon.** Figure 8.7b shows a molecule of tRNA carrying its amino acid to a ribosome. It can do this if its anticodon can pair with the mRNA codon in the ribosome. The tRNA bases pair with the mRNA bases A=U, C=G, G=C and U=A.

Figure 8.7c shows how a ribosome removes the amino acid from the tRNA molecule and links it to others. By moving along the mRNA three bases at a time, the ribosome will link amino acids to form a protein. Because the genetic code has been read correctly, the protein will have the required amino acid sequence.

Q 5 A tRNA molecule has the anticodon UGA. What amino acid will it carry?

Replication of DNA

New cells must have the same DNA code as their parents otherwise they would not be able to make the proteins needed by the cell. Before cells divide, their DNA is copied. This process is known as **DNA replication.**

When DNA is copied, the double helix of each molecule unwinds. The weak hydrogen bonds between DNA bases easily break down. The molecule then becomes two separate chains of nucleotides with unpaired bases. The bases do not stay unpaired for long. Free DNA nucleotides pair with the unpaired bases in each of the separate DNA chains. As you might expect, the pairing is specific (see Table 8.4).

Unpaired base on a single chain of DNA	Base carried on free DNA nucleotide
Adenine	Thymine
Cytosine	Guanine
Guanine	Cytosine
Thymine	Adenine

Table 8.4 DNA replication involves base pairing. After the DNA molecule separates into two chains of nucleotides, free DNA nucleotides pair with the exposed bases on each DNA chain.

Figure 8.8 shows the process of DNA replication. Figure 8.8a shows a DNA helix that has begun to separate at one end. In Figure 8.8b, free DNA nucleotides have paired with exposed bases on each of the chains of the DNA molecule. Figure 8.8c shows how the newly paired nucleotides are joined together. The enzyme **DNA polymerase** controls the joining up of these nucleotides. Notice in Figure 8.8c that each chain in the old molecule builds up an exact copy of its previous partner. Two new DNA molecules are being formed that are exact copies of the old ones. Because one of the chains in each new molecule was present in the old molecule, this process is called **semi-conservative replication.**

Q 6 Suggest why the accurate replication of DNA is important to a cell.

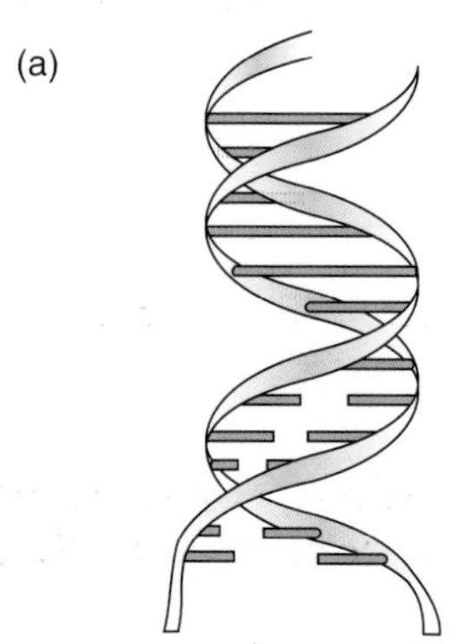

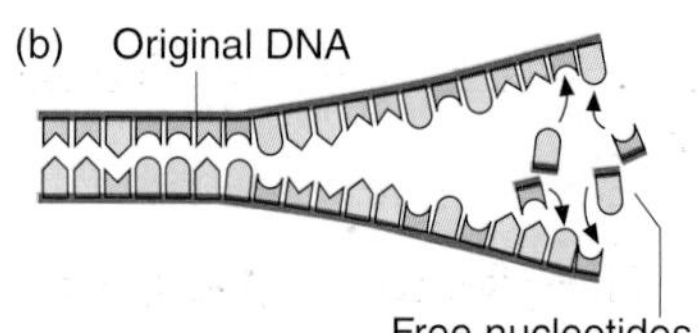

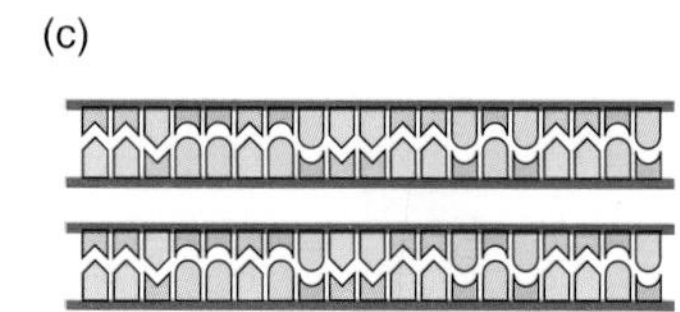

Figure 8.8
The process of DNA replication. (a) As the weak hydrogen bonds between bases break, a molecule of DNA begins to unwind where it is to be copied. (b) DNA nucleotides pair with the bases that are exposed as the DNA unwinds. (c) In each new molecule, the enzyme DNA polymerase joins the free nucleotides together to form a new strand that is an exact copy of the old.

DNA and inheritance

The sequence of bases in DNA controls the order of amino acids in proteins that are made by a cell's ribosomes. Without the correct sequence of bases, a cell will not be able to make a particular protein. This is important when we remember that many of the proteins made in a cell are enzymes. These enzymes control the chemical reactions that occur in a cell. Reactions in cells are referred to as **cell metabolism**. The sequence below shows how several enzymes can control a series of chemical reactions. Such a sequence is called a **metabolic pathway**:

$$\text{substance A} \xrightarrow{\text{enzyme 1}} \text{substance B} \xrightarrow{\text{enzyme 2}} \text{substance C} \xrightarrow{\text{enzyme 3}} \text{substance D}$$

The length of DNA that carries the code for a particular protein is called a **gene**. If the gene for a particular protein is present in a cell's DNA, the cell will be able to make that protein. If the protein is an enzyme, it will have an effect on the cell's metabolism. Often the effect of an enzyme on an organism can be observed or detected chemically. Properties of an organism that can be detected are called its **phenotype**. Hair colour is part of the phenotype that is easily detected in humans. Blood group is part of the phenotype that is less easy to detect than hair colour. Like all aspects of our phenotype, these features are influenced by metabolic pathways. Specific enzymes control these pathways. We can now see the link between DNA and phenotype (Figure 8.9).

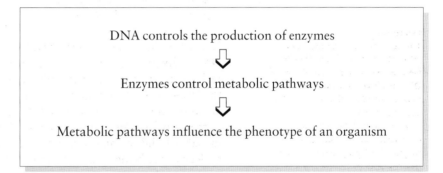

DNA controls the production of enzymes

Enzymes control metabolic pathways

Metabolic pathways influence the phenotype of an organism

Figure 8.9
By controlling the production of enzymes, DNA affects an organism's phenotype.

Note that the environment also affects phenotype. This is easy to imagine if you think of hair colour. Although inherited, many people use bleaches or dyes to change their hair colour. You can become blonde even if your genetic code is for brown hair!

key term

A **gene** is a length of DNA that carries the genetic code for a single protein.

Q 7 **Phenylketonuria (PKU) is a human disorder in which the amino acid phenylalanine cannot be broken down into tyrosine. Suggest a cause for PKU.**

Summary

- Nucleic acids carry the genetic code that cells use to make proteins. The code determines the sequence of amino acids in a protein.
- Nucleic acids are long chains of smaller units called nucleotides. Each nucleotide has a five-carbon sugar, a phosphate group and an organic base.
- Deoxyribonucleic acid (DNA) is made of two chains of nucleotides: all its nucleotides carry the five-carbon sugar deoxyribose. Ribonucleic acid (RNA) is made from a single chain of nucleotides: all its nucleotides carry the five-carbon sugar ribose.
- There are two groups of organic bases: purines and pyrimidines. Adenine and guanine are purine bases whereas cytosine, thymine and uracil are pyrimidine bases. Adenine, cytosine, guanine and thymine are found in DNA. Adenine, cytosine, guanine and uracil are found in RNA.
- Provided they have the complementary base sequence, strands of nucleic acid can be held together by hydrogen bonds between bases. These base pairs are highly specific. In DNA adenine always pairs with thymine and cytosine always pairs with guanine. RNA forms similar base pairs, except that adenine pairs with uracil.
- Hydrogen bonds between base pairs hold together the two complementary strands in a DNA molecule. They also hold parts of RNA molecules together where the single strand has folded back on itself.
- Protein synthesis can be considered in two stages: transcription and translation. During transcription, the base sequence of part of a DNA molecule is copied by a molecule of messenger RNA (mRNA). Each group of three bases on the mRNA molecule, called a codon, codes for a particular amino acid. This is the genetic code.
- Introns are lengths of nucleic acid that carry a nonsense code. Although copied from DNA to mRNA during transcription, these introns are removed from the mRNA before it is used to make proteins.
- During translation, the base sequence of mRNA is 'read' by ribosomes and used to assemble a sequence of amino acids. Transfer RNA (tRNA) molecules are involved in this process. They have an anticodon, complementary to each codon, that enables them to carry amino acids to the ribosomes.
- Specific 'start' codes ensure that the genetic code is non-overlapping. This means that it can only be read in a unique sequence of three bases at a time.

- DNA is copied so that it can be passed from cell to cell during cell division. During semi-conservative replication, the two strands in the parent DNA break away from each other. By base pairing, free nucleotides assemble along the exposed single DNA strands. Because the base pairing is specific, each new strand is an exact copy of the old partner strand. DNA polymerase joins together the free nucleotides.
- The length of DNA that carries a code for one protein is called a gene.
- Because many proteins are enzymes, the DNA determines the metabolic reactions that occur in cells. In this way, DNA contributes to the phenotype of organisms.
- Spontaneous changes, called mutations sometimes occur in DNA. These mutations often disrupt metabolic pathways. Cystic fibrosis and phenylketonuria are two examples of disorders caused by defective genes.

Examination questions

1 (a) (i) State how the individual nucleotides are held together in a single strand of DNA.

(1 mark)

(ii) Name the type of bond that holds together the two strands of nucleotides in a DNA molecule.

(1 mark)

(b) The table shows the sequence of bases on part of a molecule of mRNA.

Base sequence on coding strand of DNA									
Base sequence on mRNA	A	G	C	U	G	U	A	C	U

Complete the table to show the base sequence of the coding strand of a molecule of DNA.

(1 mark)

(c) (i) A particular strand of mRNA is 450 bases long. How many amino acids would be present in the protein coded by this mRNA?

(1 mark)

(ii) Explain your answer.

(2 mark)

2 The table shows the proportion in bases in the nucleic acid of two animals.

Nucleic acid	Adenine	Cytosine	Guanine	Thymine
DNA in gene of animal A	26%			
mRNA resulting from gene in animal A	29%	22%	18%	0%

(a) (i) Complete the table to show the percentage of bases in the DNA of this animal.

(1 mark)

(ii) Explain how your knowledge of DNA structure enabled you to calculate your answer.

(2 marks)

(b) The percentage composition of mRNA is not the same as the DNA gene from which it is made. Use your knowledge of mRNA formation to explain:

(i) the result for thymine;

(2 marks)

(ii) the result for adenine.

(1 mark)

3 (a) Draw a diagram to show the structure of a single mRNA nucleotide.

(2 marks)

(b) Describe how mRNA contains a code for protein structure.

(4 marks)

Assignment

Although our knowledge of biology has grown enormously over the last 50 years, each new discovery only adds a little to our overall understanding. In order to interpret our discoveries and gain a better understanding, we often construct models. These models can then be used to make predictions that can be tested further. You will have already come across models in your biology course. One of the simplest is the "lock and key" model used to explain the way in which enzymes work. Testing this model suggested improvements that could be made and as a result we now think that induced fit provides a better explanation of the way in which enzymes work.

Our knowledge of the structure of DNA is the result of the work of many biologists including James Watson and Francis Crick working in Cambridge. They reviewed a lot of information and used it to produce a molecular model of DNA which they used to test their predictions. In this assignment you will also produce a model of DNA. Although it will be constructed in a simple way, it should enable you to make predictions about its structure and properties. You can make the model and answer Questions 1 to 4 before you start to study the topics of DNA and protein synthesis but you may wish to leave Question 5 until you have found out rather more about this important molecule.

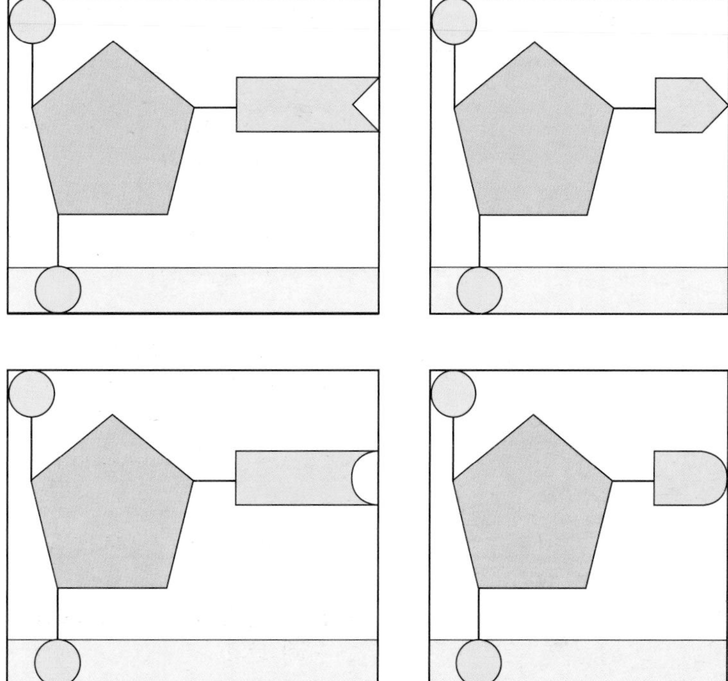

Figure 8.10
The four nucleotides which form a DNA molecule. Each nucleotide consists of a five-carbon sugar, deoxyribose, a phosphate group and a base. A different base is found in each of these nucleotides.

- Figure 8.10 shows the four basic units or nucleotides which form a DNA molecule. Before you start making your model, you will need about 10 copies of each of these nucleotides. The simplest approach is probably to cut them out from photocopies.
- Now produce a single polynucleotide chain. Mix your nucleotides together and pick out 20 at random. Stick these together so that the lower phosphate group of one overlaps the upper phosphate group of the next. This means that the top of the second nucleotide should completely overlap the shaded area at the bottom of the first nucleotide.

1 Explain why this polynucleotide chain can be described as a polymer.

2 (a) The first nucleotide that you picked out could have been any one of the four different types. If you selected two nucleotides at random from your initial collection, how many different combinations would it have been possible to have selected? Remember, it is possible that the two nucleotides might be the same.

(b) Use your answer to Question 2 (a) and your knowledge of protein structure to explain how DNA is suited to its role of coding for different proteins.

3 Watson and Crick suggested that a DNA molecule consisted of two polynucleotide chains parallel to each other. Use your model to predict:

(a) which nucleotide bases in one chain will fit with which bases in the other;
(b) how the two chains will be orientated with respect to each other.

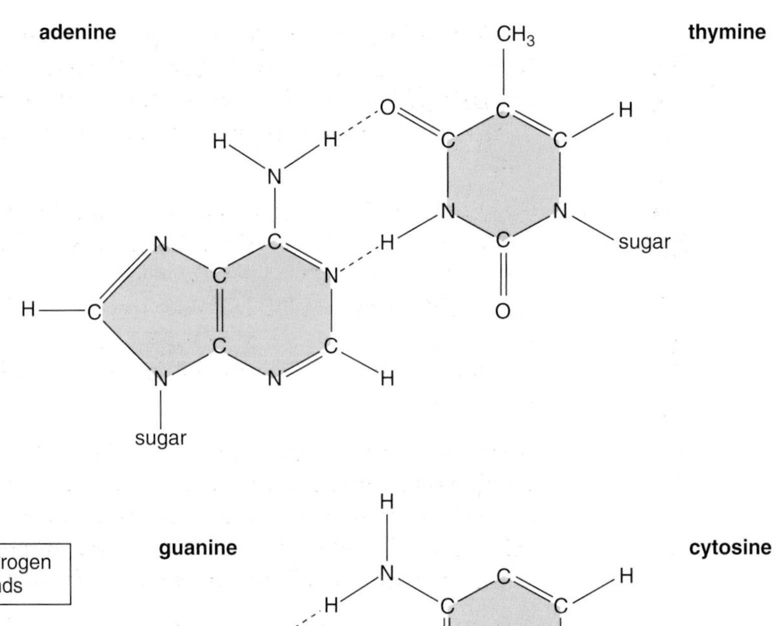

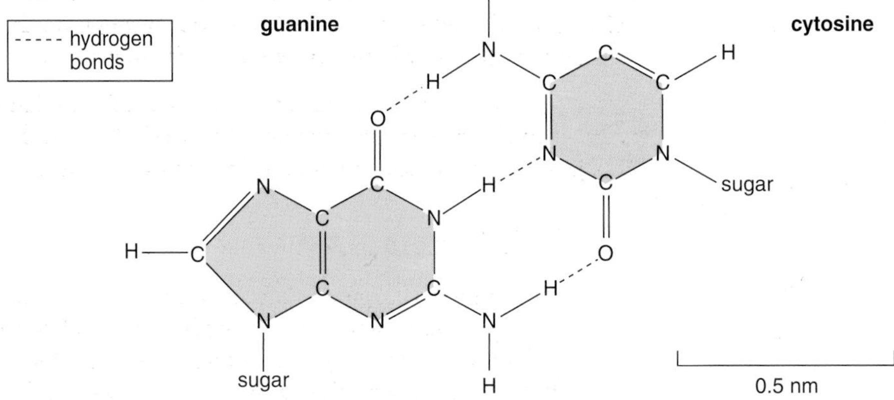

Figure 8.11
Molecular structure of the four different bases found in a DNA molecule.

4 Figure 8.11 shows the molecular structures of the four different bases found in a molecule of DNA. Use your answer to Questions 3(a) and Figure 8.11 to explain why the bases in the two polynucleotide chains of a molecule of DNA can only combine in a specific way.

5 It is important to remember that what you have constructed is only a model and it differs considerably from a real DNA molecule. In what ways does your model differ from our present ideas about the structure of a DNA molecule?

The Cell Cycle

In 1951, scientists at the Johns Hopkins University in the USA were trying to grow human cells outside the body. Such cells would enable them to study cell physiology. The scientists would also be able to study diseases without having to experiment directly on humans. At first, all the cell cultures they started died within a few weeks.

Finally, one cell culture proved successful. These cells were code named HeLa, from the first two letters of the donor's first and second name, Henrietta Lacks. The reason that HeLa cells grew so well in the laboratory was that they were cancer cells. As the HeLa cells continued to divide rapidly in the laboratory, so they did within Henrietta's body. Six months after the first diagnosis, cancer cells had spread through Henrietta's body. Two months later, at the age of only 31, she was dead.

Although Henrietta had died, some of her cells lived on in the laboratory at Johns Hopkins University. Subcultures of these cells were sent to research biologists in other laboratories who, in turn, passed them on to others. In a few years, HeLa cells were cultured in research laboratories throughout the world. Some were even sent into space aboard the *Discoverer XVII* satellite.

Every year, research that is published in hundreds of scientific papers is based on work with HeLa cells. The uncontrollable cell division that killed Henrietta some 50 years ago continues unabated in laboratory cultures around the world.

Cell division

We saw in Chapter 1 that all organisms are made of cells. New cells can only be made when existing cells divide. Taken together, these two ideas form the cell theory. Although all cells have the potential to divide, many cells in a eukaryotic organism lose this ability. For example:

- cell division replaces the entire lining of your small intestine every five days
- your liver cells divide only to repair damage, and then stop dividing
- your nerve cells do not divide.

There are two types of cell division in eukaryotic cells: mitosis and meiosis. In this chapter you will learn about only one of them: mitosis. Cell division in prokaryotic cells does not involve either mitosis or meiosis.

key term

The **cell theory** states that all organisms are made of cells and that new cells can only arise by the division of existing cells.

Mitosis

During mitosis, a parent cell divides to produce two daughter cells. Each daughter cell contains some of the cytoplasm of the parent cell, including its organelles. The most important feature of mitosis is that the daughter cells each contain the same genetic code as the parent cell. In other words, mitosis produces cells that are genetically identical to each other and to the parent cell. The parent cell can do this because it has copied its own chromosomes prior to cell division.

key term

Genetic information is passed from cell to cell during cell division. **Mitosis** produces daughter cells that are genetically identical to each other and to the parent cell.

Q 1 **Name the process by which a parent cell copies its chromosomes prior to mitosis.**

Mitosis consists of two processes: division of the nucleus (**karyokinesis**) and division of the cytoplasm (**cytokinesis**). Division of the cytoplasm usually occurs straight after division of the nucleus. Sometimes division of the cytoplasm is delayed. This leads to cells that have more than one nucleus. Cells in the muscles that move your skeleton are a rare example of human cells that have many nuclei.

Although mitosis is a continuous process, it is usually described as a series of stages. These stages are prophase, metaphase, anaphase and telophase. The time between divisions is called interphase. Table 9.1 summarises the main events that take place during each stage of mitosis. This gives us an overview of mitosis. We will then look at each stage in more detail.

Stage of mitosis	Main events that occur
Prophase	Chromosomes start to coil, becoming shorter and fatter Nuclear envelope disappears A network of protein fibres (the spindle) forms in the cell
Metaphase	Chromosomes line up on the equator of the spindle Fibres of the spindle attach to a region of each chromosome called the centromere
Anaphase	Spindle fibres contract and pull the two copies of each chromosome to opposite ends of the spindle
Telophase	The two sets of chromosomes form new nuclei During this process the chromosomes become long and thin again, and new nuclear envelopes form around the nuclei The cytoplasm usually divides to form two new cells
Interphase	The daughter cells grow more cytoplasm and get on with their normal activities The cells make copies of their chromosomes (by DNA replication) ready for their next division

Table 9.1 A summary of events that occur during the stages of mitosis.

Prophase

Figure 9.1
A cell during interphase, i.e. prior to mitosis occurring.

Chromosomes are normally long and thin. They are so thin that they cannot be resolved by an optical microscope. This is why we cannot see them in the cell shown in Figure 9.1.

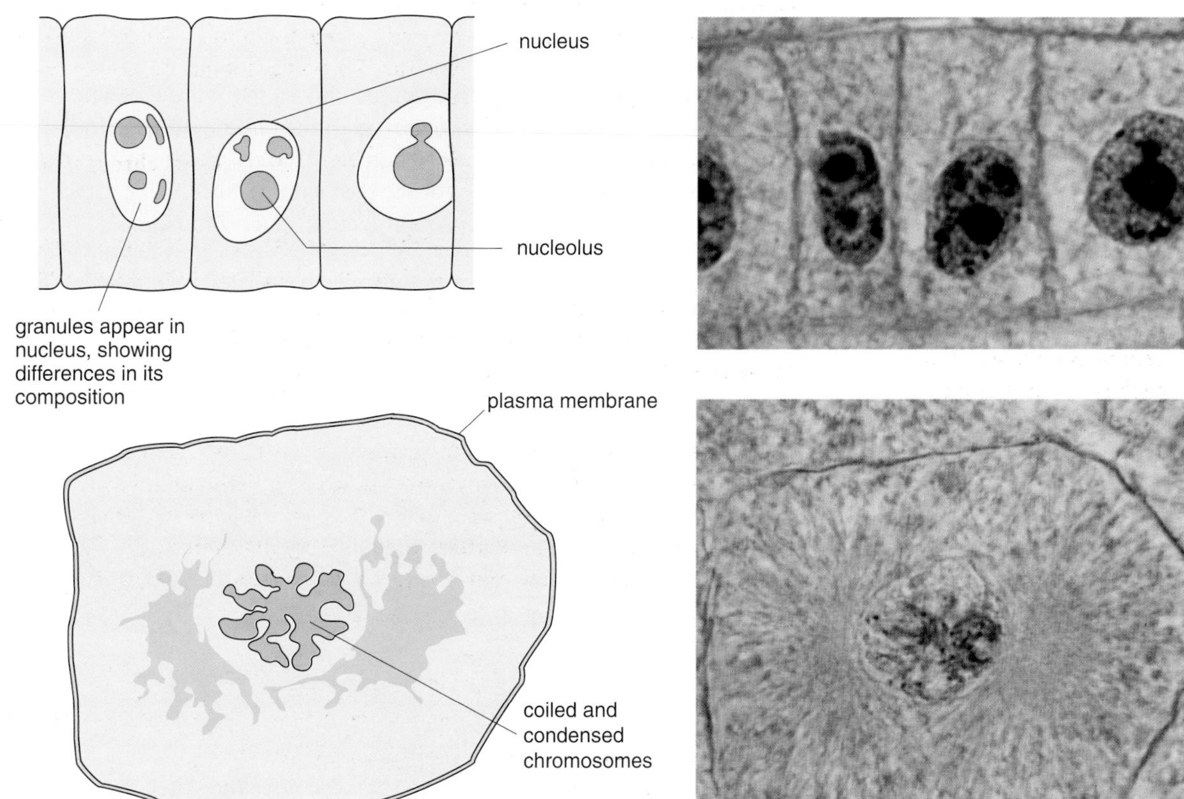

Figure 9.2 (above and right)
Cell during prophase of mitosis.

Figure 9.3 (below)
Electronmicrograph of two replicated chromosomes.

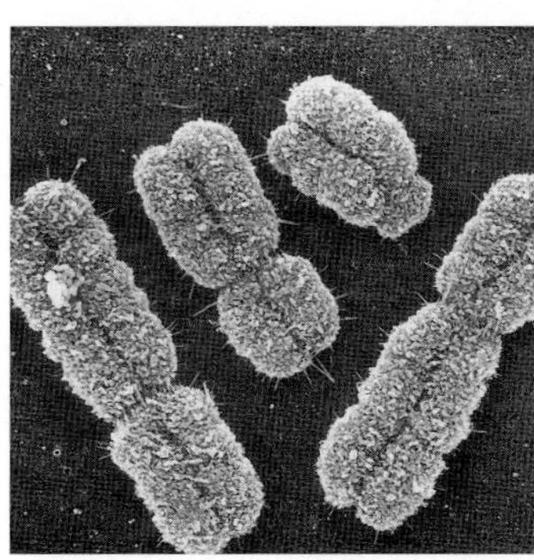

During prophase, chromosomes get shorter and fatter. They do this by coiling on themselves, a process called **condensation**. As the chromosomes condense, they become visible using an optical microscope. Figure 9.2 shows a cell during prophase. You can see its chromosomes. Each chromosome has two strands. These strands are the two copies of the original chromosome. They are held together somewhere along their length by a structure called a **centromere**. As long as the two copies of a chromosome are held together by a centromere, they are each called a **chromatid**. The chromatids of a single pair are called **sister chromatids**.

Figure 9.3 is an electronmicrograph of replicated chromosomes. In this photograph you can see the sister chromatids and centromeres very clearly.

key term

Chromatids are the two copies of a single chromosome held together by a centromere. They are made by DNA replication and are exact copies of the original chromosome in the parent cell.

Q 2 **What has happened to the nucleus of the cell shown in Figure 9.2?**

3 **Figure 9.4 shows us a two-dimensional picture of the spindle. Suggest what its three-dimensional shape will be.**

You can also see the spindle in Figure 9.2. This spindle is made from fibres of protein that radiate from two poles. During prophase, fibres from the spindle become attached to each side of the centromeres holding pairs of sister chromatids together.

Metaphase

In Figure 9.2, the chromosomes are dispersed throughout the cell. Compare this with the cell in metaphase shown in Figure 9.4. The chromosomes in Figure 9.4 have moved to the middle of the spindle. This distinguishes a cell in metaphase: its chromosomes are lined up at the equator of the spindle. We cannot see that spindle fibres have attached to the centromeres of the sister chromatids of each chromosome.

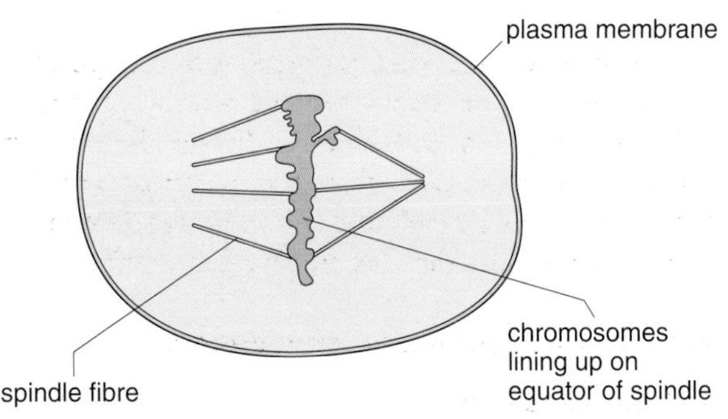

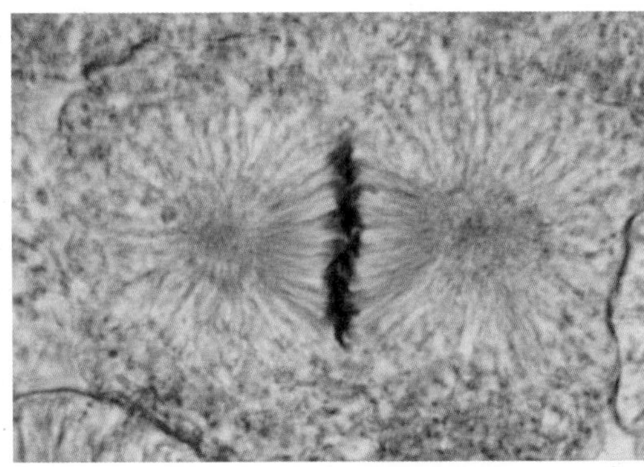

Figure 9.4
A cell during metaphase of mitosis.

Anaphase

Two important events occur during anaphase:

- the centromeres divide into two: this separates the sister chromatids in each chromosome
- the fibres of the spindle contract: this pulls the two sister chromatids apart in opposite directions, eventually reaching opposite poles of the spindle.

You cannot see the movement of chromatids in Figure 9.5. However, you can see its effects: the chromatids are now separate and are no longer at the equator of the spindle.

Figure 9.5
A cell during anaphase of mitosis.

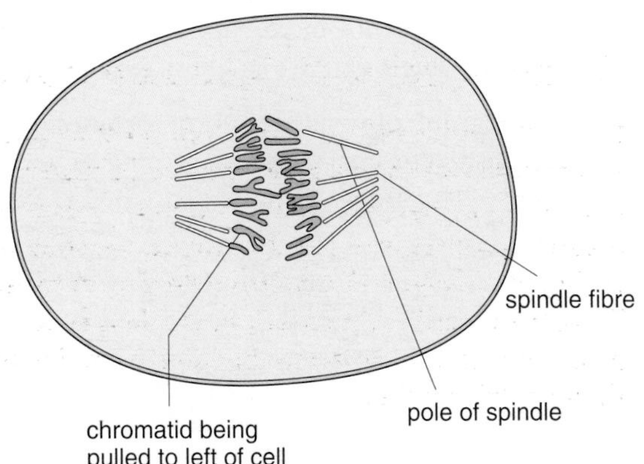

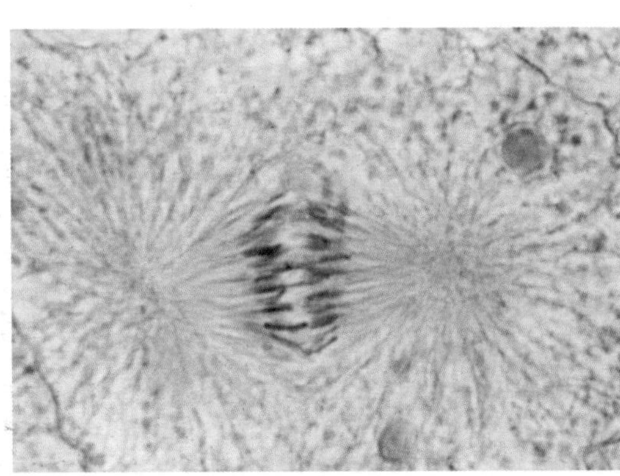

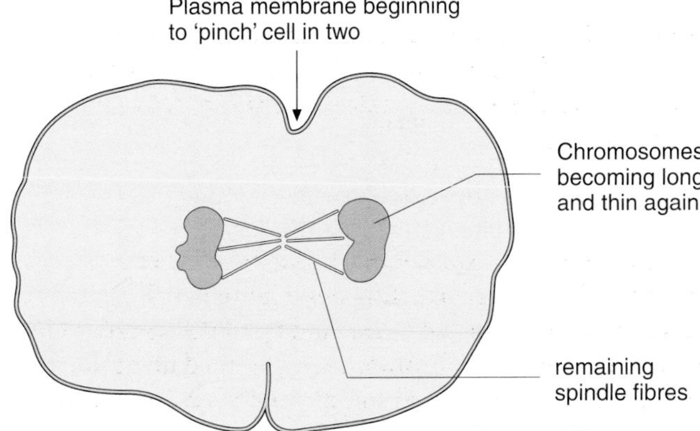

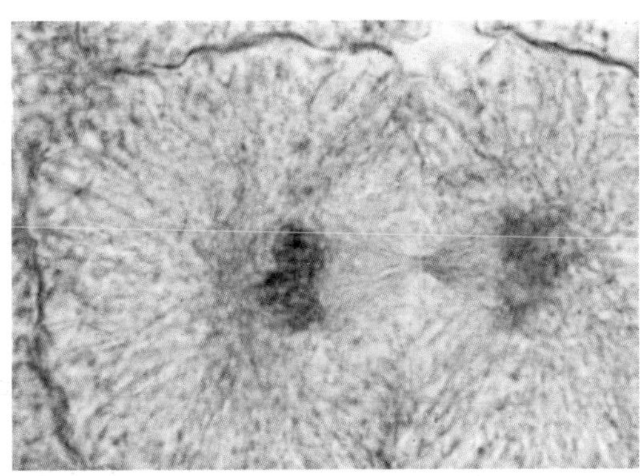

Figure 9.6
A cell during telophase of mitosis.

Telophase

One chromatid from each chromosome reaches the poles of the spindle. Around each pole there is now an exact copy of each chromosome that was present in the nucleus of the parent cell. Nuclear division is complete. This is telophase, shown in Figure 9.6. Once they have reached the poles of the spindle, the chromosomes (as we can now call them again) revert to their original form. This means that they become long and thin.

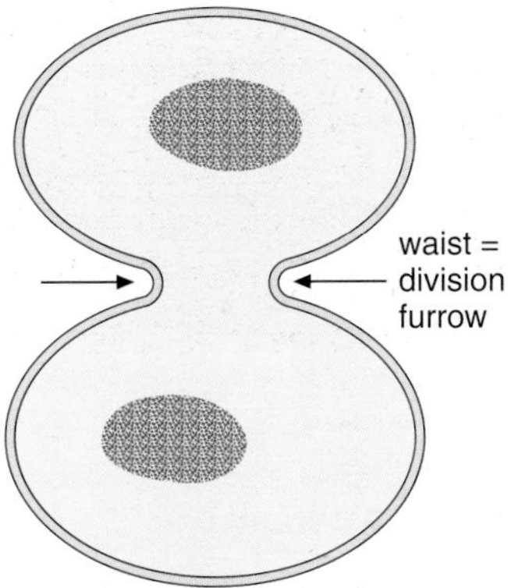

Q 4 **Suggest one advantage to explain why chromosomes:**

(a) **are normally long and thin**
(b) **become short and fat prior to mitosis.**

Cytokinesis

Division of the cytoplasm normally follows telophase fairly quickly. Cells without a cell wall usually pinch themselves in two. Figure 9.7 shows how a 'waist' forms in the middle of the cell. Eventually, cell surface membrane from one side of the cell joins that of the opposite side of the cell and the two new cells separate.

Cells with a cell wall cannot pinch themselves in two. Instead, new membrane forms in the middle of the cell. This new membrane secretes material for the production of two cell walls, one for each new cell. Secretion of new cell walls continues outwards from the middle until the new walls join the old. The two cells are now separate from each other.

Figure 9.7 (above)
Division of the cytoplasm (cytokinesis) in an animal cell.

Figure 9.8 (below)
Cells from the root tip of an onion. Many of these cells are dividing by mitosis.

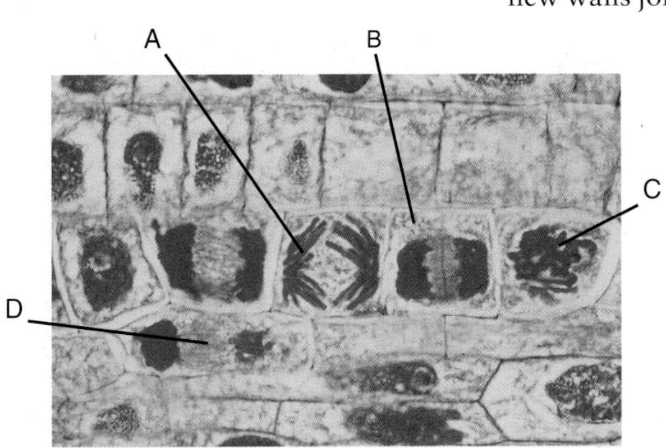

Studying mitosis in plant tissues

Figure 9.8 shows cells from the root tip of an onion. Plant tissue was chosen because it can be used safely in a school or college laboratory. A root tip was chosen because its cells divide rapidly. You can see that many of the cells in Figure 9.8 were dividing by mitosis. Table 9.2 describes the steps you would take to produce a temporary mount of onion cells, like that shown in Figure 9.8.

Steps in procedure	Explanation
Use a scalpel to cut about 3 mm from the tip of a growing root Put the cut root tip onto a watch glass	Dividing cells are found only at the very end of the growing root Behind the root tip, cells have elongated and differentiated into mature, non-dividing cells
Wearing gloves, add 1 mol dm^{-3} acetic acid (also known as ethanoic acid) to the root tip Put the watch glass onto a hotplate and warm for about 5 minutes (do not let the acid boil)	The warm acid helps to soften the tissues so that cells can be separated more easily
Use forceps to put the root tip onto a glass slide	We always mount specimens on a glass slide before viewing them with an optical microscope
Add a few drops of aceto-orcein stain to the root tip on the slide	The stain is taken up by chromosomes, making them easier to see
Gently break up the root tip using a mounted needle You should attempt to spread the root tip out thinly, rather than stir the cells from the root	This separates cells
Put a coverslip over the root and cover the coverslip with a piece of filter paper Press down on the coverslip gently with your thumb (if you press too hard, the coverslip will break)	The aim here is to produce a layer that is only one cell thick This makes individual cells easier to see
Use an optical microscope to examine the slide for cells dividing by mitosis	You will now see onion cells that have their chromosomes stained a reddish purple colour

Table 9.2 Steps in the production of a root tip squash. This process enables us to see cells dividing by mitosis.

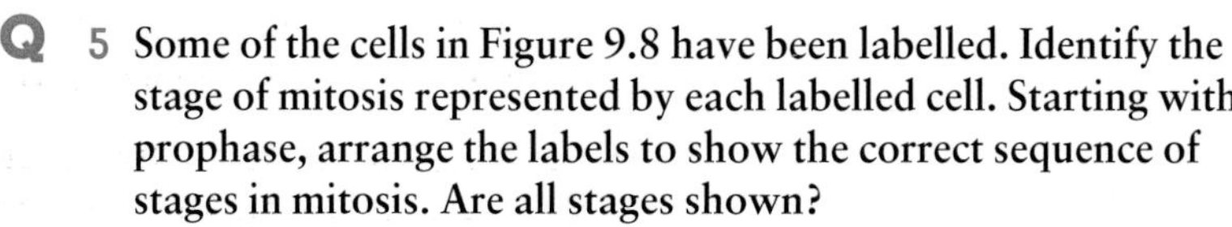

Q 5 Some of the cells in Figure 9.8 have been labelled. Identify the stage of mitosis represented by each labelled cell. Starting with prophase, arrange the labels to show the correct sequence of stages in mitosis. Are all stages shown?

The cell cycle

Eukaryotic cells have a well-defined cycle. In rapidly dividing cells, the cycle usually lasts for several hours and ends with mitosis. Figure 9.9 represents the cell cycle. Each phase of the cycle involves specific cell activities:

Q 6 Which phase, or phases, of the cell cycle corresponds to interphase?

- G1: in which cells prepare for DNA replication
- S: DNA replication occurs
- G2: a relatively short gap before mitosis
- M: mitosis.

You can see from Figure 9.9 that mitosis is a relatively short part of the cycle.

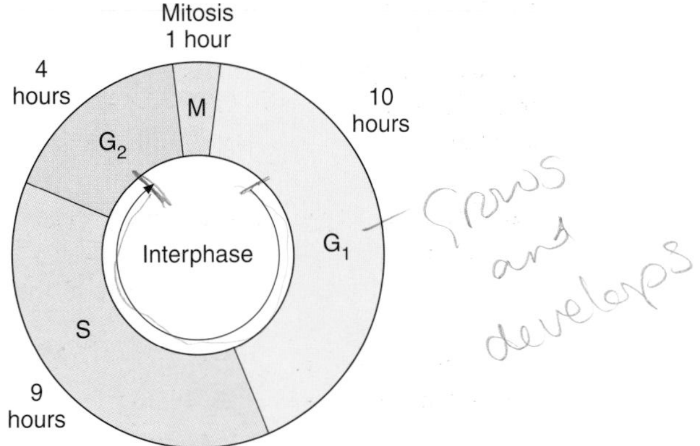

Figure 9.9
A typical cell cycle. The actual length of each phase varies from one organism to another.

Meiosis

In the last chapter, Figure 8.1 showed chromosome 22 from a human cell. In normal human cells, there are two copies of this and every other chromosome. One copy comes from the mother's egg and one copy from the father's sperm. Mitosis produces daughter cells that contain both copies of each chromosome.

Meiosis is a second type of cell division. Unlike mitosis, it separates the two copies of each chromosome into separate daughter cells. You will learn more about meiosis if you study Module 5 of the A2 specification. For AS, all you need to know is that:

- meiosis produces gametes. In humans, these are egg cells and sperm cells
- meiosis halves the number of chromosomes. Human egg cells and sperm cells have 23 chromosomes, whereas a normal human cell has two copies of each of these 23 chromosomes, i.e. 46 chromosomes. A cell with two copies of each chromosome is called a diploid cell. A cell with only one copy of each chromosome is called a haploid cell
- meiosis produces haploid daughter cells that are genetically different from each other and from the parent cell.

Q 9 A human zygote is formed when an egg cell and a sperm cell fuse at fertilisation. How many chromosomes will a human zygote contain?

Summary

- Cells can be made only when existing cells divide.
- There are two types of cell division in eukaryotic cells: mitosis and meiosis. Cell division in prokaryotic cells (such as bacteria) does not involve mitosis or meiosis.
- Mitosis produces cells that are genetically identical to each other and to the parent cell.
- Mitosis consists of several stages: prophase, metaphase, anaphase and telophase. Interphase is the period between mitotic divisions.
- In prophase, chromosomes condense to become shorter and fatter. We can see them using an optical microscope. Each chromosome has already been copied. The two copies, called sister chromatids, are held together by a centromere. The nucleus disappears and a spindle of protein fibres forms during prophase.
- In metaphase, the chromosomes line up on the equator of the spindle.
- In anaphase, each centromere divides, separating sister chromatids. Contraction of fibres in the spindle pulls the two sister chromatids of each pair to opposite poles of the spindle.
- In telophase, the separated chromatids (now called chromosomes again) uncoil to become long and thin again. Division of the cytoplasm (cytokinesis) usually occurs. In cells without a cell wall this is done by pinching the cytoplasm in the middle of the cell. In cells with a cell wall, new cell wall is secreted and separates the two new cells.
- Interphase is the time between cell divisions.
- Dividing onion cells can be viewed by preparing a temporary mount of squashed root tip, stained using aceto-orcein.
- Rapidly dividing cells undergo a cell cycle. The G1, S and G2 phases of this cycle represent interphase. The M phase of the cycle represents mitosis.
- DNA replication occurs in the G1 phase of the cell cycle.

Examination questions

1 (a) Name the phase in the cell cycle in which the following events take place.

(i) DNA replicates

(ii) Copies of each chromosome move to the poles of the cell.

(2 marks)

Figure 9.10

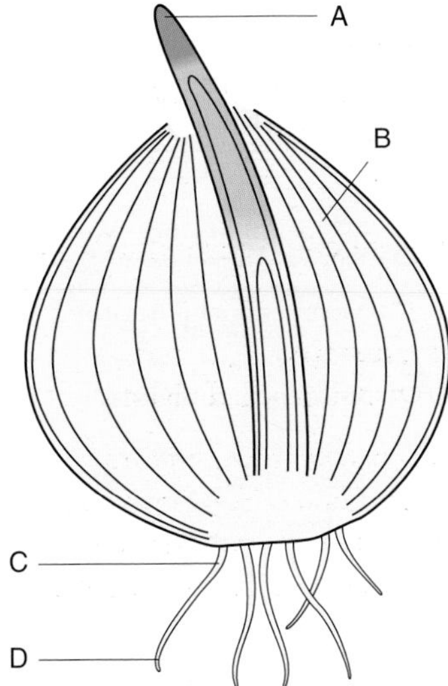

(b) Figure 9.10 shows a section through an onion bulb that is starting to grow.

(i) Which of the parts labelled A to D of this bulb would you use to prepare a slide showing mitosis?
(ii) Name the stain you would use to make this preparation.
(iii) Describe the effect this stain would have on the cells.

(3 marks)

2 Figure 9.11 shows the main stages in the cell cycle.

(a) At what stage in the cell cycle do the following events take place?

(i) Shortening of the chromosomes.
(ii) Replication of DNA.

(2 marks)

(b) The root tip of a growing plant was used to prepare a slide to study mitosis.

(i) Name one stain that you could use to make this preparation.

(1 mark)

(ii) The number of cells at each stage of the cell cycle were counted. The table shows the number of cells at each stage of the cell cycle.

Interphase
Prophase
Metaphase
Anaphase
Telophase
Division of cytoplasm

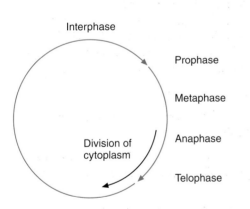

Figure 9.11

Stage of cell cycle	Number of cells seen on slide
Interphase	147
Prophase	68
Metaphase	10
Anaphase	4
Telophase	15

Calculate the percentage of time in the cell cycle spent in metaphase. Show your working.

(2 marks)

3 During the twentieth century, several experiments were carried out to discover the chemical nature of the genetic material.

(a) Alfred Mirsky found that all the body cells of any species contain the same amount of DNA, which is double the amount found in egg and sperm cells. However the same was also true of protein. What conclusion can you make from this evidence about the chemical nature of the genetic material?

(1 mark)

(b) In 1928, Frederick Griffith was interested in developing a form of vaccine against a fatal form of pneumonia, caused by the bacterium *Pneumococcus*. This bacterium has two forms: one with an outer capsule and one without the outer capsule. He treated the two forms of these bacteria in different ways and then injected them into mice. Some of his results are shown in Table 1.

Type of *Pneumococcus* injected into mice	Effect of injection
Live, non-capsulated	Mice remained healthy
Live, capsulated	Mice died of pneumonia
Heat-killed, capsulated	Mice remained healthy
Heat-killed, capsulated and live non-capsulated	Mice died of pneumonia

(i) Which type of *Pneumococcus* causes pneumonia? Explain your answer.

(1 mark)

(ii) Suggest why the mice died of pneumonia after being injected with heat-killed, capsulated *Pneumococcus* and live, non-capsulated *Pneumococcus*.

(2 marks)

(c) Bacteriophages are a type of virus that infect bacteria. The T_2 bacteriophage has an outer protein coat surrounding its DNA core. When the T_2 bacteriophage infects the bacterium *Escherichia coli*, it injects its own DNA into the bacterial cell but leaves its protein coat on the surface of the bacterium. In 1952, Hershey and Chase performed an experiment in which they labelled the outer protein coat of T_2 bacteriophage with radioactive sulphur and labelled its DNA with radioactive phosphorous. They then infected cells of *Escherichia coli* with these T_2 bacteriophages. Table 2 summarises their results.

Part of T_2 bacteriophage labelled	Site of radioactivity in *Escherichia coli* after infection with labelled bacteriophage
Outer coat	Outside bacterial cell
DNA core	Inside bacterial cell

(i) What was the source of radioactivity inside the cells of *Escherichia coli*?

(1 mark)

(ii) Explain what these results suggest about the chemical nature of the genetic material.

(3 marks)

Assignment

There are many complex techniques that can be used to investigate the processes which occur in living organisms. Biologists can also find out a lot about cell biology by using simple instruments such as light microscopes and interpreting what they see very carefully. This assignment will provide you with an opportunity to interpret a photograph showing cells undergoing mitosis. The photograph was taken through an ordinary light microscope.

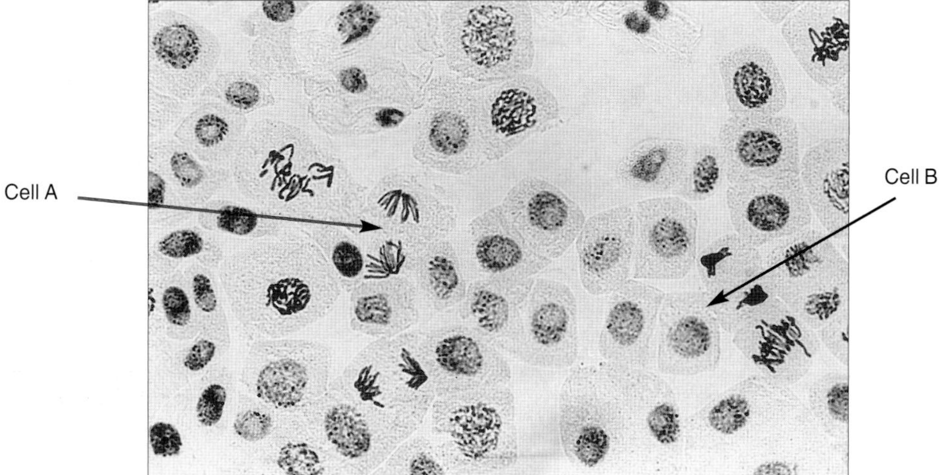

Figure 9.13
Cells at various stages in mitosis.

Look at the photograph in Figure 9.13. Some of the cells which you can see are dividing.

1 What is the evidence from the photograph that:

(a) these are animal cells?

(1 mark)

(b) the nuclei labelled A are from a cell in anophase?

(1 mark)

2 (a) Measure the diameter of cell B (use the horizontal width) in millimetres.

(b) The magnification of this photograph is ×1000. Use this and the measurement from Question 2(a) to calculate the real diameter of the cell. Give your answer in micrometres (μm). Show your working.

(2 marks)

3 Cell B is a cell which is in interphase. It is just about to divide. The nuclei labelled A are in telophase. The nucleus of cell B contains 14 chromosomes and 10 units of DNA. How many chromosomes and how much DNA would you expect to find in each of the nuclei labelled A.

(2 marks)

4 Like other animal cells, human cells also undergo mitosis. Suggest whether or not you would expect to see cells undergoing mitosis in slides of each of the following. Give a reason for your answer in each case.

(a) red blood cells;
(b) cells from the lining of the intestine;
(c) cells from the surface of the skin.

(3 marks)

Table 9.3 shows the number of cells in interphase and in the various stages of mitosis in this photograph

Stage	Number of cells	Percentage of cells
Interphase		
Prophase		
Metaphase		
Anaphase		
Telophase		
Total		

Table 9.3 Number of cells in interphase and stages of mitosis.

5 (a) Use information from the photograph to complete column 2 of this table as accurately as you can. Do not worry if you cannot decide exactly what stage every cell is in.

(1 mark)

(b) Calculate the percentage of cells in each of the stages. Write your answers in the third column of the table.

(1 mark)

6 (a) Which stage of mitosis would you expect to take longest? Explain how you arrived at your answer.

(1 mark)

(b) It takes 24 hours for a cell from this animal to go from the beginning of prophase to the end of telophase. Assume that the percentage of cells in a particular stage of mitosis is proportional to the time taken for the cell to pass through that stage. Use the information in the table to estimate the lengths of the various stages of mitosis. Present your results as a suitable scale drawing.

(2 marks)

Gene Technology

Pollution is a world-wide problem. Gene technology has been used to produce bacteria that reduce pollution in two ways:

- by removing existing pollutants
- by introducing cleaner industrial processes.

Spillages from oil tankers are a major source of pollution at sea. Although detergents can be used to remove the oil, these detergents are very harmful to marine life. Naturally occurring 'oil-eating' bacteria are able to oxidise hydrocarbons into carbon dioxide. The genes that enable them to do this are often carried on plasmids. This has enabled the transfer of these genes into marine bacteria. As a result, genetically engineered bacteria can now be used to deal with oil spills.

Glucose-6-phosphate dehydrogenase is an enzyme that is used in medicine to determine the amount of glucose in blood serum. Commercial quantities of this enzyme are produced by culturing naturally occurring bacteria and then harvesting the enzyme. Some years ago, a company in Germany used gene technology to transfer the appropriate gene to a different bacterium. This genetically engineered bacterium produces the enzyme in a new industrial process that produces one thousand times less pollution than the original.

Overview of gene technology

In this chapter we will learn about the way in which fast-growing cells can be used to make a human protein. Because the process involves combining human DNA with the DNA from another organism, it is called recombinant DNA technology. Figure 10.1 gives an overview of the process. We will examine each step in this flowchart in the rest of this chapter.

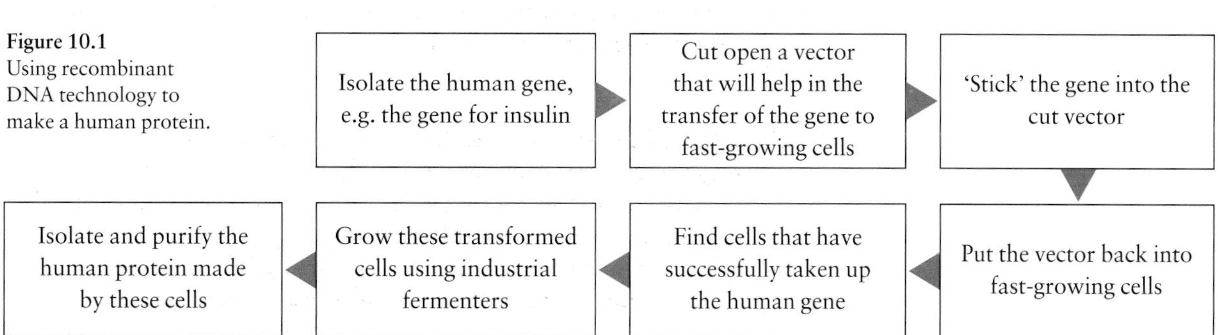

Figure 10.1 Using recombinant DNA technology to make a human protein.

Isolating the required gene

Except for egg cells and sperm cells, every cell in the human body contains 46 chromosomes. Each chromosome can contain several thousand genes. The first task is to isolate the gene that we want to use. There are three ways in which this can be done and these are described below. The first two ways depend on base pairing. If you have forgotten this process, it would be a good idea to refresh your memory by revising Chapter 8 before reading the rest of this chapter.

key term

Recombinant DNA technology involves the combination of DNA from one organism with DNA from another organism. Often this involves inserting human DNA into the DNA of another organism. When these genetically engineered organisms are cultured, they produce human protein.

Working backwards from the protein

This is, perhaps, the easiest of the three methods. First we must work out the amino acid sequence of the protein we wish to make, e.g. insulin. Once we know this, we can use the genetic code in Table 8.3 to work out the base sequence that would code for this protein. We then make DNA with that base sequence. The result is an artificial gene made from **complementary DNA (cDNA)**.

Using messenger RNA

key term

Reverse transcriptase is an enzyme that controls the formation of cDNA from mRNA.

When cells make a protein, they first transcribe its gene into a molecule of mRNA (see Chapter 8). Messenger RNA molecules carrying the code for insulin are common in the cytoplasm of insulin-producing cells. If we can find molecules of insulin mRNA, we can use them to make artificial insulin genes. To do this, we must use an enzyme called **reverse transcriptase**. This enzyme speeds up the production of cDNA from mRNA:

$$\text{mRNA} + \text{DNA nucleotides} \xrightarrow{\text{reverse transcriptase}} \text{cDNA} + \text{mRNA}$$

Using DNA probes to find our gene

To find a single gene amongst the many millions of genes in a human cell is a difficult task. To help this process, we first make a **DNA probe**. A DNA probe is a short, single strand of DNA that carries part of the base sequence of the gene we are looking for. The DNA probe is labelled with a radioactive or fluorescent marker. In the right conditions, molecules of the DNA probe will attach to the complementary sequence of bases in DNA extracted from human cells. The radioactive or fluorescent marker shows us where the target gene is.

Q 1 **What would be the base sequence of a DNA probe used to find the sequence ATC GAC CCT AGA?**

Cutting the gene out of its DNA chain

Once we have found the target gene in this way, we must remove it from its chromosome. Enzymes called **restriction endonucleases** (restriction enzymes for short) control this process. There are many different restriction enzymes. Each cuts DNA at a different base sequence, called a recognition sequence. Figure 10.2 shows the action of one restriction

key term

Restriction endonucleases are enzymes that 'cut' DNA at specific base sequences within their molecules. They are made naturally by bacteria, which use them to destroy the DNA from the bacteriophages that infect them. These enzymes are usually called **restriction enzymes** for short.

enzyme. Notice that the recognition sequence for each enzyme is palindromic, i.e. it reads the same in both directions. Figure 10.2 shows that one restriction enzyme, *Sma*I, cuts DNA vertically. This produces two fragments of DNA with blunt ends. The other restriction enzyme cuts DNA horizontally as well as vertically. This produces protruding ends. These protruding ends will form base pairs with any piece of DNA with the complementary sequence. Because of this, they are known as cohesive or **'sticky' ends**.

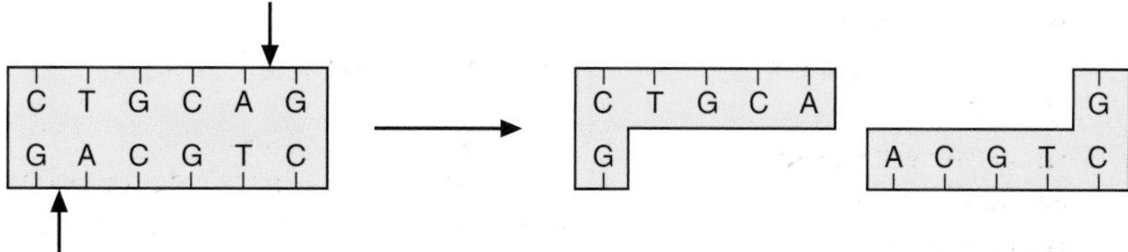

The restriction enzyme *Pst* I cuts DNA at the points shown by the arrows to form 'sticky ends'

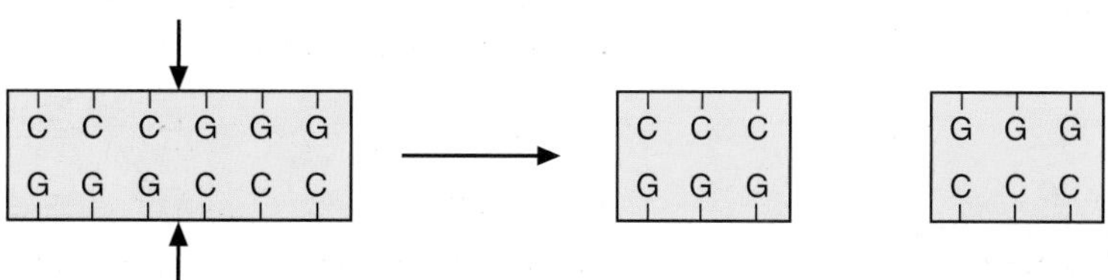

The restriction enzyme *Sma* I cuts DNA at the points shown by the arrows to form blunt ends

Figure 10.2
The action of two different restriction enzymes. Note that the recognition sequence of each enzyme is six pairs long. Recognition sequences for restriction enzymes are usually four to eight base pairs long. Note also that each recognition sequence is palindromic, i.e. it reads the same in both directions.

Q 2 **Identify the role of:**

(a) **reverse transcriptase**
(b) **restriction endonuclease.**

Splicing a gene into a vector

The section above describes how we can isolate the gene that codes for a human protein we want to make. We next need to insert it into the type of cell we will use to make the protein. The easiest way to do this is to use a **vector**. Often we use a **plasmid** as a vector. Plasmids are short, circular strands of DNA that are found in some bacteria. We could also use bacteriophages (viruses that infect bacteria) as vectors.

key term

A **vector** is gene carrier. It will carry a human gene into the cell of a bacterium or yeast that will be used to make human protein. **Plasmids,** circular strands of DNA found in some bacteria, are useful as vectors if bacteria are to be used to make human protein.

To use a plasmid as a vector, we must cut it using the same restriction enzyme that we used to cut the human gene. Using the same restriction enzyme is very important. We want to produce complementary sticky ends on the human gene and the plasmid. Only then will they pair together.

Q 3 Why must the sticky ends of a target gene and a plasmid be cut using the same restriction endonuclease?

Cut genes and cut plasmids are mixed together. Under the right conditions, the sticky ends of the gene and the sticky ends of the plasmid join together. This repair process is known as ligation. It is controlled by an enzyme called **ligase**. Figure 10.3 shows that the new molecule produced is a circular DNA molecule. Because it contains the genes that were in the original plasmid and the human gene, it is called **recombinant DNA**.

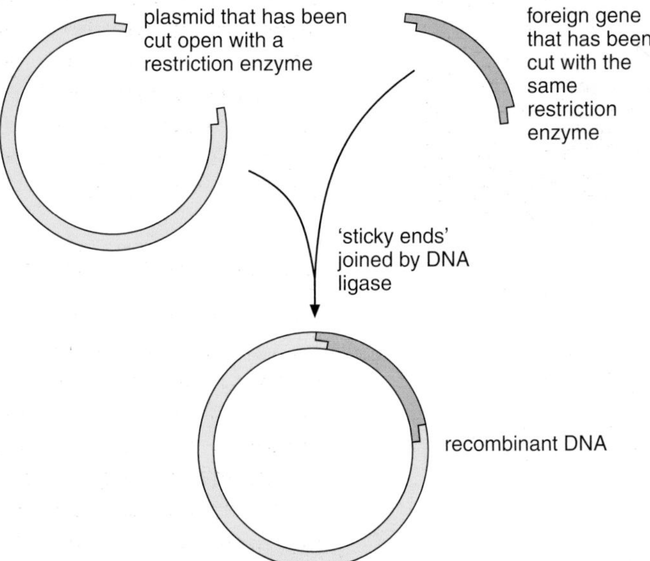

Figure 10.3
'Sticky ends' of cut DNA will join together.

Introduction of recombinant DNA to the host cell

The recombinant DNA must now be introduced to a host cell. Several successful methods have been found by trial and error. One involves soaking bacteria in an ice-cold calcium chloride solution containing the recombinant DNA plasmids. After incubation at 42°C for two minutes, many bacterial cells take up the recombinant plasmids. Because these bacteria contain DNA from another species (humans) they are **transgenic organisms**.

Finding the genetically modified bacteria

Whatever method is used, not all the bacteria will take up the recombinant plasmid. Since the bacteria are useful to us only if they have the recombinant plasmid, we need to find them. Genetic markers are used in this process.

Q 4 The β-lactamase gene (called *amp*^{r}) confers resistance to the antibiotic ampicillin. Suggest why the *amp*^{r} gene might be inserted into a plasmid that we are hoping will carry the gene for insulin production into a bacterial cell.

The most common genetic marker is a gene for antibiotic resistance. Using antibiotic resistance as a genetic marker involves splicing two genes into plasmids: one for the human protein we want to make and a second for antibiotic resistance. If a bacterial cell has successfully taken up the plasmid with the human gene, it will also be resistant to the antibiotic. Cells that have not taken up the plasmid will be susceptible to the antibiotic. When bacteria are grown on plates of agar, they form distinct colonies. If the agar plate contains an antibiotic, only those bacteria with a gene for resistance will form colonies.

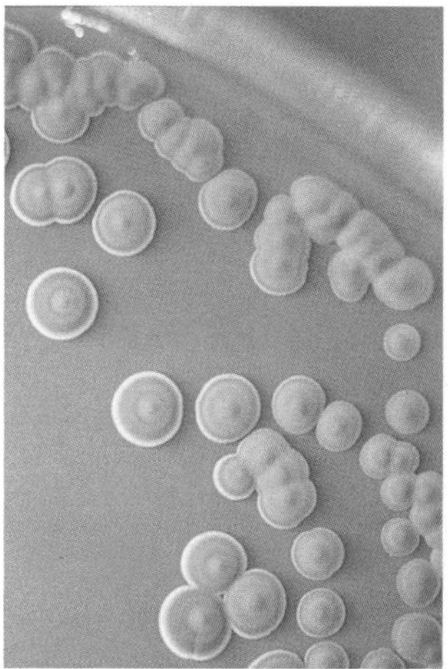

Figure 10.4
When cultured on agar plates, individual bacteria divide to form colonies. Each colony you can see on this agar plate contains millions of cells. The cells in each colony were produced in only a few hours by the division of a single cell.

Culturing host cells

The above processes sound easy. They are not. Obtaining bacterial cells that have successfully taken up the modified plasmid is a hit-and-miss process. In the majority of cases, it will turn out to be unsuccessful (only about 0.0025% of treated bacteria successfully take up the plasmid). Even when bacteria pass the screening process, they do not always use the human gene they now contain. However, those that do are used to make pure cultures. These cultures are grown in large tanks called fermenters. Conditions inside the fermenters are carefully controlled to ensure the optimal growth of the transgenic bacteria. As they grow, the bacteria inside the fermenter make the target human protein. At appropriate intervals, the culture solution is drawn off and the human protein isolated from it.

Uses of recombinant DNA technology

Humans have used organisms for thousands of years. Plants were used for shelter, as tools, as food and as a source of medicine. Yeasts were used to make bread and alcohol. Bacteria were used to make cheese, yoghurt and vinegar. Animals were used for food and clothing as well as for working with humans. As farmers, humans took care of these organisms to ensure that they gained a large crop or kept animals that were strong. However, humans also genetically changed these organisms. Only those organisms that showed the best characteristics were allowed to breed. As a result, crop plants were bred that gave a large crop. Animals were bred to give a high yield of meat, milk or wool. This process is called **artificial selection**. It is called 'artificial' because it is humans, and not nature, that determine which characteristics dominate the population. The animals shown in Figure 10.5 show some results of artificial selection. The sheep have been bred over the centuries to produce more wool than their natural ancestors. The female cattle have been bred to have a high yield of milk. The sheep dog is only one of a variety of breeds of dog.

Figure 10.5
Breeds of animals produced by artificial selection. (a) A merino sheep with short legs, to stop it escaping from pens, and long hair, for a high yield of wool. (b) Fresian cattle have a very high milk yield. (c) The sheep dog has been bred to round up sheep without harming them.

(a)

(b)

(c)

Recombinant DNA technology allows us much greater control over genetic manipulation. Table 10.1 (overleaf) summarises some of the ways that recombinant DNA technology has been beneficial to humans.

Application of recombinant DNA technology	Explanation
Genetically engineered microorganisms	Human genes can be inserted into bacteria, which are then grown in fermenters The bacteria produce a human protein Large amounts of insulin and human growth hormone can be produced cheaply in this way
Genetically modified plants	Desirable genes can be transferred from an organism to a crop plant Potato plants that are resistant to attack by a virus and maize (corn) plants that are resistant to drought have been produced in this way
Genetically modified animals	Human proteins, such as haemoglobin and blood-clotting factors, are already produced in the milk of transgenic cows, goats and sheep
Human gene therapy	Gene therapy involves inserting a 'normal' gene into an organism's body to correct a genetic disorder Severe combined immune deficiency (SCID) leaves some people with almost no functioning immune system Inserting copies of a gene coding for the enzyme adenosine deaminase (ADA) into the blood cells of sufferers has been used to 'cure' the symptoms of SCID
Mapping human chromosomes (the Human Genome Project)	Scientists in over 1000 laboratories around the world are contributing to the Human Genome Project Their aim is to create a map of all human chromosomes, identifying the precise location of every gene This will help to develop new gene therapy treatments

Table 10.1 Some applications of recombinant DNA technology that have benefited humans.

Q 5 **Explain what is meant by *transgenic cows* in Table 10.1.**

Moral and ethical concerns about recombinant DNA technology

Over the past few years, there have been a number of demonstrations about the growth of genetically modified organisms (GMOs). National newspapers have covered these demonstrations, and have also carried articles debating the use of GMOs. Some stories seem designed to upset readers. For example, one newspaper in the UK ran an article suggesting that if you ate meat from an animal with an inserted human chromosome, you would be a cannibal. You should be able to distinguish these stories from rational concerns about GMOs.

Over 70% of the land area of the UK is under some form of agricultural production. None of this is currently used for the commercial growth of GMOs. Research into the possible use of genetically modified crop plants is, however, underway in many countries. Table 10.2 shows some of the ways in which genetic engineering could, in principle, be used to produce novel crops. However, there is international agreement that genetically modified crops should pass through a safety assessment procedure before they can be grown. This involves careful research to answer the following questions:

- How does the introduced gene affect the genetically engineered plant?
- Is there evidence that the introduced gene affects the toxicity or allergenic properties of the genetically engineered plant?
- Is it likely that there will be unintended effects on other, useful organisms within the environment?
- Is the genetically engineered plant likely to become a weed or to invade natural habitats?
- Can the introduced gene be transferred to other plants (e.g. by pollination) or animals and, if so, what would be the likely consequences?

Examples of ways in which genetic modification could make crops more useful
Improve resistance to pests
Improve resistance to bacterial, fungal or viral diseases
Improve resistance to herbicides, so that crop plants are unaffected by herbicides used to control weeds
Remove allergens from crops that currently contain them, e.g. rice
Enable plants to resist cold or drought
Produce salt-tolerant varieties of crops – important in parts of the Indian subcontinent and sub-Saharan Africa if global warming causes a rise in sea level
Improve storage time, e.g. by slowing the sprouting of potatoes
Enable plants to produce pharmaceutical substances, such as edible vaccines
Enhance the production of vitamins and anti-cancer substances by plants

Table 10.2 In principle, it is possible to produce genetically engineered crops with novel properties.

Despite the safety checks, many people, including scientists and politicians, are concerned about the risks of recombinant DNA technology. Table 10.3 (overleaf) summarises some of their concerns.

Moral or ethical concern	Explanation
Mutation of transgenic bacteria or viruses	If they mutate, these transgenic organisms could become new pathogens The danger is that we might not be able to control these new pathogens
Genetically modified crops could 'escape'	If pollen or seeds were carried from test plots, they might result in genetically modified populations elsewhere Since the genetic modification could involve resistance to herbicides, these plants could become 'superweeds' that we could not control
Transgenic organisms might set up an evolutionary process that could harm the environment	Some crop plants have been given a gene enabling them to produce a pesticide Through natural selection, this might speed up the evolution of pesticide-resistant insects
Populations of transgenic organisms could upset the balance of nature	Populations of transgenic salmon have been produced in which individuals grow rapidly These transgenic fish could compete for food with other fish species The latter could become extinct and natural food webs could change
Objection to specific transgenic organisms	Many religious groups could not use products from specific organisms, e.g. to Hindus, cows are sacred animals and to Jews and Muslims, pigs are unclean The use of products from these organisms might be unacceptable to people from these religions
Eugenics	Genetic engineering could be used to insert genes into humans (or their eggs and sperm) In some cases, for example SCID, this process could prevent an early death However, it could be used to give individuals characteristics that are considered to be desirable This is unacceptable to many people and it also reminds people of some of the programmes that have been used throughout modern history to eradicate less powerful ethnic groups
Screening could lead to discrimination against individuals	When fetuses are screened for genetic disorders, parents often face a dilemma about aborting affected fetuses It will become possible to screen adults for genes that predispose them to genetic disorders This might lead to insurance companies discriminating against people with these disorders, even though they have perfect health

Table 10.3 Many people are concerned about the use of recombinant DNA technology. This table summarises their concerns.

Q 6 **Cystic fibrosis results from a gene mutation in humans. One symptom is the production of thick mucus by cells lining the lung. This results in severe discomfort and an increased risk of infection. Suggest how eugenics could be used to help sufferers of this inherited disorder.**

Extension box

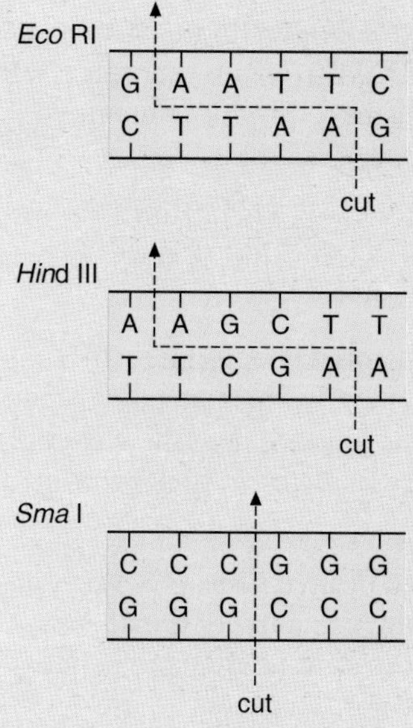

Figure 10.6
The base sequences recognised by three different restriction enzymes: *Eco* RI, *Hind* III and *Sma* I. The sequences recognised by many restriction enzymes are six base pairs long and the sequence is the same whichever way round it is read.

It may surprise you to know that humans are not the only organisms to suffer from viral infections. Bacteria can also be infected by viruses.

Viruses have a very simple structure. They consist essentially of a protein coat surrounding a piece of nucleic acid which codes for the proteins in this coat. When a virus in which the nucleic acid is DNA infects a bacterial cell, its DNA enters the bacterium and takes over the host cell's metabolism. As a result, the bacterium makes new virus particles instead of its own proteins.

Many bacteria produce enzymes which recognise foreign DNA like that from viruses and disable it by cutting it into short pieces. These enzymes are called restriction enzymes because their activity is restricted to foreign DNA. Each enzyme recognises a specific sequence, often six nucleotides long, and only cuts the DNA at this site. Figure 10.6 shows some different restriction endonucleases and the nucleotides sequences where they cut the DNA. More than a hundred different sorts of restriction enzyme have now been isolated and purified. They get their names from the bacteria from which they are obtained. *Eco* RI, for example, was the first restriction enzyme (R I) isolated from *Escherichia coli* (*Eco*).

If you look at Figure 10.6, you will see that *Sma* I cuts the length of DNA in a different way to the other two. It cuts straight through the DNA producing two blunt ends. The other two enzymes produce staggered cuts sometimes called "sticky ends". Enzymes which produce staggered cuts are particularly useful in genetic engineering enabling an introduced piece of DNA to join to the DNA in a plasmid.

Different restriction enzymes which cut the DNA at different points provide genetic engineers with a very useful set of tools. We can illustrate their value by looking at an example. Figure 10.7 shows a length of DNA. Suppose we want to isolate the gene that is represented by the base sequence, shown in red, from this length. If we use the enzyme *Eco* RI, it will cut this sequence in two places, effectively cutting out the gene we want. If we use *Hind* III on the other hand, the bases that it recognises are within the part of the sequence that makes up the gene. It would not be suitable to use this enzyme.

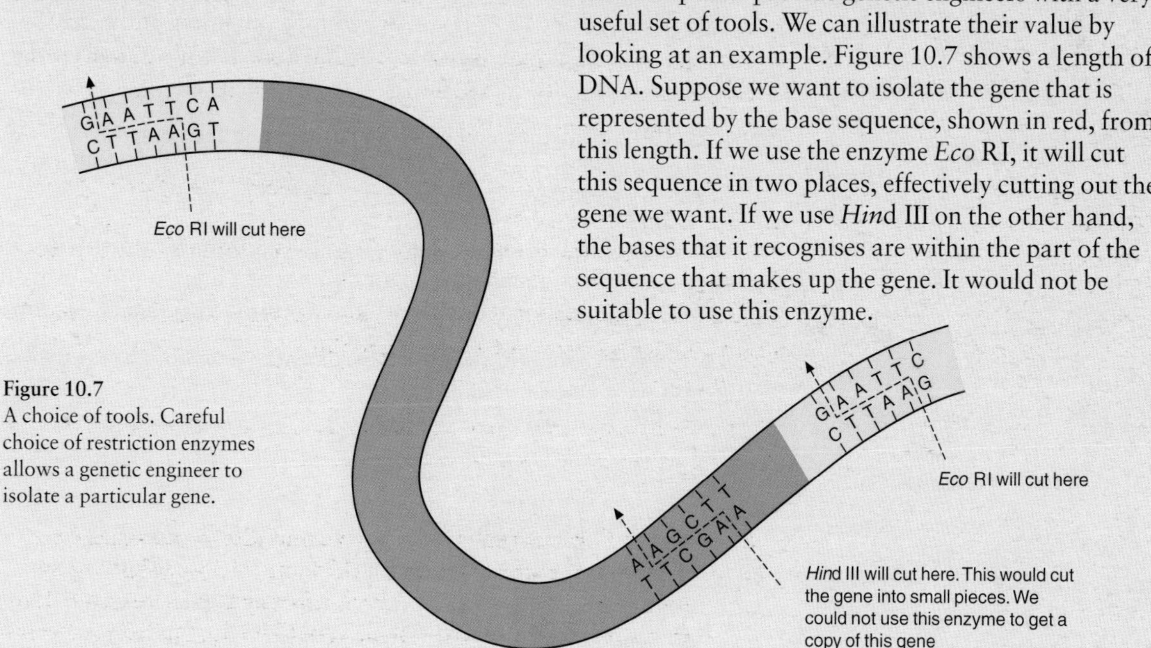

Figure 10.7
A choice of tools. Careful choice of restriction enzymes allows a genetic engineer to isolate a particular gene.

Summary

- Gene technology, also called recombinant DNA technology, involves combining DNA from one species with DNA from another species. It can be used to enable non-human organisms to produce human proteins, such as insulin.
- A desired gene can be made from its mRNA, obtained from cells, in other words by reversing the process of transcription. A reverse transcriptase enzyme is used in this process.
- Alternatively, a desired gene can be 'cut' from DNA using restriction endonuclease enzymes. When these enzymes produce staggered cuts, 'sticky ends' are formed on the DNA. The advantage of these sticky ends is that they will bind by base pairing to identical sticky ends in a vector.
- DNA probes can be used to help us find a desired gene. DNA probes have a base sequence that is complementary to parts of the desired gene. They also have radioactive, or fluorescent, markers to help us locate them once they have bound to the desired gene.
- Once isolated, a desired gene is spliced into a vector. A vector is a gene carrier that helps us to get the desired gene into the target organism.
- If the target organism is a bacterial cell, plasmids are often used as vectors. Plasmids are small circular DNA molecules found in some bacterial cells. The desired gene and the plasmid are cut using the same restriction endonuclease enzyme, and then stuck together. A ligase enzyme helps the gene and the plasmid to stick together, or anneal.
- Usually, a gene for antibiotic resistance is also spliced into bacterial plasmids. Growing bacteria on a medium containing antibiotic enables us to identify which bacteria have successfully taken up the transgenic plasmid.
- Successfully transformed bacteria are grown in fermenters. The human protein they make, e.g. insulin, is drawn off and purified from the culture medium.
- Recombinant DNA technology can be used for the treatment of some human disorders. It can also be used to harvest human proteins from the milk of cows, goats or sheep and to improve the quality of crops.
- Many people, including scientists, are concerned about the use of recombinant DNA technology.

Examination questions

1 Figure 10.8 summarises stages involved in genetic engineering.

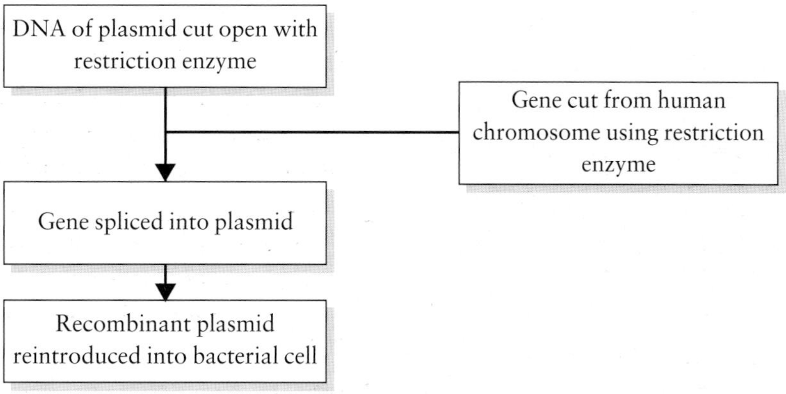

Figure 10.8

DNA of plasmid cut open with restriction enzyme

Gene cut from human chromosome using restriction enzyme

↓

Gene spliced into plasmid

↓

Recombinant plasmid reintroduced into bacterial cell

(a) Explain why the same restriction enzyme may be used to cut the plasmid and the human gene.

(2 marks)

(b) Name the enzyme that is used to splice the human gene into the plasmid.

(1 mark)

(c) Describe **one** method that can be used to discover bacterial cells that have successfully taken up the recombinant plasmid.

(3 marks)

2 Clover is a crop plant that is fed to sheep. Genetic engineering has been used to develop strains of clover that have high concentrations of proteins rich in sulphur-containing amino acids. A piece of recombinant DNA was produced which contained three genes. The recombinant DNA, shown in Figure 10.9, was inserted into clover plants.

Figure 10.9

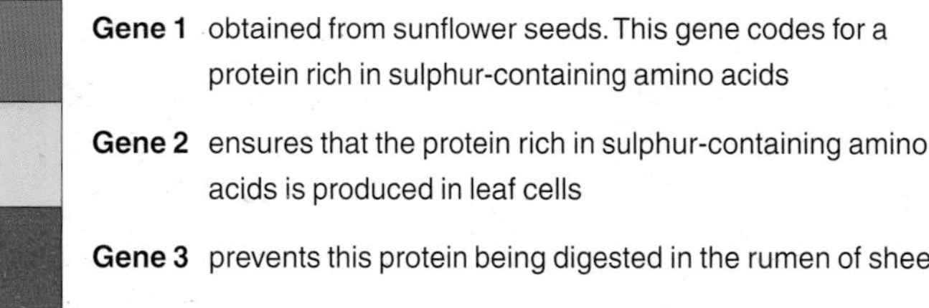
Gene 1 obtained from sunflower seeds. This gene codes for a protein rich in sulphur-containing amino acids
Gene 2 ensures that the protein rich in sulphur-containing amino acids is produced in leaf cells
Gene 3 prevents this protein being digested in the rumen of sheep

(a) Describe how enzymes could be used to remove Gene 1 from sunflower seeds and incorporate it into this recombinant DNA.

(3 marks)

(b) It is hoped that feeding sheep on clover containing this piece of DNA will make them produce more wool. This is because wool is rich in sulphur-containing amino acids.Suggest why it is necessary to add genes **2** and **3** to make this possible.

(3 marks)

Assignment

Figure 10.10
Mussels are shellfish. They produce a protein glue which helps them to stick firmly to rocks on the seashore. This glue even sticks under water!

Different organisms produce different proteins. There are many different species of living organism and there are probably even more different proteins. The genes which encode some of these proteins have been isolated and used to produce genetically modified organisms. These organisms make specific proteins in large and useful amounts.

This assignment is based on a passage about a protein which is produced by mussels. This protein helps to attach these animals to rocks on the seashore.

The questions which follow should help you to understand a little more about how particular proteins are adapted to different functions as well as about some of the problems encountered by scientists in introducing genes from one species to another. A good understanding of the material in the passage will require you to draw on topics from various parts of your AS course. Showing how different topics link together is an important skill and one which you will need to develop if you continue with your biology course.

Read the following passage:

> Walk along the seashore at low tide and you will see a variety of different organisms. Many of these organisms are sedentary and stay in one place. One of the major problems that they face is that they must avoid being washed away by pounding waves. This is where proteins such as mussel adhesive protein come in.
>
> Mussel adhesive protein has an unusual primary structure. It contains a sequence of ten amino acids which is repeated many times. Unlike proteins such as enzymes and the carriers in plasma membranes, mussel adhesive protein does not have a distinct tertiary structure. It is always changing shape. This allows the protein molecules to squeeze into the small cracks and crevices which cover rock faces. Like man-made glues, mussel adhesive protein starts off as a liquid and sets into a solid. It is the repeated amino acid sequences which allow the protein glue to set firm. Each of the repeated sequences contains the amino acid, tyrosine. Chemical bonds form between the tyrosine in one sequence and the tyrosine in another. Not only do these bonds form between different parts of the same molecule, but they also form between different molecules. The chemical bonds bind the protein molecules firmly to each other and to the surface to which the mussel is attached.
>
> Tests have shown that mussel adhesive protein does not result in the production of antibodies when introduced into the human body. Because of this, it could have many uses in medicine. But, before we can begin to consider ways of using it, we need a method of producing suitably large amounts. Attempts have been made to use genetically modified bacteria to produce mussel adhesive protein but researchers have encountered a problem. Bacteria are able to synthesise the protein, but only in

an inactive form. It is thought that the reason for this is that, in a mussel cell, chemical changes take place in the endoplasmic reticulum which convert the inactive molecule to an active one. Scientists believe, however, that we can overcome this problem by using transgenic tobacco plants to produce mussel adhesive protein. Tobacco is a good choice. Not only will the necessary chemical changes take place but tobacco is a non-food plant and it has no close wild relatives in Europe.

Use your own knowledge and information in the passage to answer the following questions:

1 (a) Describe how the structure of an enzyme molecule differs from the structure of mussel adhesive protein.

(2 marks)

(b) How is the structure of each of these molecules related to its function?

(3 marks)

2 (a) If mussel adhesive protein is to be used in medicine, it is important that it does not result in the production of antibodies when introduced into the human body. Explain why.

(1 mark)

(b) Suggest two specific uses for mussel adhesive protein in medicine.

(2 marks)

3 Figure 10.11 shows the sequence of amino acids which is repeated many times in each mussel adhesive protein molecule.

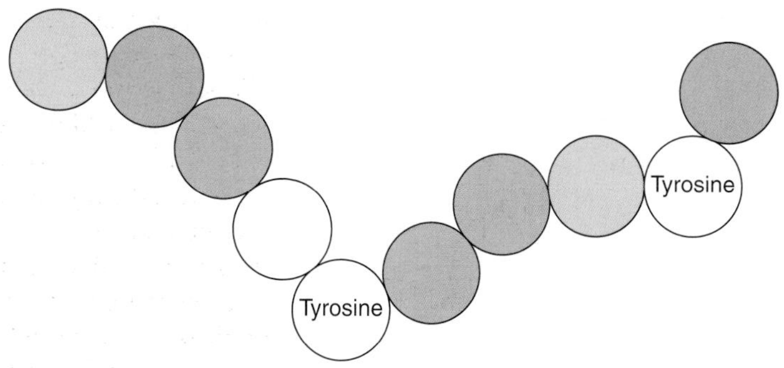

Figure 10.11
This sequence of amino acids is repeated many times in each mussel adhesive protein molecule. The different colours represent different amino acids. Only tyrosine has been named.

(a) How many mRNA nucleotides are necessary to encode this sequence?

(1 mark)

(b) What is the minimum number of different sorts of tRNA necessary to bring the amino acids to the ribosomes to produce this sequence of amino acids?

(1 mark)

(c) (i) Explain why we can work out a sequence of mRNA bases which encodes mussel adhesive protein if we know its amino acid sequence.

(1 mark)

(ii) The sequence of mRNA bases for the adhesive protein actually found in mussel cells may differ from the sequence we have worked out. Explain why.

(1 mark)

4 (a) The process which converts inactive mussel adhesive protein into its active form is called **post-translational modification.** Explain why.

(2 marks)

(b) Explain how the organelles present in a bacterial cell result in its being:

(i) able to synthesise mussel adhesive protein;

(1 mark)

(ii) unable to produce the active form of the protein.

(1 mark)

(c) Why would you expect transgenic tobacco plants to be able to produce active mussel adhesive protein?

(1 mark)

5 Suggest why a plant chosen to produce mussel adhesive protein should:

(a) be a non-food plant;

(2 marks)

(b) have no close wild relatives in the area where it is to be grown.

(2 marks)

Immunology and Forensic Biology

Although antibodies introduced into humans are useful in treating diseases, protection against infection can be given by vaccines that stimulate the immune system to develop antibodies before exposure to the disease. Most conventional vaccines use dead or disabled viruses whose complex protein coats include numerous antigens, that is molecules that stimulate the production of antibodies.

The use of conventional viruses as vaccines can be hazardous if, once disabled, they recover the ability to become infectious. They are also difficult to mass produce and store.

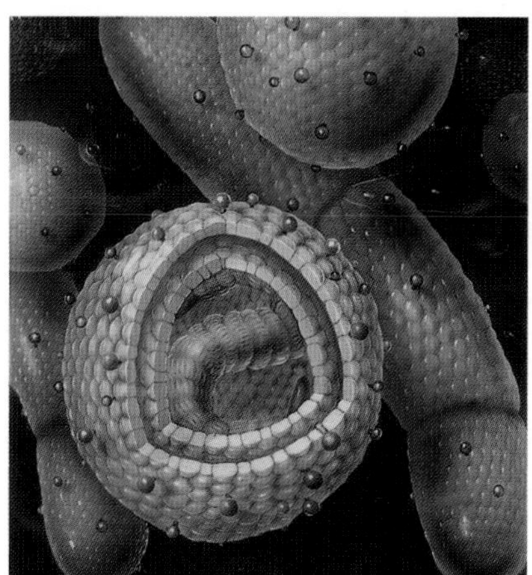

Figure 11.1
The hepatitis B virus showing surface antigen.

Figure 11.2
This could be the modern method of taking your oral vaccine.

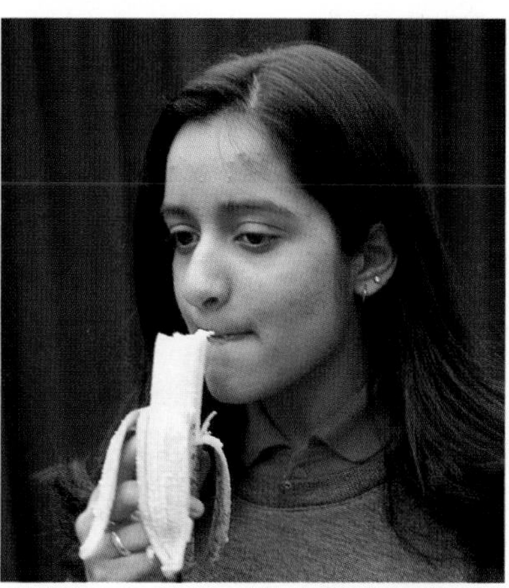

Scientists have developed genetically engineered plants, known as transgenic plants, which can produce individual antigens. However, they do not produce the same range of antigens a virus will present and therefore do not stimulate the full immune response. This may not provide the same degree of protection overall, but these antigens are potentially easier to mass produce and store. Some plant material can simply be eaten, so the antigens could be a lot easier to administer.

The most encouraging early experiments in 'plantigen' production involved the use of tobacco plants genetically engineered using a bacterium called *Agrobacterium*. This tobacco plant was able to produce hepatitis B antigen, but in this case, as tobacco plants are rarely eaten, the antigen still had to be extracted from the leaves and purified.

Transgenic plants have now been produced that contain vaccines against illnesses such as rabies, cholera, diarrhoea and tooth decay. The initial plants involved include tobacco, maize, soybean, cowpea and spinach.

Experimenters have now succeeded in producing transgenic potato and banana plants, thus producing foodstuffs that could provide palatable oral vaccines against hepatitis B. Since cooking destroys the antigens and potatoes are usually cooked before eating, experimenters are concentrating on the banana plant.

The banana is the most promising potential carrier of oral vaccines as it needs no preparation, can be eaten raw and comes with its own, natural packaging. The possibility of using vaccine-carrying bananas against a range of diseases is an attractive one.

key term

Immunity is the body's ability to identify and protect itself against disease.

Principles of immunity

There are two parts to the immune system:

- the non-specific response
- the specific response.

Non-specific response

Many pathogens do not harm us because we have physical, chemical and cellular defences that prevent them from entering the body. Should they enter, we are able to prevent them from spreading through the body. These defences, present from birth, form the non-specific response. This system does not distinguish between different pathogens and gives the same response each time any pathogen attacks.

Specific response

If pathogenic viruses, bacteria or other microorganisms penetrate natural barriers such as the skin and mucus membranes, they enter the bloodstream and encounter the body's immune system. This system consists mainly of white blood cells circulating in the blood and lymph, together with those contained in the spleen, liver, lymph nodes and tissues.

The specific system gives a highly effective, long-lasting immunity against anything the body recognises as foreign. It responds to specific microorganisms and enhances the activity of the non-specific system.

It is important that some terms are fully understood so that the processes involved in the immune system are clear.

Antigen

This is a molecule that triggers an immune response. Although antigens are usually proteins, other types of molecule such as polysaccharides, lipids and even nucleic acids can act as antigens. A **self-antigen** is a molecule found on the surface of your own cells to which you do not respond; however, it will cause a response if introduced into another human. It has a specific shape and is present on every cell. It helps with self-recognition. A **non-self-antigen** is a molecule found on cells entering your body that are not yours, e.g. bacteria, viruses or even another person's cells. This will produce an immune response in your body.

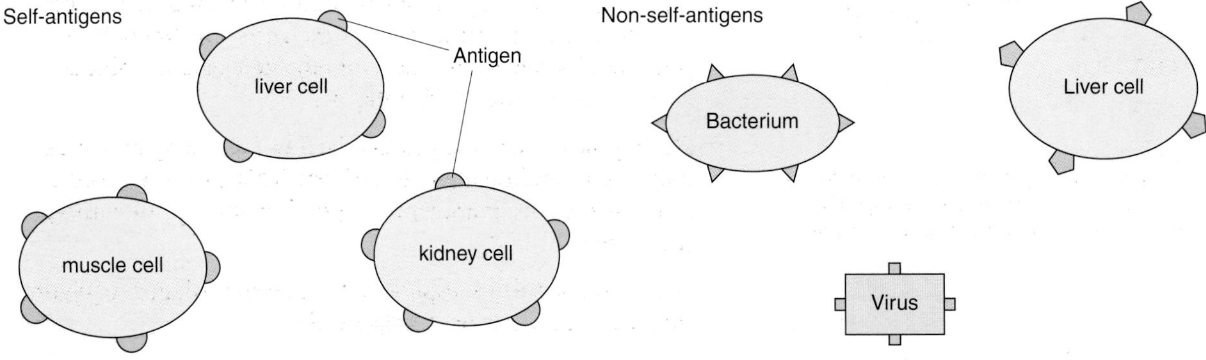

Figure 11.3
Antigens present on cells within your own body (self-antigens) and those present on foreign cells (non-self-antigens).

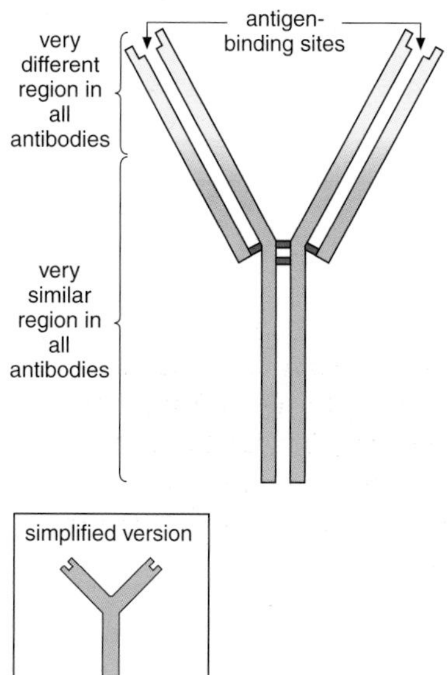

Figure 11.4
The structure of an antibody, showing its unique antigen-binding site.

Antibody

This is a term which describes simply what the molecule does. It is a molecule secreted by certain B-lymphocytes in response to stimulation by the appropriate antigen. Antibodies are types of proteins known as immunoglobulins. They have a unique antigen-binding site.

The body's specific response is based on its ability to recognise and respond to non-self-antigens. The response depends on the activity of various types of white blood cells. The main types of cells involved in defence are:

- phagocytes, e.g. macrophages
- lymphocytes.

Q 1 **How will different genetic codes affect the shape of the protein molecules in different organisms?**

Phagocytes

Phagocytes are continually produced by bone marrow throughout life. They are stored there but can leave in the blood to be distributed around the body. Macrophages circulate in the blood and pass into organs such as the lungs, kidney and lymph nodes, where they tend to remain. They are large long-lived cells that remove foreign matter from organs. They play a crucial role in initiating the specific immune response.

Lymphocytes

There are two groups of lymphocytes:

- B-lymphocytes (called B-cells)
- T-lymphocytes (called T-cells).

Both types of cell must go through a maturing process before they can take part in the specific immune response. This takes place before birth and results in the production of many different types of each group of lymphocyte.

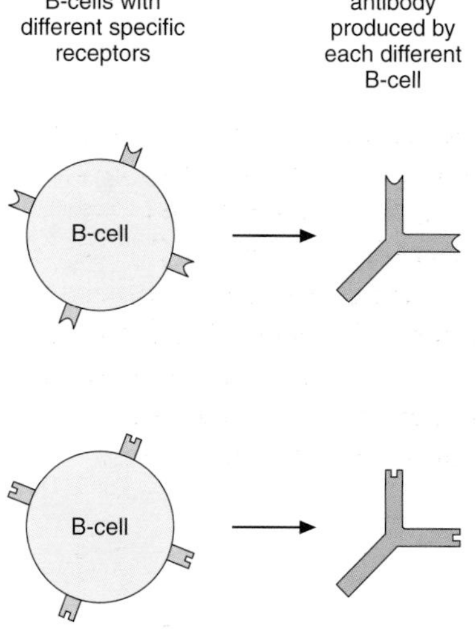

Figure 11.5
B-lymphocytes are present in a variety of forms, each having the ability to produce a specific antibody related to the receptors on its surface.

B-lymphocytes

Each B-cell is genetically programmed to develop just one type of specific receptor on its cell surface membrane. These receptors have the same shape as the antibody that will be produced by the B-cell. The antibody molecules are able to combine specifically with antigens.

During the maturation process the B-cells develop to give as many as 10 million possible variants. This gives the specific immune system the ability to respond to any type of pathogen that enters the body.

B-cells are unable to respond to the presence of an antigen on their own: they rely on T-helper cells.

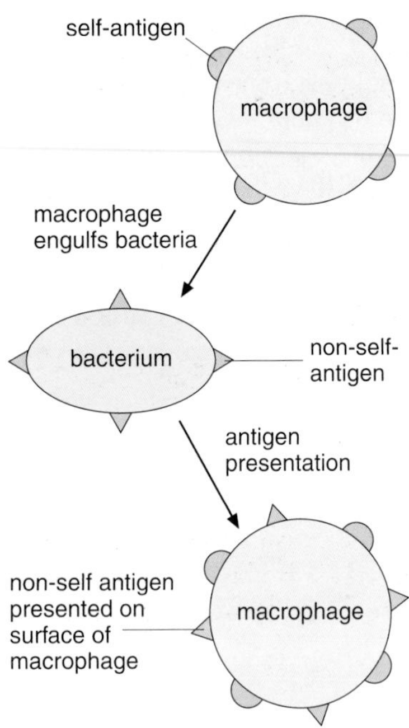

Figure 11.6
When a non-self-antigen on a bacterium is present, the macrophage will engulf the bacterium and present the non-self-antigen on its surface.

T-lymphocytes

T-cells develop specific surface receptors called T-cell receptors. A very important sub-group of T-cells are T-helper cells, which have an active role to play in the specific immune response.

Humoral immunity

This is an immune response to a specific pathogen, such as a bacterium. When bacteria enter the bloodstream they circulate around the body until they eventually reach the lymph nodes. Here they are ingested by the macrophages. The macrophages cut out the surface molecules of bacterial cell walls and membranes. These pieces are the antigens of the bacteria and the macrophages display them on their cell surface membranes. This is called **antigen presentation**.

Inside the lymph nodes the macrophage joins with specific T-helper cells and B-cells that have receptors in their membrane complementary in shape to the antigen presented by the macrophage. During antigen presentation the macrophage selects the T-helper cells and B-cells that have membrane receptors that are complementary in shape to the antigens it has exposed. This is known as **clonal selection**. The selected T-helper cells are induced to secrete cytokines, which activate the selected B-cells to divide to form a number of identical cells called **clones**.

Clonal expansion produces large numbers of identical B-lymphocytes with the ability to destroy this particular pathogen. Bacteria divide so rapidly in the ideal conditions found within the host's body that unless huge numbers of B-lymphocytes are produced the bacteria could cause damage.

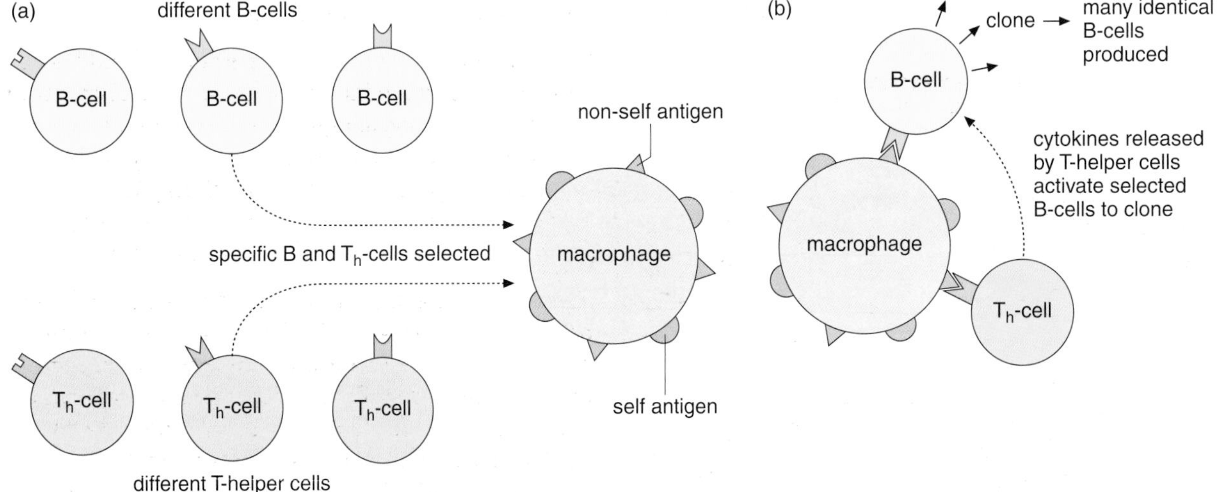

Figure 11.7
(a) Specific B-lymphocytes and T-helper cells are selected, depending on their specific surface receptors.
(b) When selected the T-helper cells secrete a chemical that stimulates the B-cells to clone.

Some of the clones of activated B-cells will differentiate into **B-lymphocyte memory cells** and some into **plasma cells.** The B-lymphocyte memory cells are stored in the lymphatic tissue, while the plasma cells very quickly produce a specific antibody at a rate of up to several thousand a second. The antibodies produced circulate in the

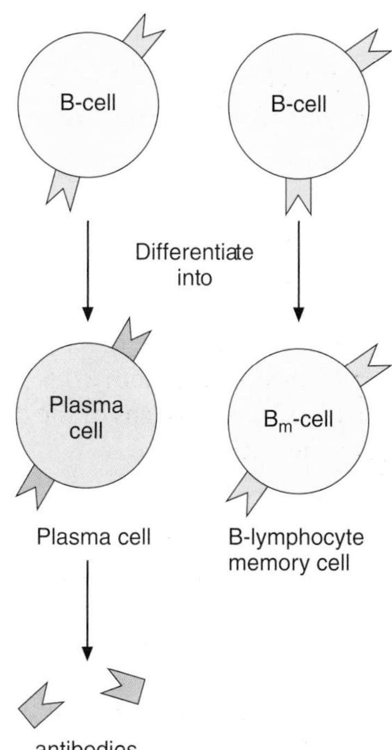

Figure 11.8 (above)
Once stimulated to clone, the B-cells differentiate into plasma cells and B-lymphocyte memory cells.

blood and lymph or secrete antibodies onto the surfaces of mucous membranes, such as those found lining the lungs. These specifically shaped antibodies bind to the antigen on the surface of the invading bacteria.

Different antibodies work in different ways:

- **agglutination** makes the pathogens clump together
- **antitoxins** neutralise the toxins produced by the bacteria
- **lysis** digests the bacterial membrane, killing the bacterium
- **opsonisation** coats the pathogen in protein that identifies them as foreign cells.

Plasma cells are short lived. After a few weeks their numbers decrease but the antibody molecules they have secreted remain in the blood for some time, maintaining immunity. Eventually, the concentration of antibodies will also decrease.

Immunological memory

When a bacterial infection occurs and an antigen is presented for the first time, clonal selection and clonal expansion take time to occur. Once B-cells have differentiated into plasma cells the specific antibodies that they secrete can be detected in the blood. This **primary response** lasts several days or weeks and then the concentration of antibody decreases as the plasma cells stop secreting them. There is little point in producing antibodies to a pathogen that is not present, so as the infection subsides plasma cells die but memory B-cells are left in the body.

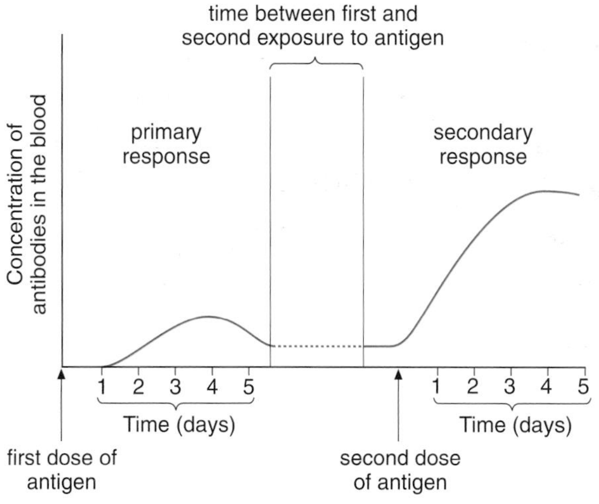

Figure 11.9
The primary and secondary responses to the same antigen.

If another infection of the same pathogen should occur the same antigen is reintroduced. There is a more rapid response, producing a higher concentration of antibody. This **secondary response** occurs quickly because of the presence of memory B-cells formed during clonal expansion in the primary response, which have remained in the lymphatic system.

When the same antigen is presented by a macrophage, the relevant memory B-cells divide rapidly to produce plasma cells. This process is repeated whenever the same antigen is identified. The secondary response is much faster because there are many memory B-cells, which can produce many plasma cells and the appropriate antibody. These destroy the pathogen before it has the chance to cause any symptoms to occur.

Memory cells are the basis of immunological memory; they last for many years, often a lifetime.

It is possible to suffer repeated infections from a single pathogen because pathogens sometimes occur in different forms, each having minor changes in the shape of the antigen, due to a possible mutation, and therefore requiring a primary response.

Each time we have an infection with a different pathogen, bearing different antigens, antigen presentation, clonal selection and clonal expansion must occur before immunity is gained.

ABO system of blood groups

key term

The **ABO system** contains four possible blood groups: A, B, AB and O.

A specialised and unique aspect of the immune system is found in human blood. There is a variety of different blood grouping systems, but the most familiar to us is the ABO system.

It is well known that if for any reason a patient has to be given a transfusion of whole blood then it is essential that the blood given is of the right group. A person of blood group A, for example, should be given blood group A.

On the surface membrane of red blood cells are a number of protein molecules, some of which are antigens. These antigens, known as **agglutinogens**, occur in two different forms: **antigen A** and **antigen B.** There are also two antibodies, known as **agglutinins**, that complement the antigens, which may be present in the blood plasma: **antibody a** and **antibody b.** These antibodies are not made in direct response to an antigen as normal antibodies are; they are present at birth regardless of any exposure to the antigen.

A person cannot have antibodies in the plasma that are specific to antigens on the red blood cells. If you have blood group A, and thus your red blood cells have antigen A, then you cannot have antibody a in your plasma or your own red blood cells would be destroyed by sticking together. However, you can quite safely have antibody b in your plasma. In fact, if you are blood group A you will have antibody b in your plasma. Table 11.1 shows the antigens and antibodies present in the different blood groups.

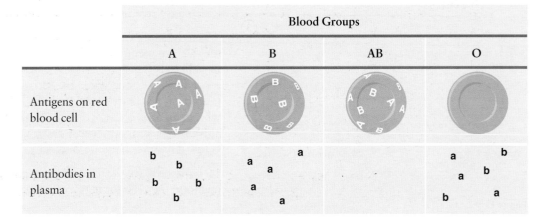

	Blood Groups			
	A	B	AB	O
Antigens on red blood cell	A A A A A A	B B B B B B	A B B A A B A B	
Antibodies in plasma	b b b b b	a a a a a		a b a b b a

Table 11.1 The antigens on red blood cells and the antibodies present in the plasma in particular types of blood.

When blood samples from two people with the same blood groups are mixed, no change occurs: the two bloods are said to be compatible. However, when two samples of different blood groups are mixed there may be important and potentially dangerous changes in the red blood cells: they may stick together to form clumps. This is caused by an antibody–antigen reaction. If clumping occurs, the bloods are said to be incompatible.

If red blood cells carrying particular antigens come into contact with plasma containing the complementary antibody, the reaction between them causes the red blood cells to **agglutinate** (stick together). This means that they can no longer function.

Blood groups have little importance for most of the population for most of the time. As long as your blood remains within your own body and you avoid major blood loss or the need for surgery, it is quite possible that no one will know your blood group. However, blood group information becomes vital when blood needs to be given to another person. When a blood transfusion is given, it is the cells of the donated blood that will be affected by any adverse reaction.

When two types of blood are mixed during a transfusion it is important to know which blood group is being donated and which blood group is receiving the donation. You can ignore the antibody given by the donor because the plasma containing the antibody is rapidly diluted by the recipient's blood and has little effect on the recipient's red blood cells.

When carrying out a blood transfusion:

- it is important to know the type of antigen on the donor's red blood cells as the wrong antigen would cause an immune response to occur
- it is important to know the type of antibody in the plasma of the recipient as the recipient can produce antibodies in large numbers as a response to an incompatible antigen.

Blood Group	Donor	A	B	AB	O
Recpient		A	B	A and B	None
A	b	✔ ③	✘	✘	✔ ②
B	a	✘	✔	✘	✔
AB	None	① ✔	✔	✔	✔
C	a and B	✘	✘	✘	✔

Table 11.2 How ABO blood types can be donated. Key: ✔ = safe transfusion, no agglutination; ✘ = transfusion not safe, agglutination will occur.

You can see three distinct patterns shown in the table:

① Individuals with blood group AB are called **universal recipients** as they can receive blood of any of the ABO groups without ill effect. This is because they are unable to produce any antibody against the A and B antigens on the donor's red blood cells.

② Individuals with blood group O are called **universal donors** as their blood can be given to people of any of the ABO groups without ill effect. This is because they have no A or B antigens on their red blood cells to stimulate an immune response.

③ Individuals of the same blood type can safely donate blood to one another. This is because they have matching antigens and antibodies.

In practice blood to be donated is always cross-matched in the laboratory with a specimen of the recipient's blood before a transfusion is given to ensure that an identical ABO blood type is used whenever possible. In the UK group O is the most common blood group (46% of the population) with group A a close second (42%). Groups B (9%) and AB (3%) are much less common.

Q 2 **Why can someone with blood group AB donate blood to someone else with the same blood group?**

3 **The saliva of several species of leech contains an anticoagulant. What are anticoagulants and why does the leech need them?**

Genetic fingerprinting or DNA profiling

One development of medical research that has proved to be extremely useful for police and forensic science work is deoxyribonucleic acid (DNA) profiling. This is a way of making a pattern from pieces of DNA cut with restriction enzymes. As everybody has different DNA, the pattern, or fingerprint, is unique to each individual. To understand the DNA profiling method it is necessary to briefly review the structure of the DNA molecule.

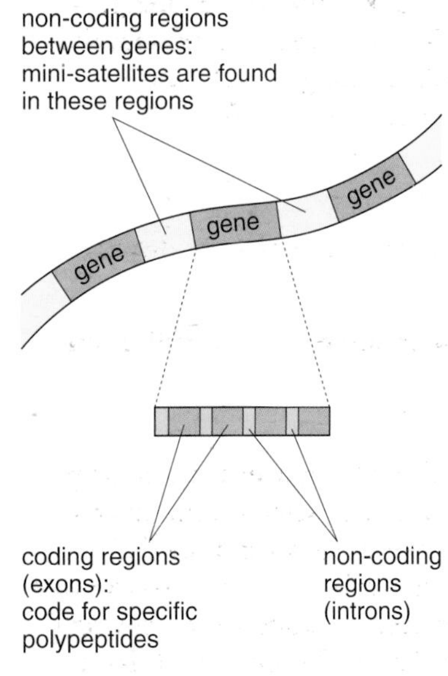

Figure 11.10
A length of DNA showing coding (exons) and non-coding (introns) regions.

DNA is a long-chain, double-helix molecule composed of building units called nucleotides (see Chapter 8). Each nucleotide consists of a sugar, a phosphate and a nitrogenous base. There are four types of nitrogenous base: adenine, guanine, cytosine and thymine. The genetic information carried by a DNA molecule is contained in the sequence of these four bases. A gene is a length of DNA that codes for a specific protein or polypeptide. In human cells, the DNA molecules are organised into chromosomes located within the nucleus. Each human chromosome contains about 4000 genes. Surprisingly, only 2% of the total DNA of a human cell consists of genes. The rest consists of non-coding sequences of bases called **introns**, which can occur between or within a gene. Each intron can be between 60 and 100 000 bases long. A single gene can harbour as many as 50 of these introns sandwiched between **exons**, which are the coding parts of the DNA molecule. The function of the non-coding introns is unknown.

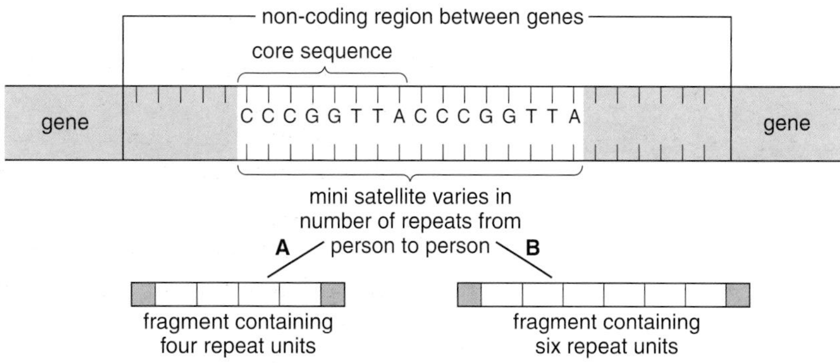

Figure 11.11
A fragment containing minisatellites in which **A** has four repeat units and **B** has the same core sequence repeated six times.

Q 4 **Faults accumulate in the non-coding regions more frequently than they do in genes. Why do you think this is?**

Minisatellites or variable number tandem repeats

Within the non-coding regions between genes, there are short sequences of bases called **core sequences** that repeat themselves over and over again, sometimes up to 100 times. These repeating regions of DNA are called minisatellites or variable number tandem repeats. Different individuals have different numbers of repeated core sequences. The greater the number of repeats, the longer the minisatellite. Different people therefore have different sized minisatellites.

DNA profiling is based on two very important observations:

- the number of repeats of a core sequence tends to vary considerably from person to person
- each individual has 50–100 different types of minisatellite made from different core sequences.

The chances of any two individuals having matching minisatellites all of the same length is minuscule, unless, of course, the two are identical twins. Some minisatellites can be similar in length and thus contain the same number of repeats. The trick in DNA profiling is to choose minisatellites that show the most variation between people.

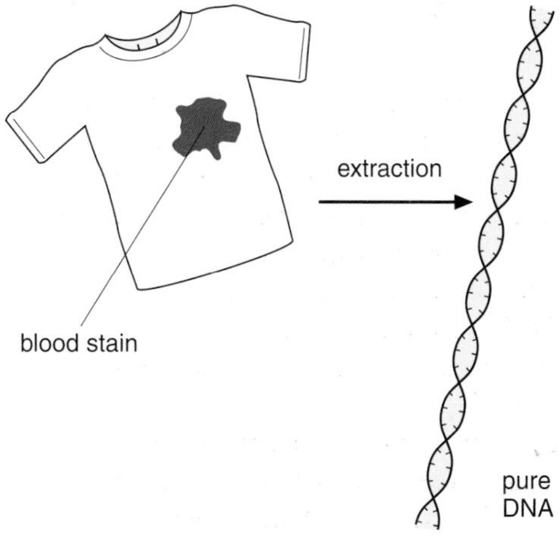

Figure 11.12
A pure sample of DNA is obtained from a specimen of blood.

Making a DNA fingerprint

The technique for making a DNA fingerprint can be divided into four main steps:

extraction → digestion → separation → hybridisation.

Extraction

A sample of tissue containing cells with a nucleus (e.g. a drop of blood, a hair root or a small sample of semen containing a few sperm cells) is taken to the laboratory where the DNA is extracted by shaking the sample in a mixture of water-saturated phenol and chloroform. The proteins precipitate out, leaving pure DNA dissolved in the water layer. The amount of tissue required is very small: 0.5 cm^3 of blood, 0.005 cm^3 of semen or one hair root.

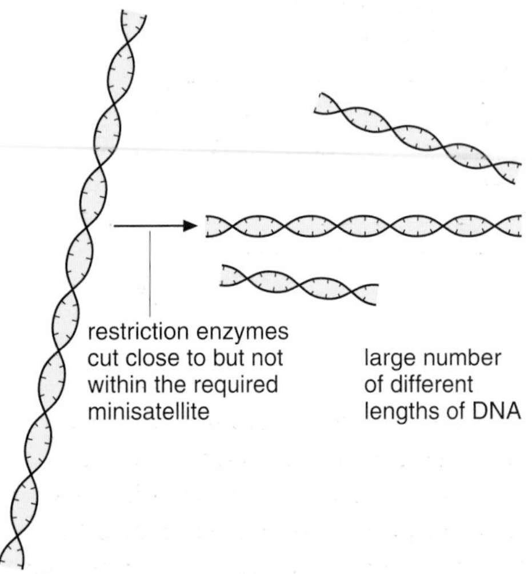

Figure 11.13
The DNA is cut using a restriction enzyme, producing a large number of different lengths of DNA.

Q 5 **Why do forensic scientists need a larger sample of blood than of semen to prepare a genetic fingerprint?**

Digestion

Certain restriction enzymes are added to the DNA to cut it. These enzymes recognise specific base sequences and so cut at specific points, close to but not within the minisatellite region, thus leaving the minisatellites intact. This process produces a number of DNA fragments of different lengths, some of which contain the minisatellite.

Separation

The DNA fragments are separated according to size by electrophoresis. This involves placing the DNA fragments in wells at one end of a block of agarose gel. An electric current is then passed through the gel. The pieces of DNA carry negative charges and so move towards the positively charged electrode. The smaller fragments move through the gel more quickly than the larger fragments and consequently travel further through the gel. The fragments are therefore separated into bands.

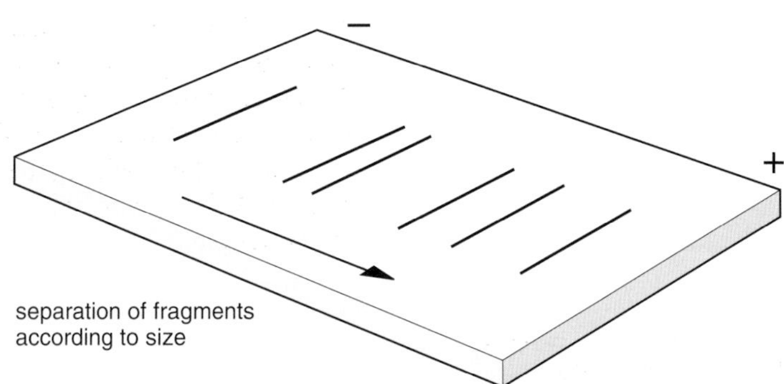

Figure 11.14 (right)
Electrophoresis separates fragments according to size. The smaller the fragment, the further it travels up the gel.

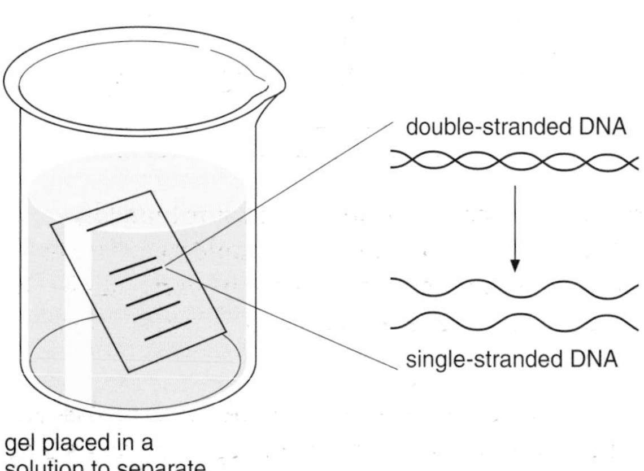

Figure 11.15
The double-stranded DNA is separated into single strands.

The pattern of bands on the gel is invisible at this stage. To show the bands containing the minisatellite a probe is added. This is a single strand of complementary DNA, which cannot attach to the existing double-strand of DNA. The next step in the process involves separating the double-stranded DNA fragments into single strands by immersing the block of gel in an alkaline solution.

After this, a technique called **Southern blotting** is used to transfer the single-stranded DNA fragments onto a nylon membrane. This involves putting a thin sheet of nylon over the gel and covering it with absorbent paper towels. The absorbent paper draws the DNA fragments up into the membrane by capillary action, with their relative positions remaining unchanged. The fragments are then fixed to the membrane by exposure to UV light.

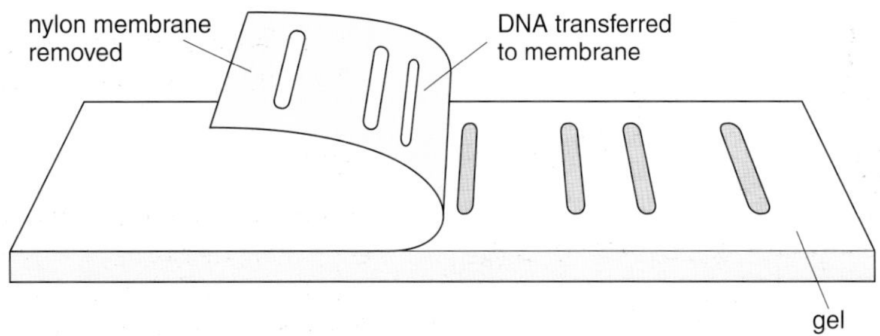

Figure 11.16
The nylon membrane is removed. The fragments of DNA that have been absorbed onto the nylon membrane are in exactly the same position as on the gel.

Hybridisation

At this point a radioactive DNA probe is used to bind onto and reveal the location of a certain type of minisatellite. The technique involves immersing the nylon membrane in a solution containing the radioactive DNA probe. DNA probes consist of a single strand of a length of DNA made up of sequences of bases complementary to the core sequence. The probes used in forensic work are commonly of a type that will bind only at one specific site or locus, and are known as single-locus probes. Excess probes are simply washed away. The process can be repeated with different radioactive probes, which bind to different core sequences and thus identify different minisatellites.

The final stage in DNA fingerprinting consists of making visible the minisatellite regions that have been picked up by the radioactive probes. This is done by putting an X-ray film over the nylon membrane. The places where radioactive probes have bound to DNA fragments will emit radiation, which will fog the film. This creates a pattern of bands known as a DNA fingerprint, which is not unlike the bar code found on items in supermarkets.

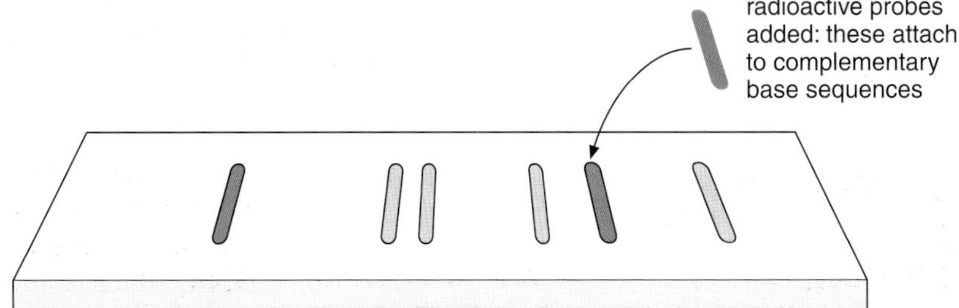

Figure 11.17
In this example the minisatellites inherited from the mother and father are of different lengths and will therefore show up as two separate strands.

Humans have two copies of most chromosomes: one contributed by the father, the other by the mother. Each individual should therefore have two of each type of minisatellite, not necessarily with the same number of repeats and therefore of different lengths. If they are of different lengths they will appear as two bands.

Forensic laboratories commonly use three or four single-locus probes on a single sample, giving a DNA print with up to six or eight bands. As only one of each chromosome is inherited from each parent, half of the bands should match those of the mother and half those of the father.

Q 6 Which of the child's DNA bands were inherited from her mother?

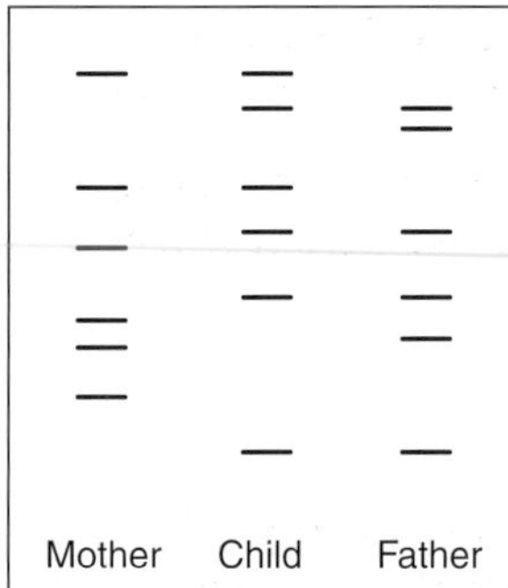

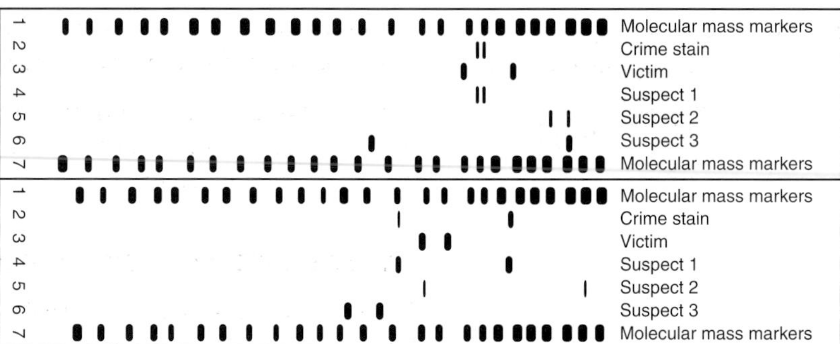

Figure 11.18 (above)
Typical DNA profile of parents and a child.

Figure 11.19 (above right)
Results of a single-locus probe. In each example the profile from a bloodstain extract is in track 2 and the profile from the victim's DNA is in track 3. Tracks 4, 5 and 6 contain the profiles of three suspects.

DNA fingerprinting can be used to identify suspects in a crime. The illustration in Figure 11.19 shows two cases in which one of a number of suspects can be linked to the scene of the crime because their DNA profile matches that of a bloodstain found on the victim.

Q 7 **Which suspect committed the crime?**

Analysing the results

Visual inspection is used first to see whether two DNA fingerprints match. If there is a visual match, an automated scanning system is used to calculate the length of the DNA fragments denoted by the bands. This involves measuring how far the fragments have travelled through the gel compared with the distance travelled by DNA of known length, called **markers**. The DNA profile is then expressed as a set of six to eight numbers.

The next step consists of calculating the odds of somebody else in the population having the same DNA fingerprint as the suspect. This typically works out to be 1 in 30 for a single band. However, by multiplying together the odds from eight bands, analysts can predict the chance of an identical DNA fingerprint being produced from two different individuals as hundreds of millions to one.

Problems with contamination

DNA fingerprints have been widely used to settle paternity disputes, to help identify rapists and murderers, and even to establish the ownership of a prized pedigree cat. Until 1989 DNA evidence was regarded as indisputable. Since then, however, scientists have questioned the reliability and accuracy of tests carried out in some forensic laboratories. The criticisms have largely been directed against the standard of practice in these laboratories and their interpretation of data, not against the underlying principle of this technique.

There are three main problems:

- forensic samples are rarely pure: a blood sample may be contaminated with DNA from bacteria and fungi
- if there was a delay in collecting the sample, the DNA could have decomposed, so that some sites normally cut by the restriction enzymes could have been lost. This could result in longer or shorter fragments being produced, forming a different banding pattern.

Q 8 Explain why DNA fingerprinting is a more useful diagnostic tool than blood typing?

- ions in the contaminants could affect the charge on DNA fragments, thus causing them to travel through the gel at a faster or slower rate.

Criticisms of DNA fingerprinting have led to a greater understanding of the strengths and weaknesses of this technique and more stringent control in forensic laboratories.

Polymerase chain reaction

In many instances there is insufficient DNA available for a genetic fingerprint test. This problem is overcome by using the **polymerase chain reaction** (PCR). Using this technique a minute quantity of whatever DNA is required can be increased a billion-fold or more.

Basically, the PCR involves the repeated replication of DNA in a test-tube, doubling the amount with each replication cycle. After 20–30 cycles enough DNA will have been produced for analysis. The procedure is very quick, each cycle taking only about 5 minutes.

In principle, all that is needed to copy DNA is the enzyme DNA polymerase and a supply of nucleotides. However, there are some problems:

- The base sequence of the DNA to be amplified must be known in advance.
- DNA polymerase cannot work on an entirely single stranded template. **DNA primers** are therefore required. These are single-stranded lengths of DNA, about 20 bases long. They are chosen to combine with the single-stranded DNA at one end of the region to be replicated, creating a double-stranded initiator section. These primers are synthesised artificially, therefore again the complementary sequence of bases must be known in advance.
- Between each cycle of DNA replication the two strands of DNA have to be separated by heat; this denatures most DNA polymerases. To avoid this, DNA polymerase from a bacterium that is adapted to high temperature is used.

Q 9 After only two cycles of replication, four copies of the double-stranded DNA exist. Calculate how much a DNA sample will have increased after:

(a) 10 cycles
(b) 25 cycles

The PCR consists of the following steps:

1. DNA is heated to 95°C to separate the double strands.
2. DNA polymerases, four different nucleotides and an excess quantity of DNA primers are added.
3. The mixture is cooled sufficiently to allow the primers to join to the DNA at either end of the region that is to be amplified. As the primers are added in excess, the original parent strands are much more likely to pair with the primers than with each other.
4. The DNA is replicated, producing two double-stranded molecules.
5. The DNA is heated again and the process repeated.

The stages in the process are carried out in a PCR machine, and are summarised in Figure 11.20 (overleaf).

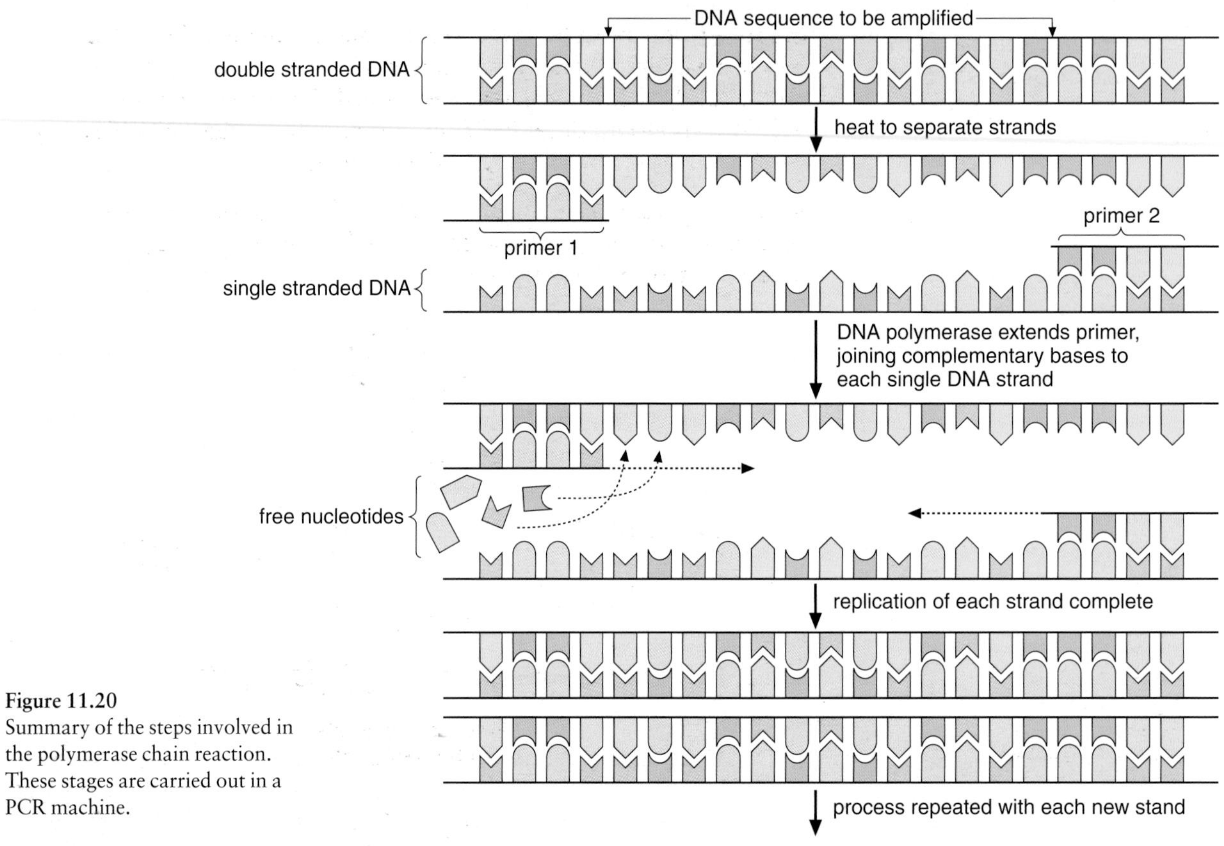

Figure 11.20
Summary of the steps involved in the polymerase chain reaction. These stages are carried out in a PCR machine.

Extension box 1

Monoclonal antibodies

All vertebrates, including humans, produce antibodies to react with antigens, which are often presented by bacteria or viruses. Human antibodies are produced by white blood cells of the immune system called B-lymphocytes. Each individual possesses millions of different types of B-lymphocytes. Each type is genetically programmed to produce just one type of antibody, which recognises the antigen that stimulated its production.

When the B-lymphocyte is activated a series of profound changes occur in the cell. The cell swells and actively divides, giving rise to a clone of daughter cells. Some of the daughter cells enlarge to become antibody-secreting plasma cells. Each plasma cell is capable of secreting more than 10 million copies of a particular type of antibody per hour.

Attempts have been made to culture plasma cells individually as a way of obtaining a large supply of a specific antibody. However, plasma cells are highly specialised and they will not divide in culture. They are consequently short lived and do not stay alive long enough to make sufficient antibodies. Fusing a B-lymphocyte with a tumour cell, usually from a mouse, has solved this difficulty. Tumour cells are cancerous cells that will actively divide indefinitely in a culture solution. The product of cell fusion between a B-lymphocyte cell and a tumour cell is called a **hybridoma.** Hybridoma cells divide actively and produce antibodies continuously. They can therefore be cloned. Since each clone of hybridoma cells produces just one type of antibody, the antibodies they produce are known as **monoclonal antibodies.**

Monoclonal antibodies are useful in biotechnology and medicine. They bind to specific molecules and can be used in a variety of ways, as described in the following sections.

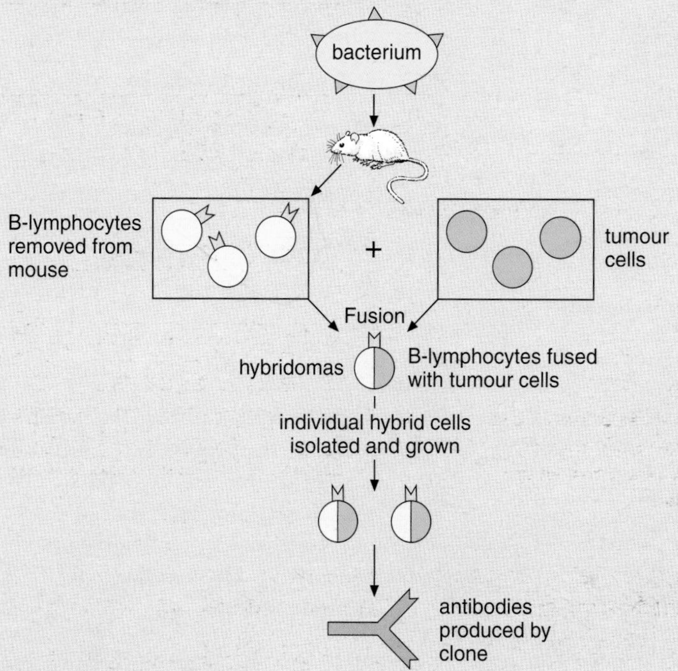

Figure 11.21 Summary of monoclonal antibody production.

Analytical reagents

Monoclonal antibodies used in a biosensor can measure minute amounts of drugs or hormones in body tissue or fluids.

Purification of proteins

Monoclonal antibodies can be used to purify a desired protein, for example interferon. The monoclonal antibodies are attached to an insoluble resin and put into a column. When the crude preparation, which includes interferon and other proteins, is poured onto the column, the interferon only attaches to the immobilised monoclonal antibodies while the unwanted proteins trickle out of the column. An acid can be used to dislodge the interferon from the antibody–resin complex before the solution is neutralised and concentrated by low-temperature evaporation.

Treatment of infectious diseases

Monoclonal antibodies can be used to treat diseases such as rabies or tetanus, where it is too late to use other methods.

Treatment of cancer

The surface molecules of cancer cells and normal cells are very similar but certain types of cancer cells make abnormal proteins called '**tumour markers**' that can be recognised as being different. Monoclonal antibodies, which recognise these tumour markers, could be attached to toxic chemicals, which they could deliver specifically to the cancer cells.

Drugs for blocking organ transplant rejection

The monoclonal antibodies in the drug act by neutralising mature white blood cells that would otherwise attack the new organ.

Detecting heart attacks

Monoclonal antibodies can detect enzymes in the blood that are released by damaged heart muscle. This is used to indicate whether or not a heart attack has occurred.

Screening

Monoclonal antibodies can screen donated blood for viruses that cause AIDS or hepatitis.

Pregnancy test

The pregnancy test works by making use of monoclonal antibodies to detect the presence of a hormone, **human chorionic gonadotrophin.** This hormone is only produced by women who are pregnant and is excreted in the urine. The test is so sensitive to minute quantities of the hormone that a positive result can be obtained within days of conceiving. The test takes only a few minutes and requires only a small amount of urine.

Extension box 2

The Rhesus system

Other than the ABO blood system, another way to type blood is known as the Rhesus system. This is based on the presence or absence of an antigen, agglutinogen-D, on the surface of the red blood cells. Of the total population, 85% possess red blood cells, which have the antigen and are said to be rhesus positive, Rh^+. The rest, which do not have this antigen, are regarded as rhesus negative, Rh^-. Unlike the antibodies for the ABO system, antibodies against agglutinogen-D, anti-D, do not occur naturally in the plasma. A person who is Rh^+ can never produce anti-D and it is only produced by Rh^- people whose blood has been contaminated by red blood cells carrying the agglutinogen-D either during a transfusion or during delivery of a baby. If blood containing the antigen enters a rhesus negative individual they will respond by producing anti-D antibodies.

This is normally not a problem as blood is very carefully matched before transfusions occur. However, because the presence of these protein antigens on the surface of the red blood cells are genetically determined, it is possible that a mother could be rhesus negative and her unborn child rhesus positive. During the later stages of pregnancy, and particularly at childbirth, some fragments of the rhesus positive cells of the fetus can cross the placenta and enter the mother's blood system. If they do, her body will produce anti-D antibodies which can pass across the placenta into the fetus. Normally, they are not produced in sufficient quantities to affect the first-born child. If she becomes pregnant again, with a rhesus positive fetus, and antigens again enters the mother's blood system, she will now produce vast numbers of the anti-D antibodies, which will cross the placenta and enter the fetal blood system. These antibodies will cause the red blood cells of the fetus to rupture and release its haemoglobin.

	Blood group	
	Rhesus positive	**Rhesus negative**
Antigen on red blood cell	D	
Antibodies in plasma	**None** Can never be produced	**None** Can be produced if antigen is present

Table 11.3 The antigens on red blood cells and the antibodies that can be potentially produced in rhesus positive and rhesus negative types of blood.

In extreme cases so many cells are destroyed that the fetus dies. This is known as haemolytic disease of the newborn. This rarely happens today because a woman who is rhesus negative is treated immediately after her first pregnancy. Within 72 hours of giving birth the mother is given an injection of anti-D antibodies. These attach themselves to the antigens on the fetal cells, which are in the mother's circulation and destroy them before the mother can be stimulated to produce her own antibodies against them.

Figure 11.22 (left)
The effect of rhesus antigens entering the blood system of a rhesus negative mother from a rhesus positive fetus.

Figure 11.23 (right)
Injecting anti-D antibodies prevents this rare immune attack.

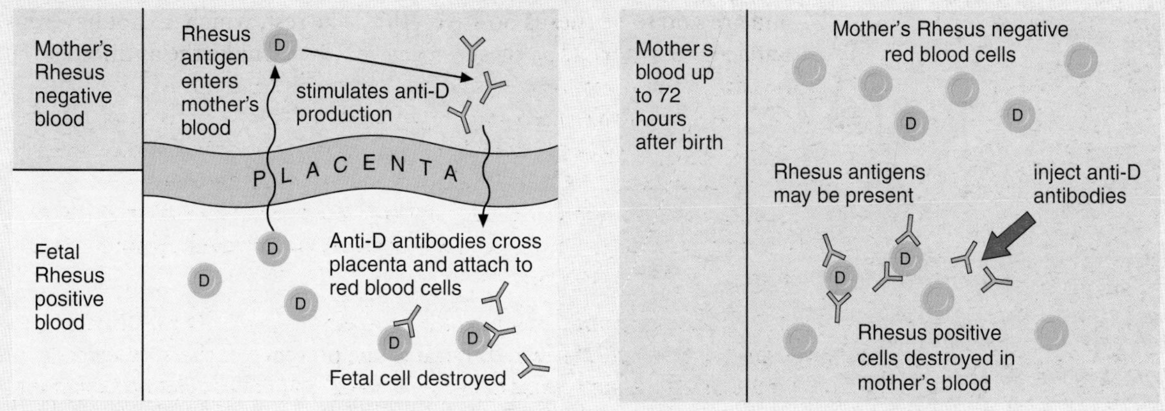

Summary

- There is a specific and non-specific immune response.
- The specific immune response acts against pathogenic microorganisms and viruses.
- Antigens, usually proteins, trigger the immune response.
- Macrophages in the lymph nodes ingest pathogens and present the antigens on their cell surface membranes.
- T-helper cells enable B-cells to respond to the presence of an antigen.
- Clonal expansion leads to large numbers of identical B-lymphocytes which can destroy a particular pathogen.
- B-lymphocyte memory cells are stored in the lymphatic tissue.
- Plasma cells very quickly produce specific antibodies which bind to the antigen on the surface of invading pathogens.
- Antibodies act by agglutination, producing antitoxins, lysis or opsonisation.
- Memory B-cells are responsible for a rapid secondary response on further infection.
- The ABO blood groups (A, B, AB and O) can be distinguished by the antigens on the red blood cells (antigen A, antigen B, both, or neither) and antibodies in the plasma.
- Everyone has a unique DNA fingerprint.
- Restriction enzymes cut DNA into fragments.
- Electrophoresis can be used to sort DNA fragments according to size.
- Radioactive DNA probes are used to locate specific DNA fragments.
- The polymerase chain reaction (PCR) can be used to greatly increase the amount of DNA in a small sample.

Examination questions

1 Explain the difference between the following pairs of terms:

(a) antigen and antibody
(b) B-lymphocyte and T-lymphocyte
(c) primary immune response and secondary immune response.

(6 marks)

2 (a) Explain why blood from a donor with blood group AB cannot be transfused successfully into someone with blood group O.

(b) By means of a simple flow diagram show how a B-lymphocyte responds when its specific antigen enters the bloodstream.

(6 marks)

3 (a) Explain why early techniques used to distinguish between blood obtained from different individuals relied on methods based on the biology of red blood cells while more modern techniques are based on white blood cells.

(3 marks)

(b) A sample of blood was tested in the laboratory. It agglutinated when it was mixed with B-antibody and it also agglutinated when it was mixed with A-antibody.

(i) What was the blood group of this blood sample? *(1 mark)*
(ii) Explain how you arrived at your answer. *(1 mark)*

(c) What is meant by 'non-coding' DNA?

(1 mark)

Assignments

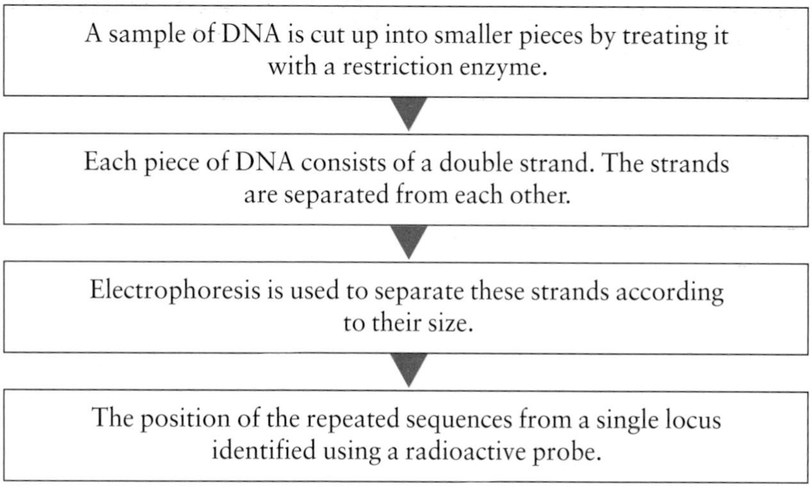

A sample of DNA is cut up into smaller pieces by treating it with a restriction enzyme.

▼

Each piece of DNA consists of a double strand. The strands are separated from each other.

▼

Electrophoresis is used to separate these strands according to their size.

▼

The position of the repeated sequences from a single locus identified using a radioactive probe.

Figure 11.24
The stages in preparing a DNA fingerprint.

In this chapter you have studied one way in which we make use of our knowledge of the structure of DNA. The technique of genetic fingerprinting, however, is not only used in helping to solve violent crimes. In this assignment we will look at how it has been adapted to help with the conservation of birds of prey.

Very strict laws apply if you want to keep a hawk or a falcon in captivity. Birds can be imported. A licence is required and this makes it easy to check up to see if a particular bird has been imported. As it is difficult to get a licence and importing birds is expensive, it might be thought better to raise the young produced by captive birds. Unfortunately this is not very easy, so people are sometimes tempted to take the young out of the nests of wild birds. This is illegal as many birds of prey are rare and protected by law. The big problem for those involved in conservation is to prove whether a bird was bred in captivity or taken from the wild. This is where genetic fingerprinting can help.

Figure 11.25
Taking the eggs or young from nests of birds of prey such as this falcon is illegal. Genetic fingerprinting enables us to prove whether a particular bird was bred in captivity or taken from the wild.

Let us look at an actual case. A Wiltshire man had a collection of birds of prey which included five peregrine falcons: an adult male, an adult female and three young birds. He claimed that the young birds were the offspring of the adult pair. The police, however, were suspicious and were of the opinion that some or all of these young birds had been taken from the wild.

The technique used to investigate this case involved the use of single-locus probes. Look at Figure 11.26. This is a diagram of a single pair of chromosomes from a peregrine falcon. Arranged along each chromosome are the genes which encode proteins. Each gene is found at a particular place or locus on the chromosome. Individual genes may exist in different forms called alleles. The alleles of a gene have slight differences in their base sequences and it is these variations which give rise to the inherited differences between individuals. The DNA between the genes is sometimes called non-coding DNA. Some of this DNA is made up of short, repeated sequences of nucleotides. The non-coding DNA found at a particular locus of one chromosome of a pair may differ from that found at the same locus on the other chromosome in the number of times the repeated sequence occurs.

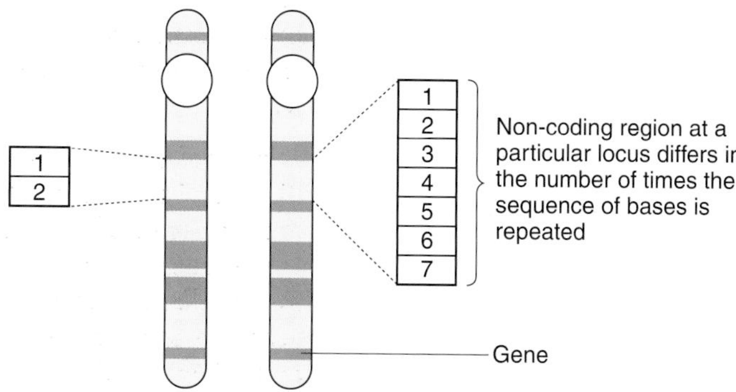

Figure 11.26
Diagram representing a single pair of chromosomes from a peregrine falcon.

1 Copy Figure 11.26. Use the information in the paragraph above to annotate your diagram to show the main features of genes and non-coding DNA.

(2 marks)

A parent will pass one of each pair of its chromosomes to each of its offspring so the non-coding sequences of DNA will be inherited in the same way that genes are inherited.

2 A female bird has a sequence of 6 repeats at a particular locus on one of its chromosomes. At the corresponding locus on the other chromosome of the pair, there is a sequence of 4 repeats. Explaining your answer in each case, answer the following questions:

(a) What proportion of the offspring of this bird would you expect to have the non-coding piece of DNA with 6 repeats?

(2 marks)

(b) A young bird has a piece of DNA with 4 repeats and a piece of DNA with 3 repeats. Could the female bird have been its mother?

(2 marks)

The main steps in carrying out this method of genetic fingerprinting are shown in the flow chart.

3 (a) There are many different restriction enzymes. Why is one particular sort of restriction enzyme is used to cut the DNA into pieces?

(2 marks)

(b) A probe is a piece of DNA with the complementary base sequence to the one being sought. Explain how a radioactive probe can be used to identify the position of a particular repeated sequence of non-coding DNA.

(2 marks)

We will now go back to our original peregrine falcon family. A test was carried out on their DNA. It produced the results shown in Figure 11.27.

Figure 11.27

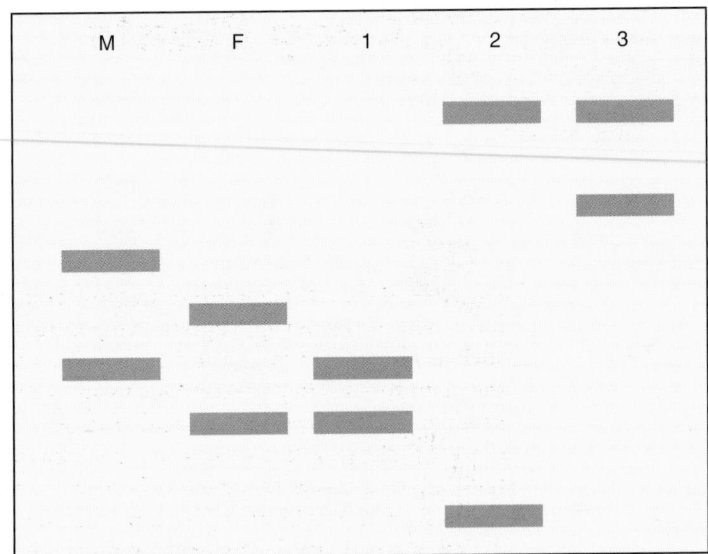

4 (a) What can you conclude from these results about the parents of the young birds? Explain your answer.

(3 marks)

(b) Suggest one reason why care has to be taken in interpreting the results of tests like this.

(2 marks)

12 Crop Plants

Plants need the right conditions to grow and produce a good harvest. They must have light, water, an appropriate temperature range and a supply of inorganic ions such as nitrate, phosphate and potassium. These ions are taken up from the soil by the plant roots. They are combined with the products of photosynthesis to produce the proteins, nucleic acids and other organic substances that the plant requires. The problem with crop plants, however, is that when they are harvested these substances are removed with them. The plants are not allowed to decompose naturally and put the inorganic ions back into the soil. The only way farmers can maintain a high yield is to make up for this loss by adding fertiliser to the soil.

Applying fertiliser is not straightforward. If the amount added is not enough, then the crop will not grow well and, when it is harvested, the yield will be poor. On the other hand, if too much is added, there are just as many problems. Adding too much fertiliser not only represents a considerable waste of money, but the surplus can be washed from the soil by rain and may eventually pollute nearby ponds and streams. The objective then is to add just the right amount. This is complicated by yet another factor. Scientific studies have shown that even in a single field the concentration of different inorganic ions differs from place to place. The right amount to add to one part of the field might be wrong for another. In one trial project in the USA information technology has been brought to the aid of the farmer in an attempt to make the addition of fertiliser more efficient. As the tractor moves through the field, a combination of satellite technology and a computer in the tractor cab allows continuous monitoring of the concentration of mineral ions in the soil. The amount of fertiliser added to the soil is then adjusted to meet these needs exactly.

The application of fertiliser to a crop is just one of the ways in which we ensure that crops produce high yields. In this chapter we look at some aspects of crop production. We start by considering the adaptations shown by cereal plants that allow them to grow in different environmental conditions. We will then go on to look at various ways in which we can alter the environment by, for example, growing plants in glasshouses or applying fertilisers and pesticides to ensure that the highest possible yield is obtained.

Figure 12.1
To produce a good harvest, fertiliser must be added to the soil, but how much should be added? Modern technology can help a farmer to add the right amount.

Cereal crops

key term

Cereals are grains produced by plants belonging to the grass family.

Cereals account for over 50% of all human energy and protein needs, and cereal plants occupy two-thirds of all cultivated land. Cereals were the earliest cultivated plants and for over 10 000 years have been the staple food for many human societies. Their importance is related to a number of features: they are relatively easy to grow, store and transport (Figure 12.2) and they have a high nutritive value. Table 12.1 shows the food values of some important cereals compared with those for some other plant foods. The figures show the amount of energy that can be obtained from them and the total amount of protein and lipid present in 100 g. Most of the energy comes from starch and other carbohydrates and so the figures in the second column are a good measure of the amount of these substances present. Note how the energy figures and those for proteins and carbohydrates that relate to cereals are much higher than the figures for the other plants. Part of the reason for this is that cereal grains contain a very low proportion of water.

Figure 12.2
Because it can be stored easily, grain forms the bulk of the world's food reserves. These grain silos contain wheat.

Food crop	Energy (kJ)	Protein (g)	Lipid (g)
Cereals			
Wheat	1420	12.0	2.0
Rice	1296	8.0	2.0
Maize	1471	10.0	4.0
Sorghum	1455	10.0	5.0
Other plant crops			
Potatoes	347	2.0	0.1
Peas	293	4.9	0.4
Lettuce	63	1.2	0.2

Table 12.1 The average energy, protein and lipid content of 100g of different crops.

Q 1 **Describe a biochemical test you could use to show that maize grains contain protein.**

A third factor that contributes to the importance of cereals is that there are many species and varieties of cereals grown in different parts of the world. Wheat and barley are found in temperate climates such as those of Europe, parts of Asia and North America. Maize grows best in hotter conditions and is an important cereal crop in many tropical areas. Rice is a crop of the wet tropics. Sorghum survives in very hot, dry conditions.

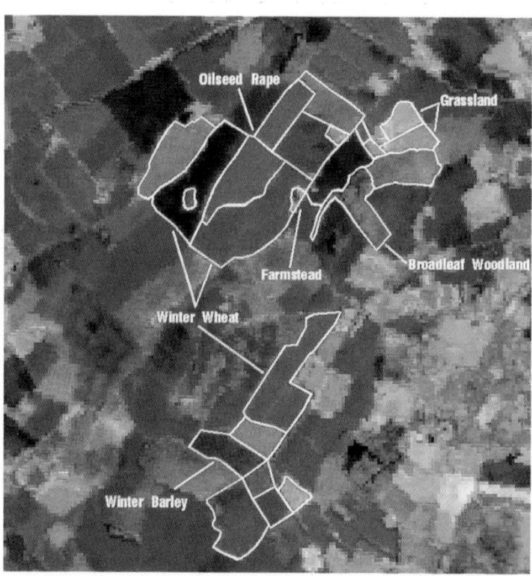

Figure 12.3
Satellite photo showing a typical distribution of cereal crops in southern England.

All of these plants have adaptations that enable them to survive and grow well in particular environmental conditions. Some of these adaptations are structural while others involve the physiology of the plant. We will now look at three species in a little more detail.

Rice

Rice is one of the most important tropical food crops and forms the main source of food for nearly half of the world's human population. There are many varieties of rice and they differ in height, in the amount of time they take to mature and in their water requirements. In various parts of its range, rice is grown in different ways but most of the rice in south-east Asia is grown in unusual conditions for a cereal plant: it is grown partly submerged in water in paddy fields (Figure 12.4).

Figure 12.4
The life cycle of rice. (a) Buffaloes are used to plough paddy fields in Sri Lanka. (b) The young rice plants are planted partly submerged in water. (c) Mature rice plants almost ready for harvesting.

(a)

(b)

(c)

The fields are flooded and then ploughed. Young rice plants are planted in the rich mud formed in these paddy fields. The oxygen concentration of this mud falls rapidly after the paddy field has been flooded. The top 10 cm or so retains some oxygen because it is able to diffuse in but below this depth **anaerobic** conditions exist and there is little or no oxygen present.

Q 2 Relatively few weeds are able to grow among rice plants. Suggest why.

Rice plants have a number of adaptations that enable them to grow well in these conditions. Look at Figure 12.5. It shows a section cut through the stem of a rice plant compared with a similar section cut through the stem of a wheat plant. The big difference between these stems is the presence of a large number of air spaces in the rice stem. These spaces allow oxygen to penetrate through to the cells of roots growing in the absence of oxygen. These root cells continue to get oxygen and are able to respire **aerobically**. Another adaptation shown by rice plants is that many of their roots are very shallow. These roots are able to make use of the oxygen that diffuses into the surface layer.

Figure 12.5
Sections through the stems of (a) wheat and (b) rice. Note the air spaces in the rice stem. These spaces allow oxygen to penetrate to cells in roots growing in anaerobic conditions.

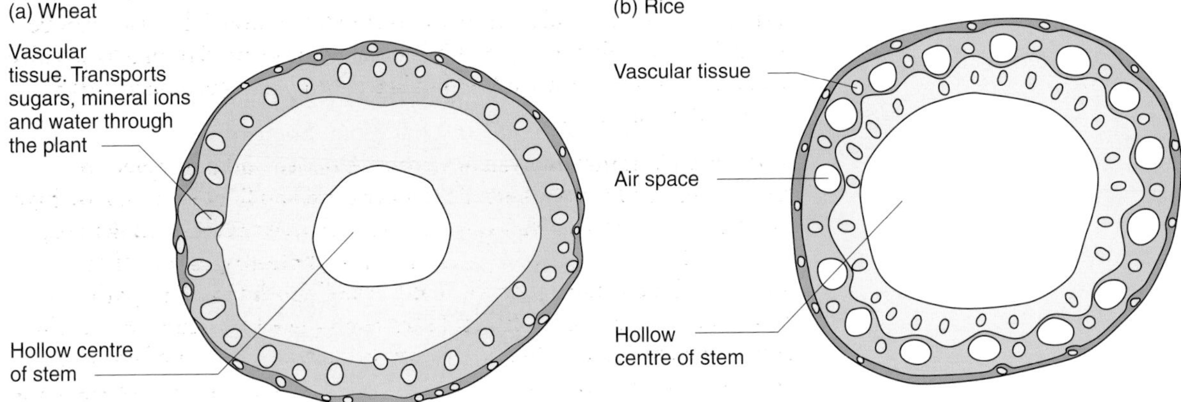

When fields in which a cereal such as wheat is growing are flooded for any length of time, the plants die. The oxygen concentration of the waterlogged soil falls rapidly. The root cells are unable to get the oxygen they need in order to respire. In these conditions they can carry on respiring without oxygen. This is called **anaerobic respiration** and results in ethanol being formed as a waste product. Unfortunately, this substance is poisonous so a plant can only respire in this way for a short time before the ethanol concentration builds up and kills it. Cells in the roots of rice plants have been shown to be extremely tolerant of ethanol, much more so than cells from the roots of other cereals. They can therefore respire anaerobically for longer periods.

Q 3 What is the main difference between aerobic and anaerobic respiration?

There are two advantages of growing rice in paddy fields. Flooding brings about chemical changes in the soil that increase the supply of soil nutrients required by the rice plants. It also reduces weeds. Rice does not grow well when it has to compete with weeds for the resources that it needs.

Maize

Maize was first cultivated in central America but it is now grown throughout the tropics, wherever the rainfall is high enough. Some varieties can also be grown in cooler parts of the world such as the southern USA and parts of Europe.

Figure 12.6
A field of maize. This crop is being grown as fodder and will be cut and fed to cattle.

Maize has many uses. Much of the maize grown in the UK is grown as a fodder crop. It is cut in early autumn and the leaves and stems are fed to cattle. In many tropical areas maize is the staple food. The grain can be roasted on the cob and eaten as it is or it can be ground to produce maize meal. This is eaten as a kind of porridge. Starch and corn oil are also produced from maize grains.

Maize is a plant that evolved in Central America and is adapted to growing in conditions where the temperature and light intensity are high. Most of the plants that grow in the UK, such as oak trees, dandelions, wheat and potatoes, have a particular type of photosynthesis that involves a biochemical pathway called the **C_3 pathway**. This pathway gets its name because the first substance produced contains three carbon atoms in its molecules. In hot, dry conditions the stomata often close, helping to reduce water loss from the plant. With the stomata closed, the concentration of carbon dioxide in the leaves falls as it is used up and is not replaced. Since carbon dioxide is essential for photosynthesis, the C_3 pathway is not very efficient under these conditions.

key term

Stomata are the pores in the leaves through which carbon dioxide enters the plant.

Maize has a slightly different way of photosynthesising, involving another biochemical pathway, the **C_4 pathway** (here, the first substance formed has four carbon atoms in its molecules). This pathway involves a biochemical process in which the carbon dioxide can be concentrated. It is much more efficient in hot, sunny conditions. More details of photosynthesis and the C_4 pathway are given in Extension box 1.

Extension box 1

Different kinds of photosynthesis

The equation often used to summarise the process of photosynthesis is

$$6H_2O + 6CO_2 + \text{light energy} \rightarrow C_6H_{12}O_6 + 6O_2$$

This is rather misleading as it suggests that photosynthesis involves a single biochemical reaction. In fact, photosynthesis is a complex pathway involving many reactions. In the **light-dependent reactions**, light energy is absorbed by chlorophyll molecules and used to produce two substances, adenosine triphosphate (ATP) and reduced nicotinamide adenine dinucleotide phosphate (NADP). In the **light-independent reactions**, these two substances are used to convert carbon dioxide to carbohydrates. This is summarised in Figure 12.7.

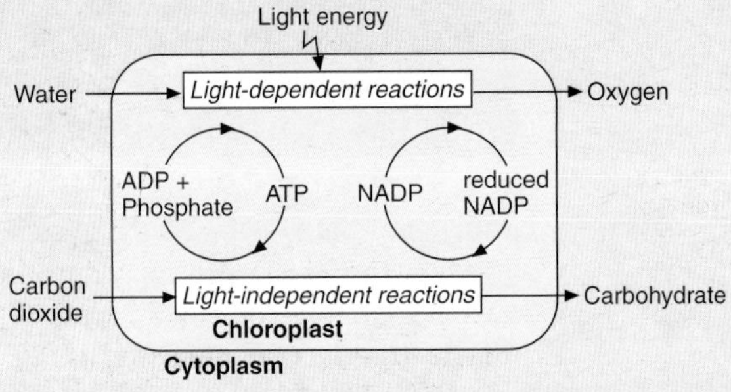

Figure 12.7
A summary of the biochemical pathways involved in photosynthesis.

Most of the plants that grow in the UK use the C_3 pathway. The first substance formed in the light-independent reaction is a molecule with three carbon atoms. The reaction that forms this substance is very slow in hot, dry conditions when the concentration of carbon dioxide in the leaves of the plant falls below its normal level.

The C_4 pathway found in maize involves a carbon dioxide-concentrating system that helps to overcome this problem. Carbon dioxide is taken into the plant and converted to a substance whose molecules contain four carbon atoms. This reaction will take place even when the carbon dioxide is present in very low concentrations. This four-carbon compound can then release its carbon dioxide into other cells in the leaf in which the C_3 pathway takes place. There is one disadvantage, however: the C_4 pathway needs a lot of energy in the form of ATP, so it is only effective at high light intensities.

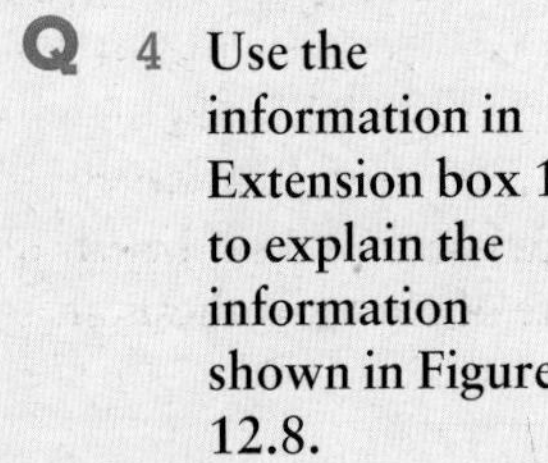

Q 4 Use the information in Extension box 1 to explain the information shown in Figure 12.8.

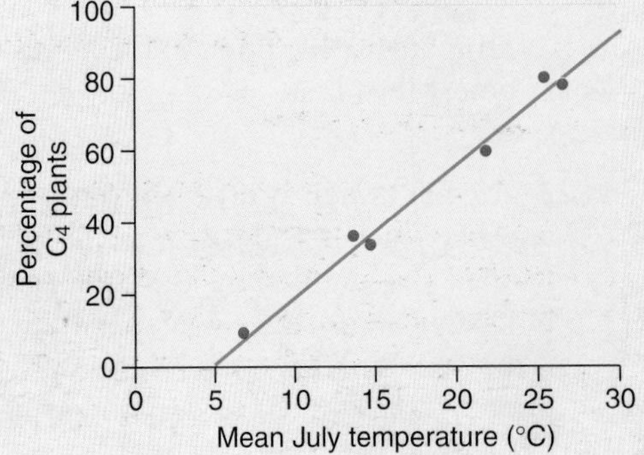

Figure 12.8
This graph shows how the proportion of C_4 plants found in North America varies with climatic conditions. The higher the mean summer temperature, the greater the percentage of C_4 plants.

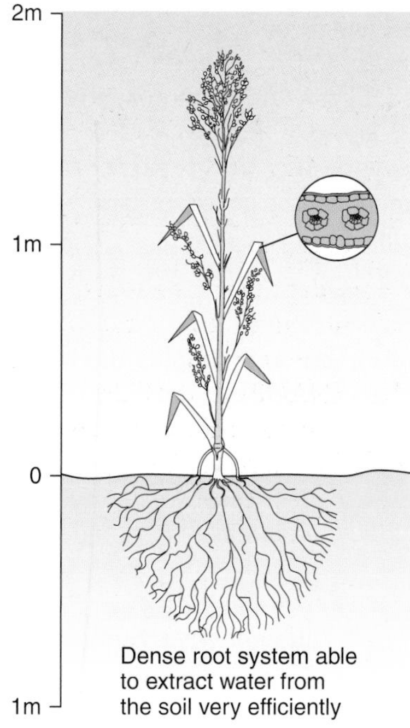
2m
1m
0
1m
Dense root system able to extract water from the soil very efficiently

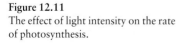

Figure 12.9 (above)
Sorghum forms the staple diet of many of the people living in the drier parts of Africa and central India.

Figure 12.10 (right)
Sorghum is often grown in areas where the rainfall is low and irregular. It has a number of adaptations that allow it to grow and produce grain in these conditions.

Sorghum

Sorghum is one of the few cultivated plants that are able to grow in hot conditions and survive long periods without rain. Because of this, it is a very important crop in the drier parts of tropical Africa and central India. The assignment at the end of this chapter shows the results of some experiments carried out by agricultural scientists into the conditions in which sorghum can grow.

Plants that live in very dry conditions have various features that enable them to survive. They are usually very efficient at extracting what little water there is from the soil and they show adaptations that reduce the rate at which water is lost from the leaves by transpiration. Figure 12.10 shows some of these adaptations.

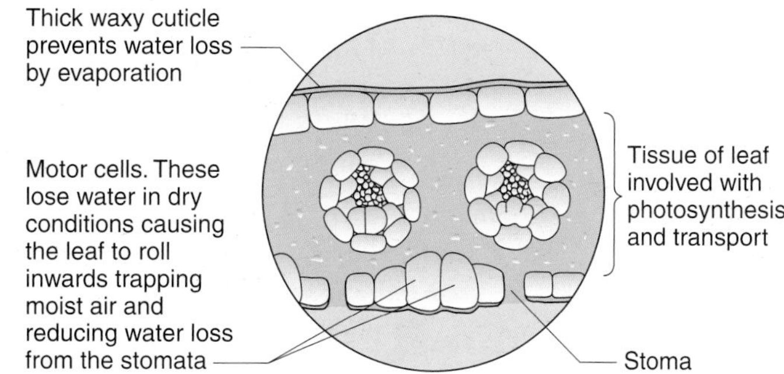
Thick waxy cuticle prevents water loss by evaporation
Motor cells. These lose water in dry conditions causing the leaf to roll inwards trapping moist air and reducing water loss from the stomata
Tissue of leaf involved with photosynthesis and transport
Stoma

Controlling the environment

Plants rely on photosynthesis to produce the carbohydrates and other substances they need for growth. The faster their rate of photosynthesis, the greater their rate of growth and, if we are considering crop plants, the higher their yield. A number of environmental factors affect the rate of photosynthesis. These include the amount of light available, the concentration of carbon dioxide in the air around the plant and the temperature. An advantage of growing crops in glasshouses is that these factors can be controlled to produce a high rate of photosynthesis and good crop yields.

Q 5 **Use your knowledge of enzymes to suggest how temperature might affect the rate of photosynthesis.**

Limiting factors

Let us consider how light affects photosynthesis. On a June day, with the sun shining, light intensity is very high. As evening approaches the light starts to fade. There will come a point when the rate of photosynthesis is affected and starts to decrease. Light intensity is said to limit the rate of photosynthesis. We can show this in a graph (see Figure 12.11).

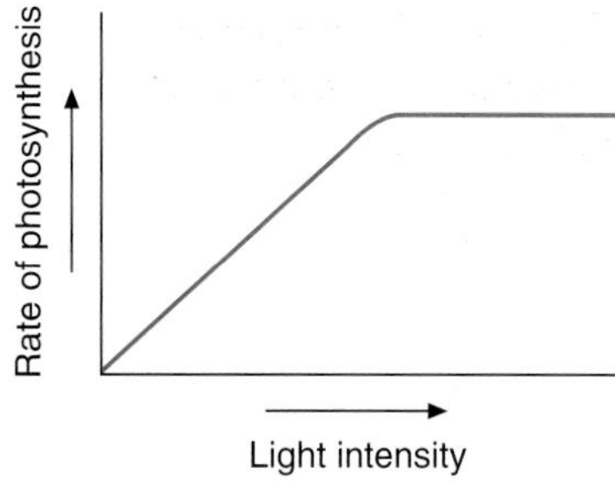
Rate of photosynthesis
Light intensity

Figure 12.11
The effect of light intensity on the rate of photosynthesis.

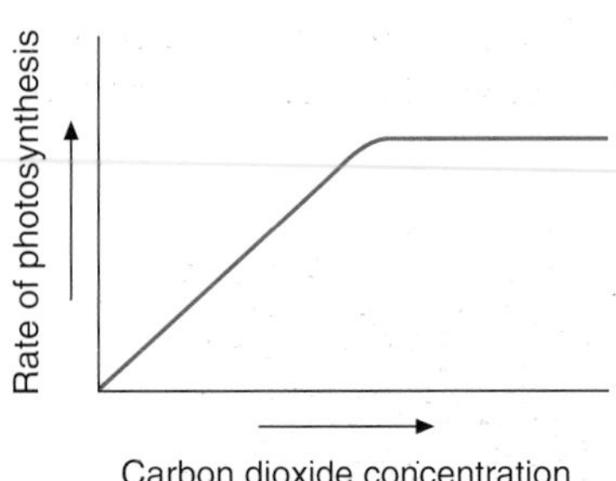

Figure 12.12
The effect of carbon dioxide concentration on the rate of photosynthesis.

Obviously, when there is no light, the plant will be unable to photosynthesise as light is the source of energy for the reactions to occur. Over the first part of the curve in Figure 12.11 the rate of photosynthesis is directly proportional to light intensity. Because of this we can describe light intensity as **limiting** the rate of photosynthesis over this part of the curve. As the light intensity continues to increase, the curve starts to flatten out. Light intensity is no longer limiting the rate of photosynthesis. The most likely explanation for this is that something else is now acting as a limiting factor.

Carbon dioxide is also essential for photosynthesis as it supplies the carbon atoms used to produce carbohydrates and other organic compounds. We might therefore expect the concentration of carbon dioxide to limit the rate of photosynthesis. When we grow plants in bright light and vary the amount of carbon dioxide present we do indeed get a very similar curve to that obtained when light intensity is varied. This is shown in Figure 12.12.

Q 6 Why is it necessary to grow plants in bright light when finding out the effect of carbon dioxide concentration on the rate of photosynthesis?

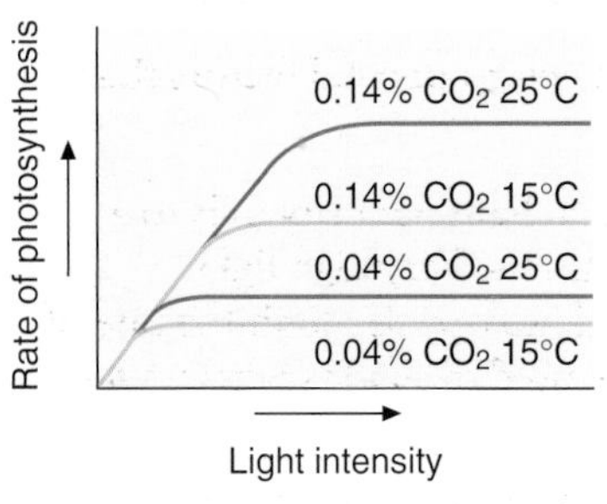

Figure 12.13
The effect of light intensity, carbon dioxide concentration and temperature on the rate of photosynthesis.

Under brightly lit conditions, the concentration of carbon dioxide is the factor that normally limits the rate of photosynthesis. Photosynthesis will not go any faster unless the concentration of carbon dioxide is increased.

A third factor also influences the rate of photosynthesis. This is temperature, which is not surprising when we consider that many of the reactions of photosynthesis are controlled by enzymes. Figure 12.13 shows the effects of all three of these variables, light intensity, carbon dioxide concentration and temperature, on the rate of photosynthesis.

We can draw a number of conclusions from looking at this graph.

- At low light intensities, such as on a winter day in the UK, light is the limiting factor. The rate of photosynthesis will only increase if light intensity increases.
- In bright conditions, carbon dioxide is usually the limiting factor. Increasing the concentration of carbon dioxide in the atmosphere from its normal level of approximately 0.04% to 0.14% has a considerable effect on the rate of photosynthesis.
- In bright conditions, increasing the temperature also increases the rate of photosynthesis. The effect is greater when carbon dioxide is at higher concentrations and therefore not limiting.

Increasing use of fossil fuels has led to an increase in the concentration of carbon dioxide in the atmosphere. Most scientists now believe that this is one of the main factors that has led to the gradual rise in the Earth's mean temperature. An increase in both carbon dioxide concentration and temperature could lead to an increase in crop yields but this may well be offset by other things such as loss of land through rising sea levels and increases in the numbers of insect pests.

Glasshouses

Large glasshouses, such as those shown in Figure 12.14, can be used to grow crops in areas where they could not otherwise be grown. Crops can also be grown throughout the year.

Figure 12.14
Many food crops, such as tomatoes, are grown in glasshouses where the environment can be regulated to give high yields.

Glasshouses can also be used to control the environmental factors that normally limit the rate of photosynthesis.

- Artificial lighting can be used if the natural light intensity falls to a level where it limits the rate of photosynthesis. Very high light intensities damage chlorophyll and will therefore also limit the rate of photosynthesis. Automatic blinds can be used to shade plants under these conditions.
- The concentration of carbon dioxide can be increased either by pumping the gas in or by using paraffin heaters. Tomato plants grow best when the carbon dioxide concentration is maintained at about 0.1%.
- The glasshouse can be heated. This prevents the night temperature falling to levels that might damage the plant but also ensures that optimum temperature conditions exist.

Q 7 **A smallholder kept his pigs in one half of a small glasshouse and grew a winter crop of lettuce in the other. He claimed to have produced a better crop of lettuce as a result! Use your knowledge of limiting factors to suggest an explanation for his claim.**

In addition to controlling light, carbon dioxide concentration and temperature, care has to be taken to supply plants with the other things they require for growth. Fertilisers must be supplied or the supply of mineral ions may limit growth and the plants will not be able to produce sufficient amounts of substances such as proteins and chlorophyll. The rate of photosynthesis will also fall if the plants have inadequate water. One thing that happens under these circumstances is that the stomata close so the supply of carbon dioxide to photosynthesising cells in the leaves is reduced.

What a grower has to consider is the cost of controlling the environment in this way. Paraffin for heaters and electricity for extra lighting cost money. These costs have to be balanced against the extra profit made by a larger yield or from an out-of-season crop.

Fertilisers

Figure 12.15
Nutrients are removed when plants are harvested. To maintain a high yield the nutrients have to be replaced either by inorganic fertiliser or, as shown here, by organic manure.

In addition to the raw materials for photosynthesis, plants also need soil nutrients. The most important of these are nitrogen, phosphorus and potassium, which are sometimes called **macronutrients** because they are required by plants in relatively large amounts. Plants also need other nutrients such iron and magnesium. These are called **micronutrients** or **trace elements** as they are only required in small amounts. These nutrients are absorbed through the plant roots and combine with the products of photosynthesis to produce substances such as proteins, nucleic acids and chlorophyll.

Under natural conditions the soil nutrients are recycled. When the plant dies, it decomposes and bacteria and other soil microorganisms break down organic molecules and release the nutrients back into the soil. When we grow a crop, however, we harvest and remove it, taking away the nutrients it contains. If we want to maintain a high yield over a period of years we must replace these nutrients. This is where fertilisers are important. When an area of rainforest, such as that shown in Figure 12.16, is cleared and burnt, many of the nutrients in the plants are returned to the soil in the ash resulting from the fires. Crops planted in these soils grow very well at first but if fertiliser is not added the yield of later crops decreases rapidly. This is shown by the figures in Table 12.2.

Q 8 **What macronutrient do plants need to absorb from the soil in order to synthesise amino acids?**

Figure 12.16
Because of their lush growth, it is sometimes thought that rainforests grow on extremely fertile soils. This is not the case. Most of the nutrients are locked up in the plants. The soil underneath is often very poor in nutrients.

	Yield of rice (tonnes per hectare)	
Crop number	Without fertiliser	With fertiliser
1	1.2	No fertiliser applied
2	1.6	3.2
3	1.2	2.4
4	0.4	1.6
5	0.3	1.4
6	0.1	1.2

Table 12.2 The effect of adding fertiliser to crops growing on land resulting from the clearing of rain forest.

The three macronutrients all help to produce a higher yield.

- Nitrogen is very important for leaf growth. Since photosynthesis takes place mostly in the leaves, this nutrient obviously plays an important part in increasing crop yield. It is also a constituent of proteins, so adding nitrogen-containing fertilisers will also affect the quality of the crop.
- Phosphorus is found in many of the organic substances the plant produces, such as nucleic acids and ATP. It is very important for cell division so it is needed by parts of the plant that are growing rapidly.
- Potassium plays an important part in both respiration and photosynthesis.

Q 9 Suggest why the yield shown in Table 12.2 fell even when fertiliser was added to the soil.

Adding these fertilisers should replace the nutrients in the soil and increase the yield. The graph in Figure 12.17 shows what happens to the yield of wheat when different amounts of nitrogen-containing fertiliser are added to the crop.

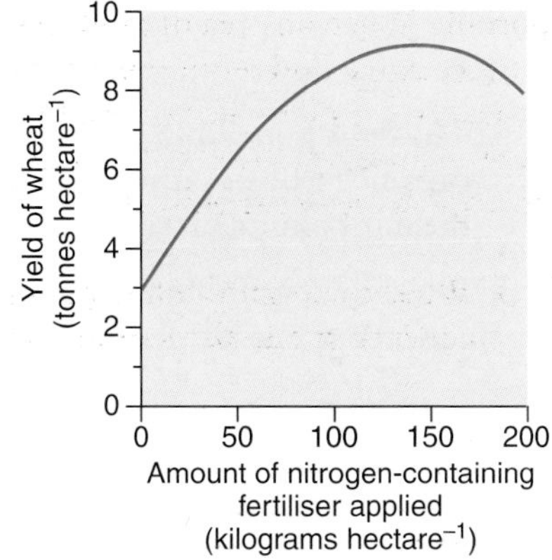

Figure 12.17
The effect of nitrogen-containing fertiliser on the yield of grain from a wheat crop.

At first, the curve rises steeply. The more fertiliser is added, the better the yield. However, there comes a point where the increase in yield gets smaller and smaller. At this stage, there is no point in adding more fertiliser because it does not increase the yield any more. This is sometimes described as the **law of diminishing returns**. There are a number of possible explanations for this. High concentrations of mineral ions in the soil can damage roots and prevent the efficient uptake of soil nutrients or we may be looking at limiting factors again. For example, if something else is limiting plant growth, such as the amount of phosphorus, it does not matter how much more nitrogen-containing fertiliser we add, we still will not get an increase in the crop yield.

There are two ways in which nitrogen, phosphorus and potassium can be added to the soil. We can use inorganic fertilisers or we can use organic fertilisers such as farmyard manure. Both contain the same macronutrients although, as Table 12.3 shows, they are present in different amounts.

	Composition as percentage of total mass		
Fertiliser	**Nitrogen**	**Phosphorus**	**Potassium**
Inorganic fertilisers			
Compound fertiliser	20	10	10
Ammonium nitrate	34	0	0
Organic fertilisers			
Cattle manure	0.6	0.1	0.5
Chicken manure	1.5	0.5	0.6
Sewage sludge	1.0	0.3	0.2

Table 12.3 The composition of some types of inorganic and organic fertiliser. Note: There are many different types of inorganic fertiliser. Some, such as ammonium nitrate, only contain one of the macronutrients required by plants. Others, called compound fertilisers, contain all three. Different compound fertilisers differ in the proportions of nitrogen, phosphorus and potassium they contain.

Q 10 The figures in Table 12.3 show that the amount of nitrogen in organic fertilisers is much less than in inorganic fertiliser. What is the most abundant substance in organic fertiliser?

The figures shown in Table 12.3 help us to make comparisons between inorganic and organic fertilisers.

Inorganic fertilisers

- are concentrated sources of macronutrients and can therefore be applied in smaller amounts – this saves on transport costs and on damage done by heavy farm machinery being driven over the soil

- are clean and lack the strong smell of many organic fertilisers
- are convenient to handle and apply so it is much easier to ensure that they are spread evenly over the crop.

Organic fertilisers

- are not 'pure' chemical substances and may contain valuable micronutrients
- add organic matter to the soil – organic matter is important because it affects the soil structure and can reduce erosion and improve water-holding properties
- may be produced as a waste product on mixed farms – applying farm manure to a plant crop is a useful way of disposing of this waste.

Extension box 2

Where do all the ions come from?

Figure 12.18
The vigorous growth of algae and water plants in this East Anglian stream is the result of pollution with nitrate and phosphate ions.

The stream shown in Figure 12.18, which runs through an area of farmland in East Anglia, is very heavily polluted. The high concentrations of nitrate and phosphate ions in the water have led to increased growth of algae and water plants. As the stream is in a rural area, little of this pollution comes from sewage. It is mainly the result of agriculture. There are two main sources of these polluting nitrates and phosphates:

- fertilisers add nutrients to the soil and not all of these are taken up by crop plants. The rain leaches soluble ions from the soil, with the result that they drain into streams and ponds.
- cattle and other livestock produce large amounts of urine and faeces. This material is not always disposed of efficiently and can lead indirectly to an increase in the concentration of nitrate and phosphate ions in streams such as that shown in Figure 12.18.

Figure 12.19 shows the monthly concentration of nitrate ions in this stream. It also shows the total volume of water present. This is indicated by the curve showing the flow.

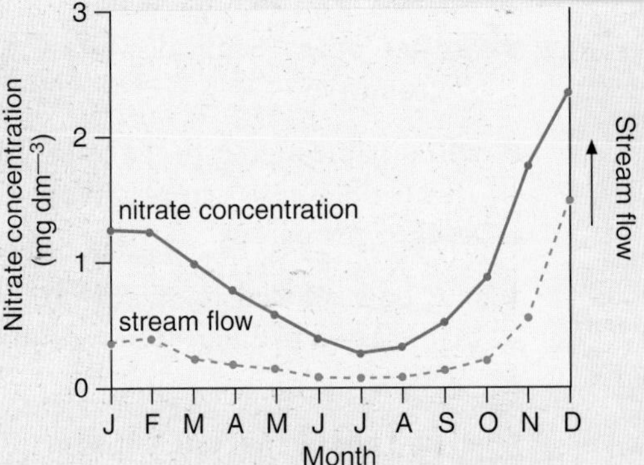

Figure 12.19
Graph showing how nitrate ion concentration and stream flow vary at different times of the year.

This is a relatively small stream so the flow provides a good indication of the amount of rain falling on the land immediately around it. During the winter months stream flow and rainfall are high. More rainwater drains into the stream during these months so more nitrates are leached from the surrounding farmland. Other factors are also involved. Crops sown in the autumn are only just beginning to grow so there is a lot of bare soil. Heavy rain results in water running off the soil, taking soil nutrients with it. It is cold and plants only grow slowly. Consequently, at this time of the year they take up a relatively small amount of mineral ions from the soil.

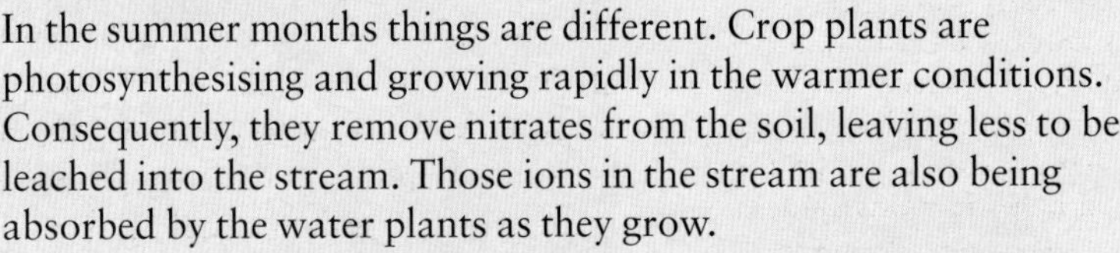

In the summer months things are different. Crop plants are photosynthesising and growing rapidly in the warmer conditions. Consequently, they remove nitrates from the soil, leaving less to be leached into the stream. Those ions in the stream are also being absorbed by the water plants as they grow.

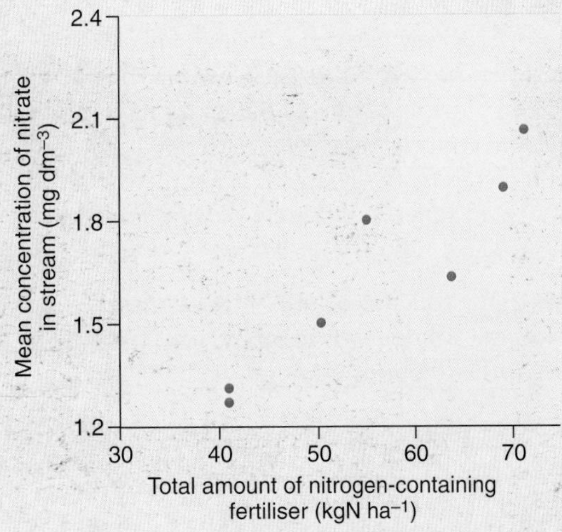

Figure 12.20
Graph showing the relationship between the mean concentration of nitrate in the stream and that added as fertiliser applied to the surrounding land.

Now look at Figure 12.20. This graph shows that there is a positive correlation between the concentration of nitrate in the stream and the amount of nitrate added in fertiliser to the surrounding land. The more nitrate added as fertiliser in a particular year, the more nitrate there is in the stream. We have to be careful about the conclusions that we draw from sets of data, however. We must be aware that just because these two things are correlated, it does not mean that the fertiliser is the only cause of pollution in the stream.

Controlling pests

Figure 12.21
Plants such as these nettles are affected by a large number of other organisms. Caterpillars eat their leaves, aphids suck their sap and other plants compete with them for sunlight, water and mineral ions in the soil.

In many ways crop plants are no different from plants growing under more natural conditions. They compete with other species of plant for mineral ions and water in the soil and for light. Insects feed on their leaves and they suffer from diseases caused by fungi and viruses. All these things reduce the growth rate of the plant and are likely to affect the amount of seed it produces. When crop plants are involved, this represents an important loss. Controlling unwanted plants, insects and fungi will increase crop yield considerably.

In the UK arable farming is based on **monoculture**. This means giving over large areas of land to the growth of a single crop. With plants growing closely together fungal diseases and insect pests can spread rapidly from plant to plant so that large areas are affected and heavy damage results in a very short time. It is therefore important to control unwanted organisms, such as the weeds, pests and things that cause disease, in order to obtain a good harvest. This will also ensure that the harvest is of a high quality.

Figure 12.22
Pests are organisms that reduce the quality or yield of a crop. Many crop pests, such as aphids (a), are insects, although slugs and snails also do a lot of damage. Diseases, such as potato blight (b), are caused by viruses, fungi and bacteria. Weeds such as wild oats (c) are simply plants growing in the wrong place. They compete with the crop and reduce its yield. In some cases, their seeds mix with harvested grain and reduce its value.

(a)

(b)

(c)

The data in Table 12.4 show some of the losses caused by pests, diseases and weeds. The figures relate to individual crops. Table 12.5 show losses in different parts of the world.

Crop	Estimated amount that could be produced (tonnes × 10^6)	Percentage of potential harvest lost to		
		Pests	Disease	Weeds
Wheat	356.8	5.1	9.5	9.8
Rice	445.7	27.5	9.0	5.7
Maize	344.9	13.0	9.6	13.0
Vegetables	279.9	8.4	11.1	8.5
Fruit	200.4	5.7	16.5	5.9

Table 12.4 Estimated losses due to pests, diseases and weeds for a number of different crops.

Q 11 Using Table 12.4:

(a) calculate the total mass of rice lost to pests, disease and parasites

(b) suggest why the percentage of fruit crops lost to weeds is low.

Area	Percentage of potential harvest lost to		
	Pests	Disease	Weeds
Europe	5.1	13.1	6.8
Africa	13.0	12.9	15.7
Asia	20.7	11.3	11.3
World-wide	13.7	11.6	9.5

Table 12.5 Estimated crop losses in different parts of the world.

Q 12 Using Table 12.5 explain why world-wide losses due to weeds are lower than those for Africa alone.

Reducing crop yields

As we saw in the earlier part of this chapter when we discussed the importance of fertiliser, plants need certain mineral ions if they are to grow well. There is a limited supply of these in soil. If weeds are growing in a wheat crop they will take some of the mineral ions that would

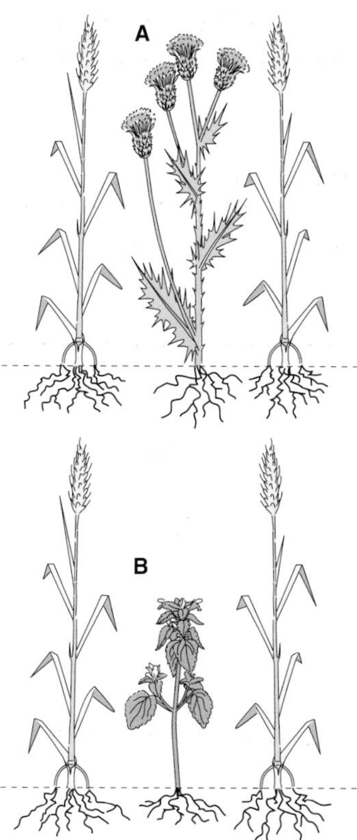

Figure 12.23 (above)
Some weeds have a greater effect on crops than others. The weed labelled A is the same height as the crop plants and its roots extend to the same depth in the soil. Its roots will compete with the roots of the crop plant for mineral ions and water, and its leaves will compete for light. Plant B is much smaller and will not have much effect on the yield of the crop.

Figure 12.24 (right)
This photograph is a scanning electron micrograph of an aphid. If the aphid inserts its needle-like mouthparts into a plant infected with a virus disease, it may suck up virus particles. It can then transfer these virus particles to another plant.

otherwise go to the wheat plants. Competition between organisms of different species for the resources they need is known as **interspecific competition.** This is in contrast to **intraspecific competition**, which is competition between organisms of the same species. In addition to mineral ions, weeds compete with crops for water and light.

Q 13 **When wheat plants are grown very close together the yield per plant is lower than when the plants are grown a greater distance apart. Explain this in terms of competition.**

Pests reduce the yield of plants in different ways. They can have a direct effect by damaging the plant leaves. Aphids such those shown in Figure 12.22a feed by inserting their mouthparts into plant cells. This often causes leaves to curl up and become distorted. Since the leaves of a plant are the places where most of the photosynthesis takes place, it follows that a plant with stunted and misshapen leaves will not photosynthesise effectively. As a result there will be smaller amounts of sugars available to be transported to other parts of the plant such as the fruits, seeds and roots.

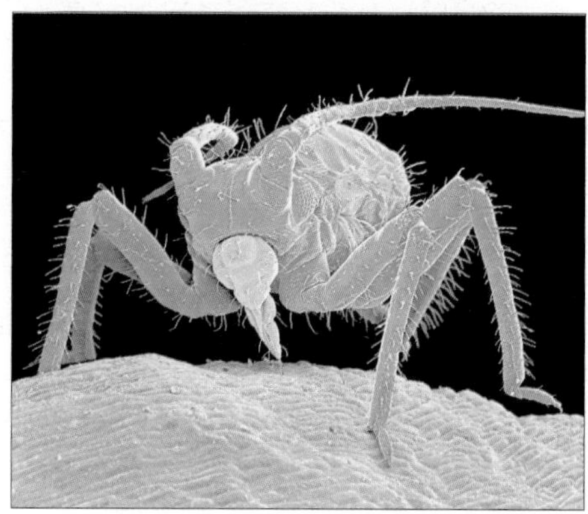

Q 14 **Two species of aphid are commonly found on the leaves of sugar beet. Black bean aphids are very common but seldom move from the plant on which they are feeding. Peach-potato aphids are not as common but often move from one plant to another. Explain why the peach-potato aphid is the more important pest.**

Controlling pests with chemicals

Pests and weeds may be controlled with chemical **pesticides.** Pesticides may be divided into **insecticides**, which are used to combat insect pests, **fungicides**, which target the fungi that cause many plant diseases, and **herbicides**, which are used to kill weeds. A variety of chemical substances are used as pesticides and they work in different ways, but we can classify them as one of the following:

- **contact pesticides:** these are pesticides that are sprayed directly onto the crop. Contact insecticides are usually absorbed by the insect through tiny gas-exchange pores, the **spiracles**, along its body. Contact herbicides and fungicides are absorbed directly through the surface. They have the advantage of being inexpensive but they often have to be reapplied. This is because their effect is usually short lived and, in addition, there are always some pests that avoid the pesticide

- **systemic pesticides:** systemic insecticides are sprayed onto the crop. They are absorbed by the leaves and transported round the plant. A sap-sucking insect such as an aphid feeding on the crop plant takes in the pesticide and is poisoned. Systemic insecticides are very effective because it is not necessary for the spray to come into contact with every insect. They are also selective in their action. Systemic herbicides are absorbed by the leaves and, because they are transported through the weed, kill all of its tissues, including underground parts such as the roots
- **residual pesticides:** these are sprayed onto the soil or used to treat seeds before the crop is planted. They remain active in the soil and will kill fungal spores, insect eggs and larvae, and weed seedlings as they germinate.

Figure 12.25
A farmer has many things to consider before spraying a crop. Spraying may increase yield by reducing the number of pests. However, the pesticide itself, and the heavy machinery used to apply it, can cause a reduction in yield. Spraying may be economically worthwhile when the crop is young and growing vigorously but not when it is ripe and ready to harvest.

Q 15 Why are systemic insecticides usually selective in their action?

A good chemical pesticide should be:

- specific: this means that it will have an immediate and lethal effect on the pest concerned but should be harmless to humans so it can be applied safely. In addition, it should not affect mammals and birds, beneficial organisms such as the natural predators of the pest, earthworms and honeybees, or vital plant processes such as photosynthesis
- chemically stable but biodegradable: this means it will have a long shelf-life but, once applied, will be broken down rapidly in the soil to a harmless substance
- cost-effective: a new pesticide has to be tested extensively before it is released for sale. This may make the development costs very high. The manufacturer needs to recover these and production costs from sales. This is further complicated because a new pesticide only remains useful for a limited time. When it is used repeatedly, the pest may develop genetic resistance so the pesticide becomes ineffective.

Biological control

Biological control does not use chemicals. It uses other organisms that are predators or parasites of the pest. There are already many examples of successful biological control and many other organisms are being investigated to see if they are suitable for using in the control of pests. Extension box 3 describes some work that has been carried out on the biological control of water hyacinth, a weed that grows rapidly, blocking rivers and lakes in many parts of the world.

Extension box 3

The water hyacinth story

The water hyacinth is a plant that floats on the water surface. Parts of it break off easily and regenerate to form new plants. In many subtropical and tropical countries it is a significant weed that completely covers many lakes and rivers. Not only does it prevent boats from using these waterways but few species of fish can survive in the oxygen-poor water under the dense carpet of plants. In areas where it has been introduced there are few herbivores that feed on it.

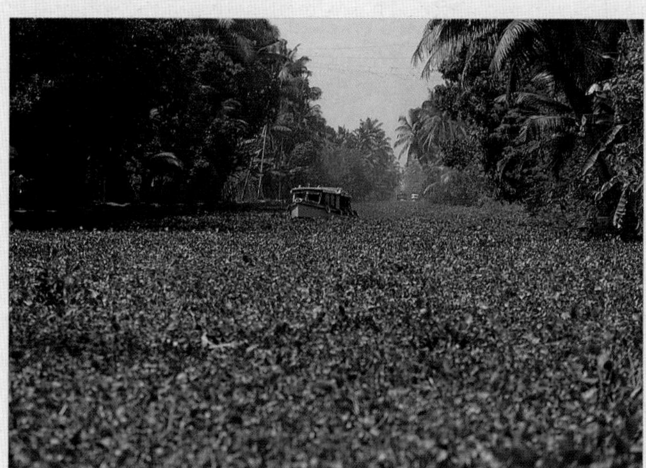

Figure 12.26
Water hyacinths might look attractive when in flower but in many parts of the world they are weeds that have major economic effects.

A series of experiments was carried out on a beetle, *Neochetina eichhorniae*, to see if it was suitable to use for the biological control of water hyacinth. Mature plants were placed individually in identical pots in a glasshouse. Two pairs of beetles were added to each pot and measurements of the plants in these experimental pots and in control pots were made at monthly intervals. Table 12.6 summarises the results.

	Mean length of longest root (cm)		Mean leaf area (cm^2)		Mean number of daughter plants produced	
Month	Experimental	Control	Experimental	Control	Experimental	Control
Aug	7	10	61	73	5	6
Sep	20	19	40	68	6	6
Oct	21	20	74	122	4	7
Nov	11	20	46	120	6	7
Dec	11	23	41	119	7	7

Table 12.6 Some results from an investigation into the effect of N. eichhorniae *on the growth of water hyacinth plants.*

The workers were able to draw two main conclusions from the data from these experiments:

- the beetle reduced the rate of growth of the roots and leaves of the water hyacinth
- the beetle did not have a noticeable effect on the number of daughter plants that were produced. It is probable that any differences between the experimental pots and the control pots were due to chance.

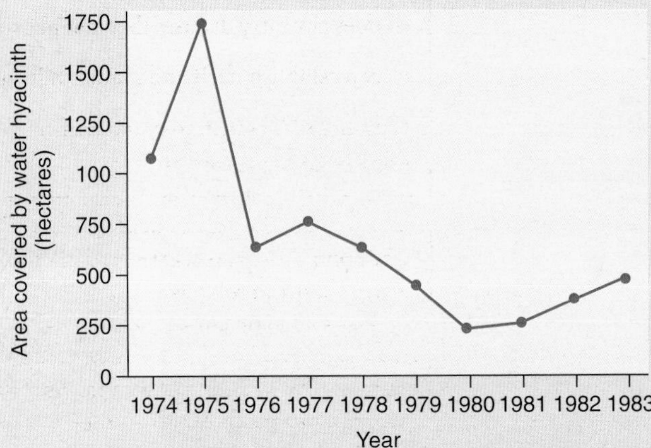

Figure 12.27
Graph showing the change in the area of a lake covered in water hyacinth following the release of the *Neochetina eichhorniae* beetle.

On the basis of these experiments, it was decided to carry out a field trial. Before releasing the beetles, however, other investigations were carried out. These showed that the beetles would be able to survive in the area concerned and that they only fed on water hyacinth

Figure 12.27 shows how the area of water covered by water hyacinth changed in the years immediately following release of the beetle.

This graph shows:

- there was an increase the area covered by water hyacinths in the year following release of the beetles – the numbers of beetles were probably too small to prevent the spread of the plant at this stage
- once the beetles had established themselves and started reproducing, the area covered by water hyacinths decreased rapidly
- the area of water covered by water hyacinth stabilised at a much lower level – whenever there was an increase in water hyacinth numbers, the beetle population increased and reduced the plant growth again.

Before a pest can be successfully controlled by another organism, extensive research has to be carried out. Only then can the biological control agent be safely released. Figure 12.28 shows some of the main steps that are carried out in setting up a biological control programme.

Figure 12.28
Flow chart showing the main steps carried out in setting up a biological control programme.

Many animals and plants are not native to the area in which they are pests. In their new environment there are often no predators or parasites to keep their numbers in check. Biologists search the area from which the pest originally came for suitable predators and parasites to use in a biological control programme.

▼

Trials are carried out to make sure that the control organism:

- will only attack the pest
- does not carry diseases that it might spread to native animals or plants
- can establish itself and maintain its numbers in its new environment.

Care has to be taken to prevent escape of the control organism before these trials are complete.

▼

The control organisms are bred or cultured in large numbers. They are distributed and released.

▼

Agricultural scientists collect the necessary information to find out if the programme has been successful.

Integrated pest management

Figure 12.29
Integrated pest control can be used in gardens. Insects such as ladybirds eat large numbers of aphids. If chemical pesticides are used sparingly, ladybirds and other predators will keep the aphid population under control.

A successful programme of biological control has a number of advantages over controlling pests with chemical pesticides:

- it is very specific and only affects the pest – no matter how carefully chemical pesticides are applied they affect other organisms
- once introduced, the control organism establishes itself and does not have to be re-introduced – chemical pesticides must be used repeatedly and in the long term this makes their use more expensive
- pests do not become resistant to organisms used for biological control – one of the reasons why we are constantly having to develop new pesticides is because pests rapidly develop resistance.

Biological control does have some disadvantages, however:

- there is often some time between introducing the control organism and a significant reduction in numbers of the pest – chemical pesticides act much more quickly
- an effective biological control programme keeps numbers of the pest at a low level – it rarely exterminates the pest completely.

A really effective control programme would make use of a variety of methods. We refer to this approach as **integrated pest management**. It often involves making use of natural parasites and predators or biological control and supporting this with the occasional use of pesticides.

Agrochemicals and the environment

We often use **agrochemicals** as a general term to refer to both fertilisers and pesticides. Large amounts of agrochemicals are used in modern farming. As we have seen in this chapter, their use results in an increase in crop yield. Unfortunately, however, the advantages of using agrochemicals are offset by the damaging impact they have on the environment.

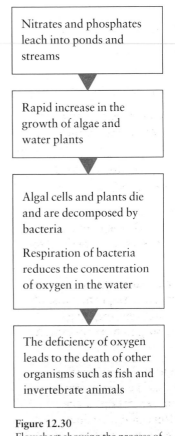

Nitrates and phosphates leach into ponds and streams

▼

Rapid increase in the growth of algae and water plants

▼

Algal cells and plants die and are decomposed by bacteria

Respiration of bacteria reduces the concentration of oxygen in the water

▼

The deficiency of oxygen leads to the death of other organisms such as fish and invertebrate animals

Figure 12.30
Flowchart showing the process of eutrophication

Fertilisers and the environment

A farmer tries to apply the right amount of fertiliser at the right time of the year. Unfortunately, it is impossible to do this. There will always be some fertiliser that will not be taken up by the crop. The mineral ions in the fertiliser will be washed or **leached** from the soil and will ultimately find their way into ponds and streams (see Extension box 1). This leads to **eutrophication**. Increasing the concentration of mineral ions, in particular of nitrates and phosphates, results in a rapid increase in the growth of algae and water plants. Light is blocked out by the resulting thick surface layer of vegetation and soon many of the algal cells die off. Bacteria in the water bring about their decomposition and increase in number. Like all living organisms, bacteria respire so the increased bacterial activity leads to a fall in the concentration of oxygen. This in turn causes the death of other organisms. We can summarise these changes in a flow chart (Figure 12.30).

Pesticides and the environment

Pesticides affect the environment in a different way from fertilisers. Many older insecticides and some of those in use today are **persistent**. This means that they are substances that are only broken down slowly so they remain in the bodies of insects for a long time after they are applied. If these insects are eaten by animals further up the food chain, they pass into the animal concerned. With each step in the food chain, they become more concentrated and more likely to build up to a lethal concentration.

Q 16 In a study of birds living in an area of farmland, the mean concentration of pesticides in the body fat of insect-eating birds was 1.1 parts per million. The concentration of pesticides in the body fat of sparrowhawks was 19.3 parts per million. Sparrowhawks feed on small birds. Explain the difference between these figures.

Figure 12.31
The sparrowhawk is one of the commoner birds of prey in the UK, but in the early 1960s numbers fell dramatically. In 1949 several hundred pairs bred in Norfolk. By 1965 a single breeding pair remained. Control of the use of persistent insecticides has led to a recovery in numbers and sparrowhawks are now common again over much of the UK.

Summary

- Cereals are an important food source with high nutritive value.
- Different kinds of cereals are adapted to grow in different environments.
- Rice is one of the most important tropical food crops.
- Rice is adapted to growing partly submerged in water by having air spaces in the stem enabling the root cells to obtain oxygen for respiration.
- Flooding favours rice growing because it increases the supply of nutrients and inhibits the growth of competing weeds.
- Maize is adapted to grow in high temperatures and light intensity. Maize photosynthesises using the C_4 pathway. This enables it to photosynthesise when the CO_2 concentration is less than optimal for the C_3 pathway, i.e. when it is hot and dry.
- Sorghum is adapted to grow hot dry climates that have long periods without rain. The leaves have a thick waxy cuticle to prevent loss of water by evaporation, and role up in dry conditions, trapping moisture and further preventing water loss. Sorghum is very efficient at extracting water from the soil.
- The growth of greenhouse crops can be improved by increasing carbon dioxide concentration, temperature and light intensity to certain maximum levels. Carbon dioxide concentration can be increased by pumping in gas or using paraffin heaters. Artificial lighting and heating can increase light intensity and temperature.
- The most important macronutrients are nitrogen, phosphorus and potassium.
- Fertilisers supply nutrients to the soil and improve crop yields. Fertilisers can be inorganic or organic.
- Weeds reduce crop yields by competing for nutrients.
- Pests reduce crop yields, for example by damaging plant leaves.
- Pesticides, fungicides and herbicides can be used to control insects, fungi and weeds respectively.
- Biological control uses predators or parasites of pests to control their numbers.
- Agrochemicals (fertilisers and pesticides) damage the environment. Minerals leached from soil contaminate fresh water leading to eutrophication.
- Persistent pesticides move through the food chain, becoming increasingly concentrated and possibly reaching lethal concentrations.

Examination questions

1 (a) Land used for crops of wheat and other cereals needs regular applications of fertilisers. Explain why fertilisers have to be added to this land each year.

(1 mark)

(b) The table shows a comparison of some features of organic and artificial fertilisers.

Feature	Organic fertilisers	Artificial fertilisers
Nutrient content	Variable	Constant
Solubility	Not immediately soluble	Soluble
Rate of release	Slow	Rapid
Concentration of ions in soil	Low	High
Bulk	Large	Small

Use the information in the table and your own knowledge to explain three advantages of using organic rather than artificial fertilisers.

(3 marks)

2 Cucumbers are often grown in glasshouses. The table shows the yield of cucumbers in a well-illuminated glasshouse under different conditions.

Temperature (°C)	Yield of cucumbers (kg per 10 plants)	
	Concentration of carbon dioxide (%)	
	0.03	0.13
20	25	25
30	59	79

(a) In which conditions was the yield greatest. Explain why?

(3 marks)

(b) Describe how a cucumber grower might economically alter the conditions inside a glasshouse during the winter months, in order to obtain the maximum yield.

(2 marks)

Assignment

This assignment will give you another opportunity to practise your data-handling skills as well as looking at some aspects of the way in which experiments are designed.

Look at Figure 12.32. It shows the results of an investigation into the effect of temperature on the time taken for different varieties of sorghum to reach maturity. Temperature was measured as the mean temperature for the 30 days immediately after planting. Maturity was determined as the day on which the plants first released pollen.

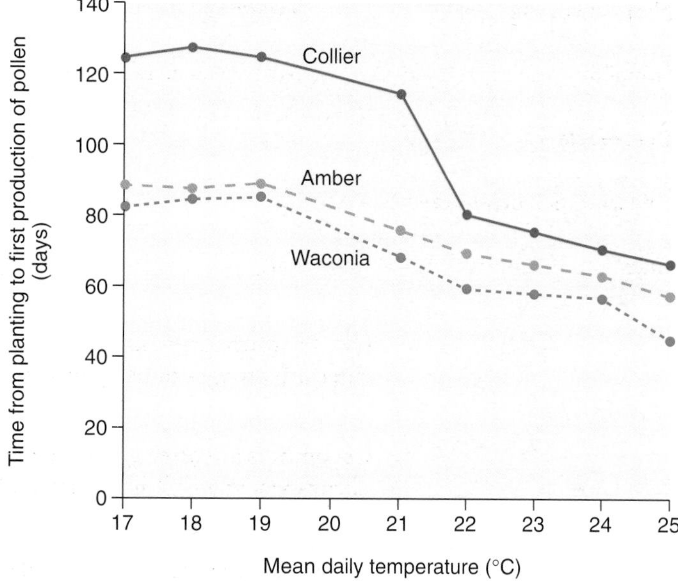

Figure 12.32
The relationship between temperature and the time taken for different varieties of sorghum to reach maturity.

1 Suggest why maturity was determined as the time when the plant first produced pollen rather than when it produced grain.

(1 mark)

2 (a) Use the graph to summarise the way in which temperature affects the time taken for sorghum to reach maturity.

(2 marks)

(b) Describe the main differences in the ways in which temperature affects the time taken for the three varieties to reach maturity.

(2 marks)

3 Explain why the data have been plotted with time on the y-axis.

(1 mark)

4 Suggest two environmental factors associated with the soil that might influence the amount of time sorghum plants take to reach maturity.

(2 marks)

In a second investigation, the water requirements of different varieties of maize and sorghum were measured. In this case, the growth of the plants was determined by measuring the increase in their dry mass. The results of this investigation are shown in Table 12.7.

Variety	Mass of water transpired by plant in producing 1 kg of dry mass (kg)	Rainfall required to produce 1 tonne of dry mass (mm hectare^{-1})
Sorghum		
Waconia	289	25.5
Amber	294	25.9
Maize		
Sherrod	336	29.6
Pride	355	31.3

Table 12.7

5 Most of the water that is transpired by a plant has been absorbed from the soil by the roots. Plants need only a small amount of water for the actual chemical reactions involved in photosynthesis. Suggest two other reasons why a plant needs water in order to produce dry matter.

(*4 marks*)

6 (a) Apart from the water that is taken up by the plant, what else may happen to the water falling as rain on the ground on which a crop of sorghum is growing?

(*1 mark*)

(b) Use this information to suggest one method by which it would be possible for a farmer to reduce the proportion of rainfall lost in this way. Explain your answer.

(*2 marks*)

7 In parts of Sudan it is extremely hot and dry for most of the year but there is a short, wet season in which a limited amount of rain falls. Use the information in this assignment to suggest the advice you would give to a Sudanese farmer concerning which type of cereal to grow. Give the reasons for your answer.

(*4 marks*)

13 Manipulating Reproduction

Deer are the largest wild land animals found in the UK. It is difficult to think of them as pests but, with no natural predators to control their numbers, they have become very common in some areas. They feed on agricultural crops and damage young trees by removing their bark. They have to be controlled in some way. Since people are very sensitive about killing these animals by shooting them, experiments have been carried out to test other methods of controlling deer populations.

Two different approaches have been investigated. In the first, female deer are trapped and hormone pellets implanted under their skin. Hormones are gradually released from these implants into the animal's blood. This prevent ovulation from taking place. No ovulation means that no egg cells are released and the deer cannot produce young.

The other approach that is being considered is the development of a vaccine that prevents fertilisation. Trials have shown that vaccinated animals produce a protein that blocks particular sites on the membranes surrounding the egg cell. These are the sites that are normally recognised by sperm cells. When they are blocked, sperm can no longer recognise and fertilise egg cells.

Figure 13.1
Fallow deer. In some parts of the UK deer are pests, damaging trees and agricultural crops. Ways have to be found to control their numbers.

These methods of controlling deer numbers rely on a knowledge of reproduction and the many different hormones that control it. In this chapter we look at other ways in which we can manipulate and control the process of reproduction. We start by considering the oestrous cycle. This is the basic reproductive cycle in a female mammal. We then go on to look at various ways in which this knowledge enables us to use hormones as contraceptives and to treat human infertility. Finally we look at how an understanding of hormones can be of use in cattle and sheep farming.

The female reproductive cycle

key term

The **oestrous cycle** is a cycle of events associated with the development and release of mature sex cells or **gametes**.

Different mammals have different patterns of reproduction. Some, such as humans, rats and mice, can breed all year round. Others have distinct breeding seasons. In the females of all these mammals, however, there is a cycle known as the **oestrous cycle**.

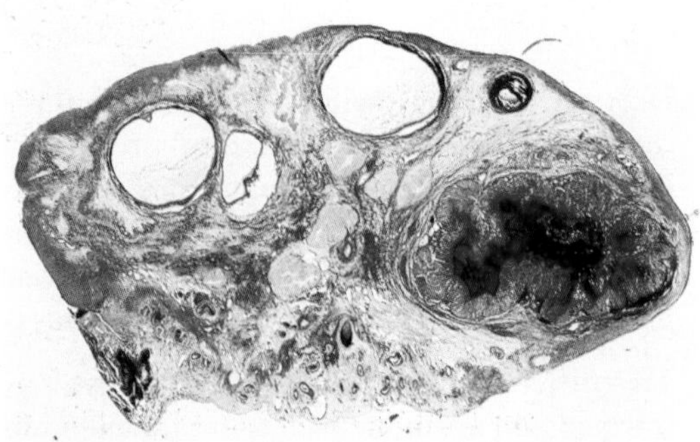

Figure 13.2
A section through an ovary as seen with a light microscope. Many immature follicles can be seen as well as a small number of follicles in the later stages of development.

Figure 13.2 shows an ovary from an adult mammal. It consists of a mass of tissue in which there are many blood vessels. These blood vessels not only supply the organ with the oxygen and nutrients it needs and remove waste products such as carbon dioxide, but also transport the hormones that are responsible for controlling the process of reproduction. Within the tissue of the ovary are **follicles** at different stages of development. Each follicle consists of a hollow ball of cells, the **follicle cells,** surrounding a developing egg cell or **oocyte.** An oestrous cycle involves the development of one or more of these follicles. The exact number that develops varies from species to species. There is usually only one in humans but in animals such as pigs and mice many more are involved. We can recognise three main stages in development (see Figure 13.3):

- **The follicular stage:** During this first part of the cycle, one or more follicles start to develop. The follicle cells surrounding the oocyte grow and divide. They produce hormones that affect other parts of the reproductive system. The follicular stage leads to the development of a mature female gamete and prepares the rest of the reproductive system for a possible pregnancy.
- **Ovulation:** The oocyte is released from the ovary and passes down the fallopian tubes towards the uterus.
- **The luteal stage:** Most of the follicle cells remain in the ovary after ovulation. They continue to develop and form a structure called the **corpus luteum.** The corpus luteum produces hormones. These hormones maintain the reproductive system in a condition that will allow implantation of a fertilised egg and maintain pregnancy.

If pregnancy takes place, this cycle is inhibited. What happens if pregnancy does not take place, however, depends on the species of mammal concerned. In humans and other mammals that do not have a distinct breeding season, the cycle begins again immediately. In other mammals, there may be a lengthy period of time before reproductive activity starts again and another oestrous cycle takes place.

Figure 13.3
The main stages in the oestrous cycle.

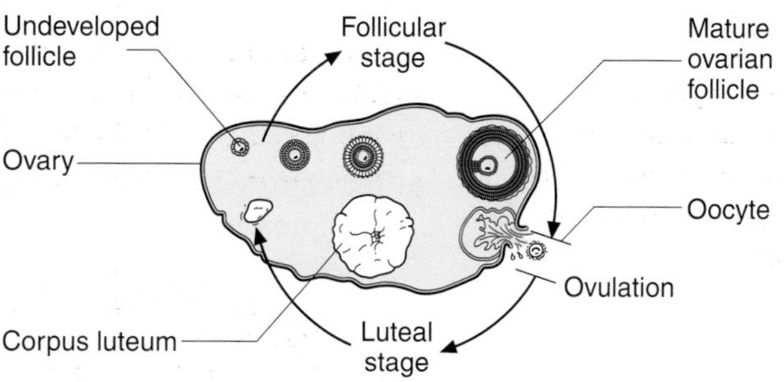

Q 1 **At which of the three stages, the follicular stage, ovulation or the luteal stage, is fertilisation likely to occur?**

The follicular stage

The pituitary gland is a small gland situated underneath the brain. At the start of the oestrous cycle, this gland secretes a hormone called **follicle-stimulating hormone** (FSH). FSH travels round the body in the blood. In the ovary, it causes the follicle cells surrounding one or more of the future egg cells to begin to grow and divide, with the result that the follicle becomes a mature **ovarian follicle**. The follicle cells secrete a hormone. This is **oestrogen** and its functions include:

- stimulating growth of the **endometrium** or lining of the uterus and its blood supply
- inhibiting further secretion of FSH by the pituitary gland
- stimulating the pituitary gland to secrete a second hormone, **luteinising hormone** (LH).

These events are summarised in Figure 13.4.

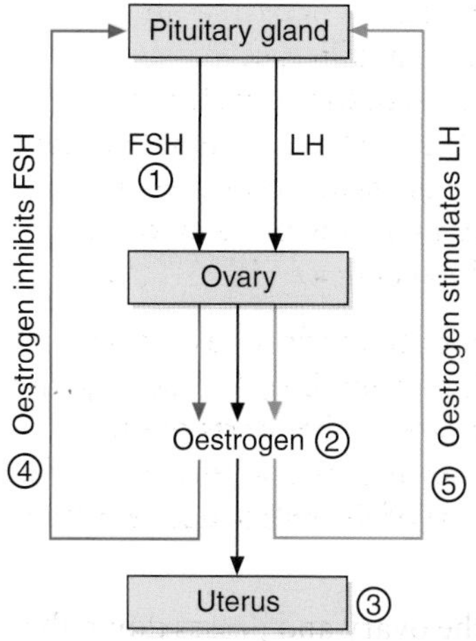

Figure 13.4
The main hormonal events associated with control of the follicular phase of the oestrous cycle. The pituitary gland secretes FSH (1), which stimulates the ovary to secrete oestrogen (2). Oestrogen stimulates the growth of the uterus lining (3). It also affects other hormones, inhibiting FSH (4) and stimulating LH (5). The increase in LH brings about ovulation.

Ovulation

As the follicular stage progresses, the developing follicle steadily increases in size and finally becomes a mature **ovarian follicle**. As it gets larger it produces more and more oestrogen, with the result that the oestrogen concentration in the blood increases. Most of the LH produced during the follicular stage of the cycle is stored in the pituitary gland. When the amount of oestrogen in the blood is high enough, it triggers the release of this LH so that there is a surge in the concentration of LH in the blood. It is this surge that brings about ovulation. The mature follicle bursts and the oocyte, together with some of the follicle cells that surround it, leaves the ovary and passes into the fallopian tube on its way down to the uterus. It is at this stage, with the oocyte in the fallopian tube, that fertilisation can take place.

The luteal phase

The surge of LH that brings about ovulation also has an effect on the follicle cells that remain in the ovary. The follicle now becomes a structure known as the corpus luteum. The corpus luteum secretes some oestrogen and a large amount of another hormone, progesterone.

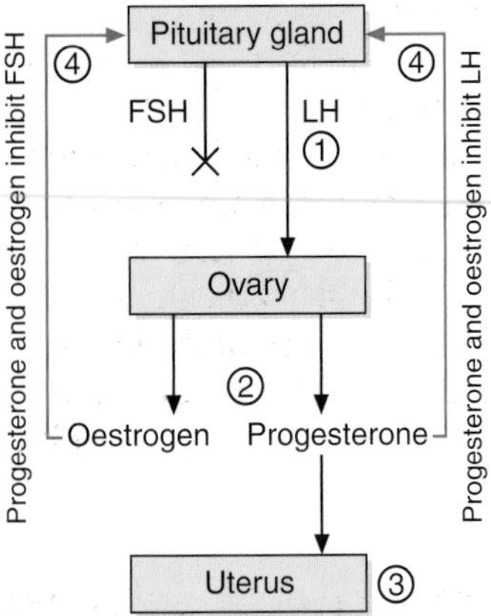

Figure 13.5
The main hormonal events associated with control of the luteal phase of the oestrous cycle. LH produced from the pituitary gland (1) brings about the development of the corpus luteum in the ovary. The corpus luteum secretes oestrogen and progesterone (2). Progesterone stimulates further growth of the uterus lining (3). Progesterone and oestrogen inhibit the production of LH and FSH by the pituitary gland (4).

Together these hormones have two important functions:

- progesterone continues to stimulate the growth of the endometrium and its blood supply. It also brings about the development of glands in this lining layer. The function of the glands is to produce a nutritive fluid
- the high concentrations of oestrogen and progesterone inhibit the production of FSH and LH. Without FSH and LH the cells of the corpus luteum start to shrivel and produce less and less oestrogen and progesterone. Since these are the hormones that inhibit the production of FSH, the cycle can start again.

These events are summarised in Figure 13.5.

This basic cycle is found in all mammals. Human females, however, have an oestrous cycle that is slightly different from that found in nearly all other mammals. As we have seen, at the end of the luteal stage, the concentration of oestrogen and progesterone fall. As a result, special spiral blood vessels in the lining of the human uterus contract and cut off the blood supply and the cells that form the uterus lining die. These spiral vessels then dilate again. The increased blood flow causes the uterus lining to be shed and lost from the body, an event known as **menstruation.** Because of menstruation we often refer to the human oestrous cycle as the **menstrual cycle.** Figure 13.6 shows the hormonal events and changes that take place in the endometrium during one complete menstrual cycle.

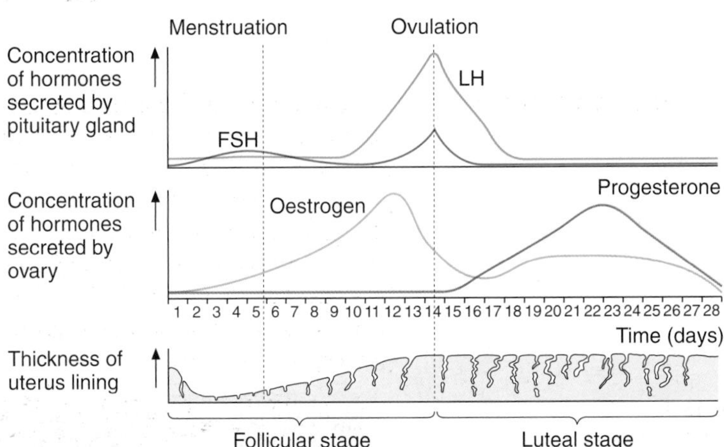

Figure 13.6
The hormonal events that control the menstrual cycle in humans are very similar to those that control the oestrous cycle in other mammals. The main difference is that at the end of the human cycle menstruation takes place and the lining of the uterus is shed.

Q 2 **Suggest how the changes that take place in the endometrium during the luteal phase would help the fertilised egg to survive and develop in the uterus.**

Oestrus

A mature female gamete only lives for a short period of time. If mating is to lead to fertilisation, it has to take place around the time of ovulation when a mature female gamete has been released into the fallopian tube. Too long before and the sperm cells will have died before the egg cell is ready; too long after and the egg cell will have died. The behaviour of many female mammals changes around the time of ovulation. The behavioural and physiological changes that occur at this stage in the oestrous cycle are known as **oestrus**. Other terms that are sometimes used include describing the animal as 'coming into season' or being 'on heat'. Oestrus marks the time when the female will allow mating to take place.

As an example, we will look at oestrus in cattle. As we saw when we looked at the way in which hormones control the oestrous cycle, there is a peak in oestrogen secretion just before ovulation and it is this that triggers the events of oestrus. The cow becomes restless, feeding less and moving around more. Early in oestrus she may attempt to mount other cows in the herd but as oestrus progresses, other cows will attempt to mount her. Instead of moving away, she will stand still. At the same time, various internal changes occur. Oestrogen causes the neck of the uterus, the cervix, to dilate and the mucus it produces to become thin and slimy so that it flows from the entrance to the vagina. These changes do not last long, about 18 hours on average.

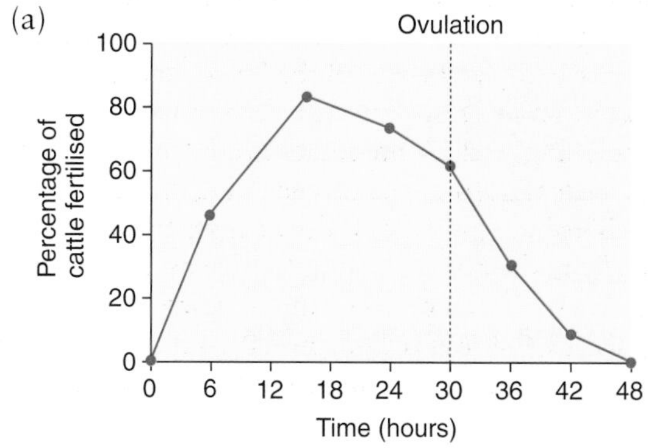

Figure 13.7
(a) The behavioural changes associated with oestrus allow a farmer to judge the time when artificial insemination is likely to be successful. (b) More cattle are fertilised if insemination takes place in the 24 hours immediately before ovulation. This is around the time when the cow is in oestrus.

Extension box 1

All female mammals have an oestrous cycle that is controlled by hormones. Different species of mammal, however, live in different conditions and have evolved adaptations in their reproductive physiology. Many of these adaptations concern the timing of the events of reproduction. In simple terms, receptors detect features such as changes in day length or temperature. This results in nerve impulses being sent to an area of the brain called the **hypothalamus.** In turn, the hypothalamus controls the secretion of the hormones that affect the process of reproduction.

Figure 13.8
A vixen and her cubs.

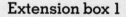

In the UK, the female fox is receptive to the male for a period of about three weeks in late January and February. If she mates and fertilisation takes place, the young are born seven to eight weeks later. This means that by the time they have been weaned and no longer rely on their mother's milk, there are plenty of young mammals and insects for the growing cubs to eat. The increase in day length in late winter brings about the secretion of a hormone called **gonadotrophin releasing hormone** (GnRH) from the hypothalamus. As its name suggests, this hormone then stimulates the release of gonadotrophins. These are hormones such as FSH, which are necessary to start the development of follicles in the ovary. The control process involves both the nervous system, which detects changes in day length, and the hormonal system, which triggers the process of ovulation.

Figure 13.9
A colony of seals off the Scottish Isles.

Seals spend most of their lives in water and many species only come out onto the land during the short breeding season, when they give birth to their young. Within a few days of birth the female seal comes into oestrus and mates. The fertilised egg travels down the oviduct into the uterus, where it starts to develop into a hollow ball of cells called a **blastocyst**. The blastocyst then stops developing and remains in the uterus in a resting state for a period of about three months. At the end of this time, it continues with its development once more. The result is that the young are fully developed and births occur at the same time each year. Control of this resting period is another aspect of reproduction that is triggered by changes in day length influencing hormone concentration.

Wildebeest are found in the savanna grasslands of Africa. Their young are very vulnerable to predators such as lions, hyenas and jackals. In parts of East Africa, wildebeest form huge migratory herds and within these herds the breeding behaviour is closely coordinated. The females in the herd mate over a very short period of time and then produce their young over a two to three week calving period. Instead of seeking seclusion when giving birth, expectant mothers gather together in calving areas early in the morning and, after a period of labour of about an hour and a half, they produce their young. Hundreds of young animals are born in a very short period of time. With so many young animals, many are able to escape the attentions of predators and survive.

Figure 13.10
A wildebeest in Southern Africa.

Synchronised breeding behaviour in wildebeest is controlled by many factors. Since the peak of mating behaviour is at the time of the full moon, it is thought that changes in the lunar cycle are detected by the nervous system. This triggers the production of the hormones, which bring about development of the reproductive organs. The precise synchronisation of breeding is helped by complex behaviour patterns.

Controlling reproduction in humans

Reproduction is the underlying cause of many of the problems affecting humans and their environment today. At an individual level, unwanted pregnancies give rise to personal problems and concerns. On a global scale, the rapid rise in world population is associated with famine, disease and pollution. There is, however, another aspect of human reproduction that we shall consider in this section. This is infertility and, although it may seem odd to refer to it in the same paragraph as overpopulation, we must bear in mind that it causes considerable distress to many childless couples.

We have a good understanding of the ways in which hormones control the menstrual cycle in a human female. We also have the chemical knowledge to produce a wide range of synthetic substances that either act as hormones or block their action. These substances can be used as contraceptives or in the treatment of infertility.

We have the ability to offer middle-aged childless couples the opportunity to have children. Is this a wise use of limited resources? Should we deny them this opportunity? Manipulating human reproduction often poses moral and ethical problems.

Hormones as contraceptives

Many room heaters are controlled by thermostats. The heater is switched on and the temperature increases. This is detected by a thermostat. As soon as the room temperature rises above a set level, the heater switches off and the room cools down again. Maintaining the temperature in this way is an example of **negative feedback**. A departure from the set level results in a series of changes that restore this set level. Negative feedback is important in controlling many things in the human body: temperature, blood glucose concentration and the levels of hormones such as those that control reproduction.

Q 3 **Explain how negative feedback can prevent body temperature from rising too high.**

In Figures 13.4 and 13.5 we saw how oestrogen and progesterone inhibited the production of FSH from the pituitary gland. The **combined contraceptive pill** is based on this example of negative feedback. The pill contains a mixture of progesterone and oestrogen. As these substances inhibit FSH, follicles are not stimulated to develop. With no mature ovarian follicles, ovulation cannot take not take place.

Although oestrogen is very efficient at inhibiting the production of FSH, it does have a number of disadvantages. In particular, it has been linked with diseases affecting the heart and circulatory system. There is some evidence that oestrogen can increase the risk of thrombosis, strokes and heart attacks. Because of this, the concentration of oestrogen in combined contraceptive pills is kept as low as possible.

Another type of hormonal contraceptive is the so-called **mini-pill**, which contains progesterone only. Progesterone on its own is not very reliable at inhibiting ovulation and there is still some uncertainty over the exact way in which it works. However, it is known to interfere with the process of cell division in the developing egg cell and can also prevent the implantation of a fertilised egg in the wall of the uterus. More information about hormones as contraceptives is included in the assignment at the end of this chapter.

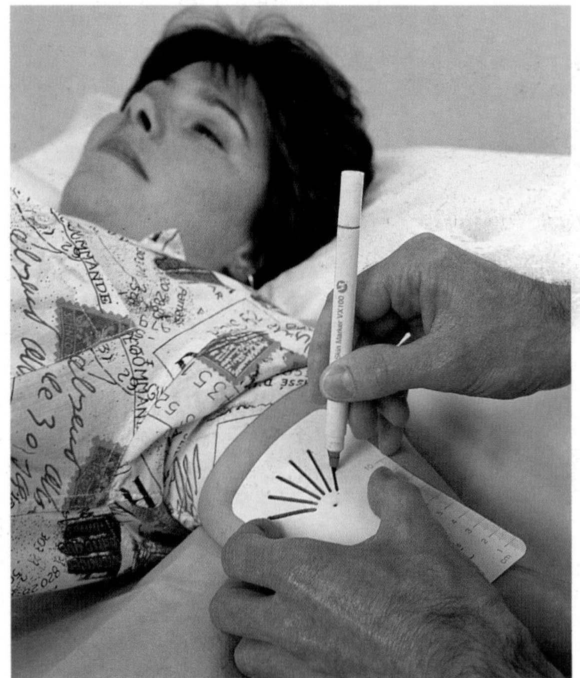

Figure 13.11
Contraceptives such as Norplant do not have to be swallowed as pills. They can be inserted under the skin, where they gradually release their hormonal contents over a period of up to five years.

Treating infertility

Infertility has many causes. At one time it was thought that it was the woman's 'fault'. Now we recognise that we should be looking at infertility as a problem for the couple rather than as the fault of one or other partner. As Table 13.1 shows, either of or both partners might be the cause of a woman failing to conceive. In this section we will look at how reproductive hormones are involved in the treatment of female infertility.

Infertile partner	Percentage of infertile couples
Female only	35–55
Male only	30–50
Both male and female	15–35

Table 13.1 The causes of infertility.

Ovulation is the key event in the menstrual cycle. Without it fertilisation cannot take place. Despite this, ovulation only takes place in about eight out of every ten cycles and this proportion is much less at the beginning and at the end of reproductive life. If a woman is unable to conceive, it is important that tests are carried out to check that she is actually ovulating. One way of doing this is to measure the concentration of progesterone in her blood. When ovulation takes taken place, a corpus luteum is formed and this secretes progesterone, so high levels of progesterone provide a good indication of ovulation.

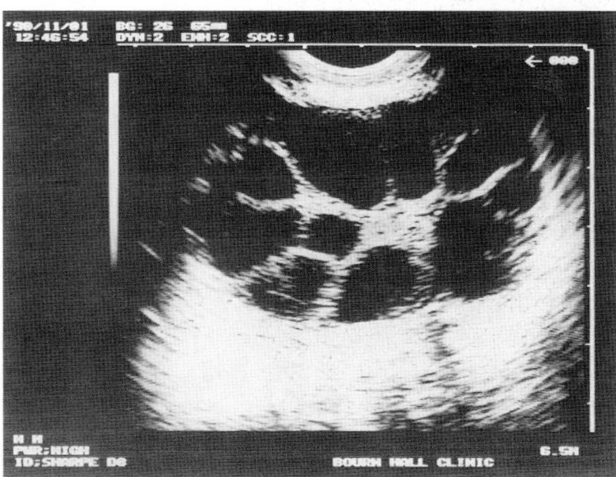

Figure 13.12
One problem with treating infertility with hormones is that there is an increased probability that several follicles might develop. This increases the risk of a multiple birth. Ultrasound scans can be used to check the number of mature follicles.

If it is found that ovulation is not taking place, various techniques can be used to overcome the problem. Many women who have apparently normal menstrual cycles but do not ovulate, produce small amounts of FSH, although this is not enough to cause a follicle to develop. They can be treated with a drug called clomiphine. To understand how it works you will need to look back at Figure 13.4. In the follicular stage of the menstrual cycle, the follicle cells secrete oestrogen. One effect of oestrogen is that it inhibits the secretion of FSH. This is fine in a fertile woman where enough FSH is produced to cause a follicle to develop but it does not help the situation when the amount of FSH is already low. Clomiphine prevents inhibition by oestrogen so more FSH is secreted. This may be enough to trigger the normal development of a follicle, which will lead to ovulation.

Extension box 2

How sex hormones work

Sex hormones are like other hormones. They are produced by specialised glands called **endocrine glands.** These glands do not have ducts but secrete the hormones they produce directly into the blood. Although hormones circulate all round the body, they only affect particular cells in certain **target organs.** Research into the biology of hormones has allowed us to answer two particular questions:

- why do hormones only affect particular cells when they are carried by the blood all round the body?
- how do hormones actually work?

We will look first at LH, one of the hormones produced by the pituitary gland. This hormone is a protein. Like many proteins, its molecules are folded in such a way that they have a tertiary structure that gives them a specific shape.

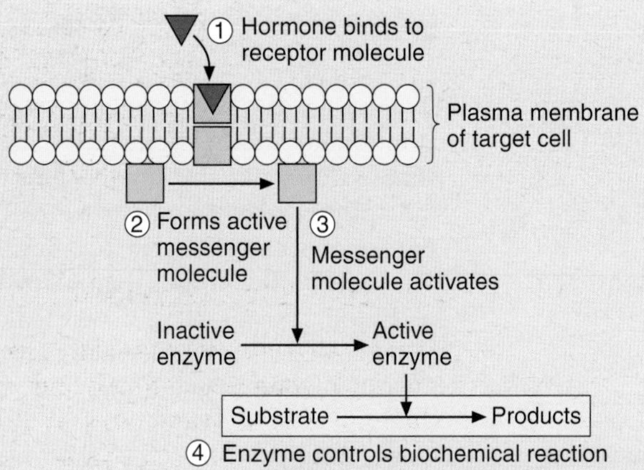
① Hormone binds to receptor molecule
Plasma membrane of target cell
② Forms active messenger molecule
③ Messenger molecule activates
Inactive enzyme
Active enzyme
Substrate
Products
④ Enzyme controls biochemical reaction

Figure 13.13
Protein hormones such as LH bind to specific receptor molecules in the cell surface membranes of target cells. This leads to the formation of a messenger molecule in the cytoplasm. The messenger molecule activates the enzymes that control the metabolism of the cell.

Q 4 Explain, in terms of receptor molecules, why some cells are affected by LH but others are not.

Figure 13.13 shows how protein hormones such as LH work. The plasma membrane of a target cell contains receptor molecules. These receptor molecules are proteins and have specific receptor sites that have a complementary shape to that of the hormone. The hormone is unable to pass through the plasma membrane and enter the cytoplasm of the cell. Instead it binds to the receptor and triggers a complex series of changes inside the cell. In simple terms, the binding of the hormone to the receptor activates a messenger molecule in the cell. This messenger molecule called cyclic AMP (cAMP) activates enzymes that control particular metabolic pathways. In this way, for example, LH produced by the pituitary gland brings about the biochemical changes in the cells of a mature ovarian follicle that lead to ovulation.

Oestrogen and progesterone are also sex hormones but they are not proteins. They are steroids and work in a different way to LH and FSH. This is summarised in Figure 13.14. Steroids are lipids and so they are able to pass through the phospholipid bilayer of the plasma membrane and enter the target cell. Once inside the cell, they bind to receptors in the cytoplasm. These receptors act as carriers for the hormone and transfer it into the nucleus of the cell. The hormone then switches on the genes responsible for the synthesis of particular proteins.

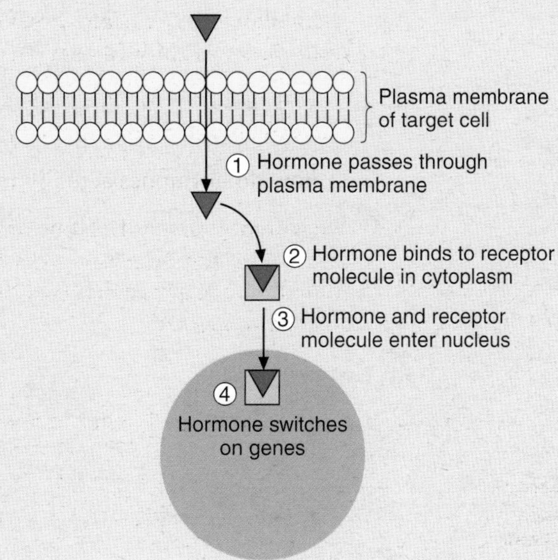
Plasma membrane of target cell
① Hormone passes through plasma membrane
② Hormone binds to receptor molecule in cytoplasm
③ Hormone and receptor molecule enter nucleus
④ Hormone switches on genes

Figure 13.14
Steroid hormones such as oestrogen penetrate the plasma membrane and bind to receptors present in large numbers in the cytoplasm of the target cell. The hormone and its receptor move to the nucleus of the cell and the hormone switches on particular genes.

Controlling reproduction in domestic animals

If we are to continue to provide large amounts of affordable food for a growing human population we must use technology to improve yields. This applies to farm animals as well as to crops. In this section we will look at three examples of how hormones can be used to increase productivity in cattle.

Transplanting embryos

About 40 weeks after an egg is fertilised, a cow gives birth. She usually has a single calf. This means that a genetically valuable cow has relatively few calves during her lifetime. However, cows produce large numbers of egg cells, many of which do not get fertilised. It ought to be possible to make use of these unfertilised eggs to increase the number of calves produced. This is where embryo transplantation comes in.

Figure 13.15
It has been estimated that if farmers used the technology of 30 years ago, the cost of food would be two to three times higher than it currently is.

An animal with the desired characteristics, such as high-yielding milk cow, is injected with a mixture of FSH and LH. As a result, follicles in her ovaries develop and she ovulates. By carefully working out the hormone dose, the cow can be made to produce a larger number of egg cells than normal. As soon as she comes into oestrus, she is artificially inseminated. The egg cells are fertilised and start to develop into embryos. Six to eight days later, these embryos are washed out of her uterus. Up to 20 embryos may be recovered this way. They are examined with a microscope and sorted out. Those that are developing normally are transferred to other cows which will act as mothers for the developing calves. It is important that the oestrous cycles of the recipient cow and the donor are synchronised or the uterus will

not be in a suitable condition for the embryos to implant and develop naturally. The process is summarised in Figure 13.16.

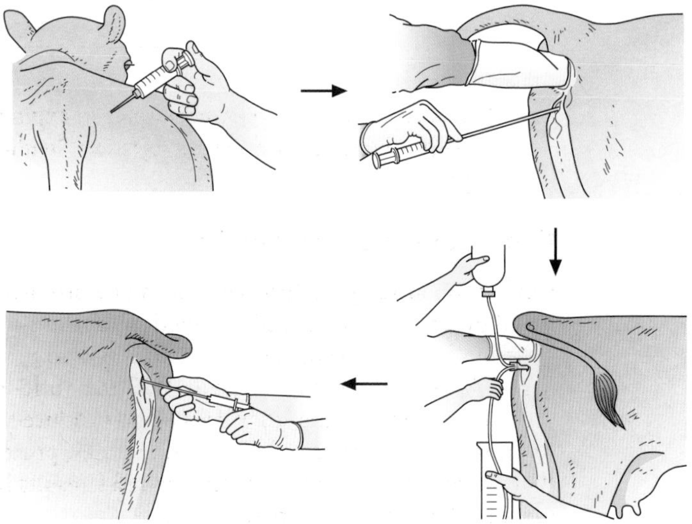

Genetically valuable donor cow injected with FSH and LH. This results in the development of a large number of follicles and egg cells
Donor cow is artificially inseminated 5 days after hormone injection
Embryos are collected 6–8 days later
After sorting, a healthy embryo is put into the uterus of a recipient cow, where it will develop normally

Figure 13.16
Embryo transplantation. Hormones are used to produce large numbers of egg cells from the ovaries of the donor cow. They may also be used to make sure that the oestrous cycles of the recipient and donor animals are synchronised.

Q 5 Cattle embryos can be stored by freezing. Draw a flow chart to show how embryo transplantation could be used to send calves from one continent to another.

Synchronising breeding

Bringing a group of animals into oestrus at the same time can be very helpful to a farmer. It reduces the time that the farmer has to spend with the animals in order to supervise mating and giving birth. If artificial insemination is used, it means that all the animals in the group can be inseminated at the same time. This can save considerably on cost.

One technique involves inserting a plastic coil containing progesterone into the animal's vagina. The progesterone is absorbed into the blood and the resulting high concentration makes sure that the corpus luteum continues to function. As long as the corpus luteum is functioning, ovulation will not take place. The coil is then removed. With no progesterone to maintain it, the corpus luteum shrivels and stops producing the hormones oestrogen and progesterone. The pituitary gland starts to produce FSH again, and ovulation takes place after a short follicular stage. Provided the coils are removed at the same time, all animals in the group will enter the follicular stage and ovulate at the same time.

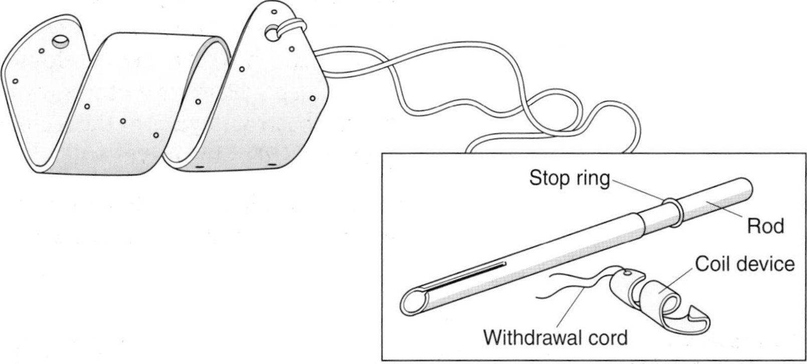

Figure 13.17
A progesterone-releasing coil. The coil is wound into a tight spiral for insertion. Once inside the vagina it springs back to its original shape and clings to the wall. The cords allow withdrawal.

Q 6 Use your knowledge of hormones to explain why, as long as the corpus luteum continues to function, ovulation does not take place.

Hormones and milk yield

Bovine somatotrophin (BST) is a hormone produced by the pituitary gland in cattle. It causes cells to divide and results in an increase in growth. We can now use genetic engineering to produce synthetic BST. When dairy cows are injected with BST, their udders show increased growth. This means that they are able to produce greater quantities of milk. In addition, more of the carbohydrates, proteins and fat in the food they eat are used for milk production. BST results in cows producing more milk over a longer period of time.

Summary

- All female mammals have an oestrous cycle with a follicular stage, ovulation, and a luteal stage.
- During the follicular stage, the follicle stimulating hormone (FSH) stimulates a follicle to become a mature ovarian follicle.
- A surge in luteinising hormone (LH) brings about ovulation.
- The same surge in LH causes the follicle remains to become the corpus luteum. The corpus luteum secretes oestrogen and progesterone.
- Oestrus refers to the changes occurring around the time of ovulation.
- For many female mammals, oestrus marks the time when they will allow mating.

- Human females can control their reproductive cycle using substances that mimic or block the action of hormones.
- The combined contraceptive pill is an example of negative feedback. It contains oestrogen and progesterone which inhibit FSH, so that follicles do not mature.
- The mini-pill contains progesterone.
- Increased knowledge of the human reproductive cycle has led to better diagnosis of and new treatments for infertility.
- Farmers maximise their animal product yields by controlling animal reproduction.
- Synchronised breeding of farm animals increases farming efficiency.

Examination questions

1 The diagram shows the production of follicles, ovulation and the formation of a corpus luteum in a human ovary. The stages in the cycle are shown clockwise round the ovary from A to G.

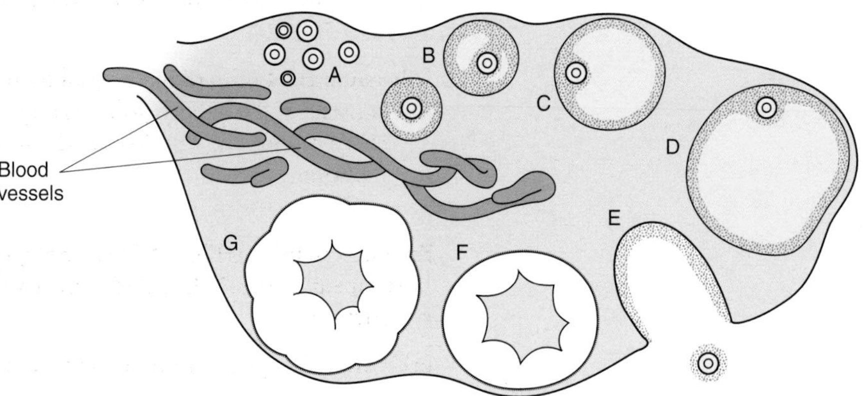

(a) Which of the lettered structures would you expect to produce the largest quantity of progesterone?

(1 mark)

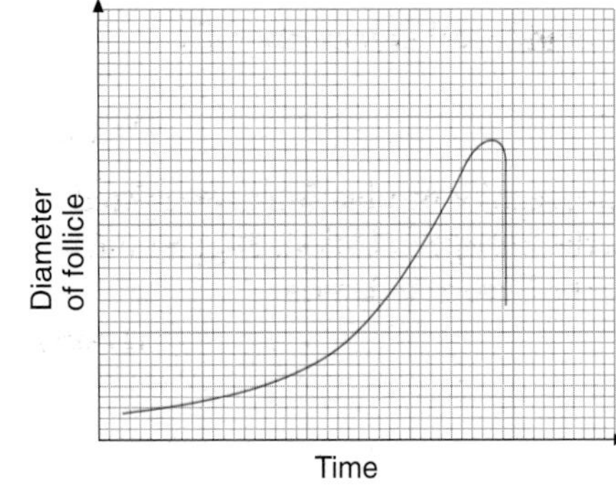

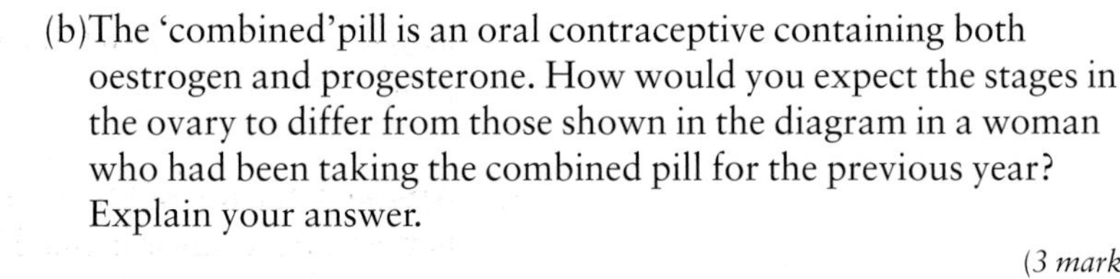

(b) The 'combined' pill is an oral contraceptive containing both oestrogen and progesterone. How would you expect the stages in the ovary to differ from those shown in the diagram in a woman who had been taking the combined pill for the previous year? Explain your answer.

(3 marks)

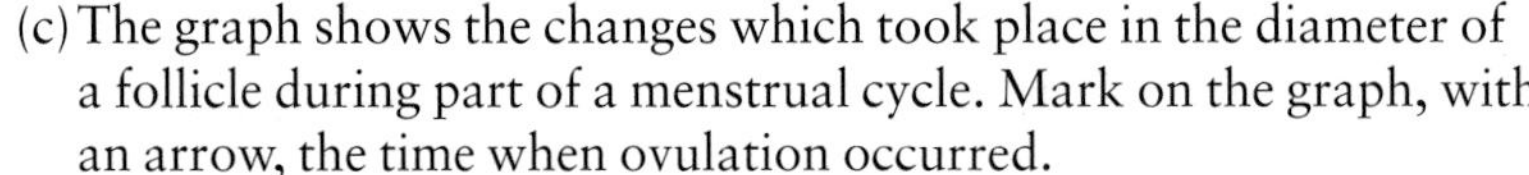

(c) The graph shows the changes which took place in the diameter of a follicle during part of a menstrual cycle. Mark on the graph, with an arrow, the time when ovulation occurred.

(1 mark)

2 In a study of three groups of women, the mean blood loss during menstruation was measured. Each of the women in group A was fitted with a standard intra-uterine device (IUD). Those in group B were fitted with an IUD which continually released small amounts of progesterone. Those in group C acted as a control. The results are shown in the table.

Group	Treatment	Mean blood loss during menstruation (cm^3)
A	Standard IUD	91
B	IUD which releases progesterone	19
C	Control group	41

(a) Suggest an explanation for each of the following observations:

(i) anaemia is often associated with the use of intra-uterine devices.

(1 mark)

(ii) the mean blood loss during menstruation of the women in group B was lower than that of the women in the control group.

(3 marks)

(b) There are 12.5 g of haemoglobin in 100 cm^3 of blood. The iron content of the haemoglobin is approximately 0.3%. Calculate the mean amount of iron lost by the women in group C. Show your working.

(2 marks)

3 Bovine somatotrophin (BST) is a polypeptide which acts as a growth hormone in cattle. When BST is injected into cows, it stimulates milk production.

(a) Name the type of monomers from which BST is made.

(1 mark)

(b) Two effects of BST are

(i) increased rate of blood flow to the udders
(ii) stimulation of mitosis.

Suggest how these two effects could together increase milk production.

(3 marks)

(c) Although the use of BST to increase milk production is permitted in the USA, its use is banned in the European Union. Suggest one reason why its use is banned in the European Union.

(1 mark)

Assignment

Hormones as contraceptives

It is possible to learn facts but this is of very little use if you do not understand them. Understanding allows you to interpret information and apply your knowledge to new situations. This assignment is based on a passage about the use of hormones as contraceptives. The questions it contains have been designed to test your understanding of the underlying biology. In addition, the last question also involves you bringing together ideas from different areas of the subject. If you continue your studies to 'A' level, you will come to appreciate that biology involves the linking together of individual topics in many different ways.

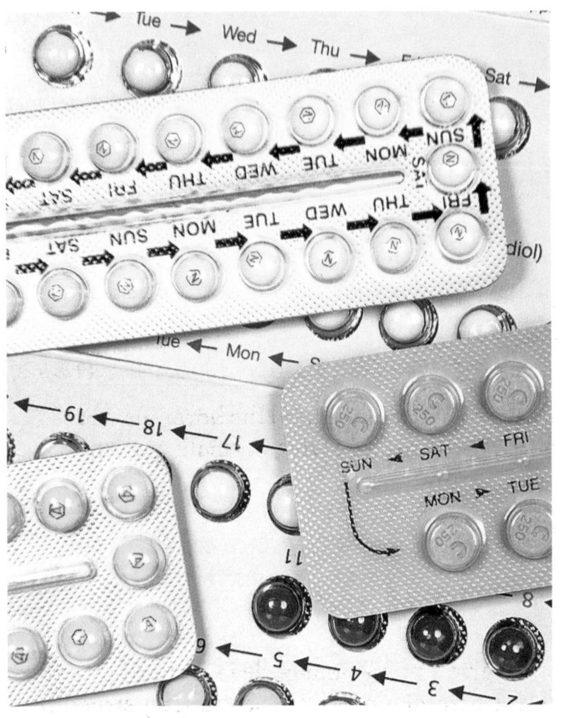

Figure 13.18
There are a number of different types of oral contraceptive, but most of them are based on the hormones that normally control the menstrual cycle.

Once ovulation has taken place, two steroid hormones are produced by the ovary. These hormones are oestrogen and progesterone. The **combined contraceptive pill** contains a mixture of these hormones. They are both lipids, as the fact that they are steroids suggests, but progesterone is not absorbed very well by the cells lining the gut. Because of this a synthetic form of this hormone is used. The pill works by inhibiting the release of FSH and LH by the pituitary gland, preventing ovulation from taking place. If the ovaries of women taking combined contraceptive pills are compared with those of women who are not taking hormonal contraceptives, a number of differences can be seen.

Oestrogen can have side effects so a **mini-pill** has been produced which contains progesterone only. It works in a different way to the combined pill as it does not normally inhibit ovulation. Instead it interferes with the process of meiosis. In rare cases, fertilisation does take place, in which case implantation is prevented from taking place. With the mini-pill, FSH is not inhibited, so oestrogen concentrations follow the same pattern as in a woman not taking oral contraceptives.

A third type of oral contraceptive is known as the **morning-after pill.** One form involves taking a single large dose of progesterone on the day following intercourse. This stimulates the lining of the uterus to develop. As it is a single dose, the concentration of progesterone in the blood then decreases rapidly. The effects this has on the uterus lining stops implantation from occurring.

A possible future development involves using a synthetic polypeptide that is very similar to gonadotrophin-releasing hormone (GnRH). This is a hormone produced by the hypothalamus. It triggers the release of FSH and LH by the pituitary gland. GnRH is a small polypeptide containing 10 amino acids. By substituting its constituent monomers, over 300 similar molecules have been produced. Some of these have been found to block GnRH receptor sites on the plasma membranes of cells in the pituitary gland. This could represent a major advance in producing a safe but reliable and cheap contraceptive.

Now answer these questions.

1 Three groups of 30-year-old women who were not pregnant were each using a different type of contraceptive. Copy the table below. Use ticks to show the stages in follicle development that might be found in the ovaries of the women in each of the three groups.

	Stage in follicle development that might be found in the ovaries		
Type of contraceptive used	**Immature follicle**	**Mature ovarian follicle**	**Corpus luteum**
No contraceptive			
Combined contraceptive pill			
Mini-pill			

(*3 marks*)

2 Explain what is meant by:

(a) a steroid
(b) a gonadotrophin.

(*2 marks*)

3 Use your knowledge of the menstrual cycle and the hormones that control it to explain how each of the following normally prevent pregnancy:

(a) the combined contraceptive pill
(b) the mini-pill
(c) the morning-after pill.

(*6 marks*)

4 Polypeptide chains fold to give a protein molecule its tertiary structure. Proteins have numerous structural functions. Their tertiary structure allows them to act as receptors and as enzymes. In addition, they form an important part of the diet. They are broken down during digestion and the amino acids produced are absorbed into the body through the wall of the intestine. Use your knowledge of proteins to explain:

(a) how the 300 or so similar molecules produced differ from GnRH

(*2 marks*)

(b) how a contraceptive based on one of these molecules could prevent GnRH from stimulating the pituitary gland to produce FSH

(*3 marks*)

(c) why a contraceptive based on one of these molecules would have to be injected rather than swallowed.

(*2 marks*)

In each answer make use of the information provided but try to add a little more detail from your own knowledge as well.

Index